IET ENERGY ENGINEERING 104

Power Transformer Condition Monitoring and Diagnosis

Other volumes in this series:

Volume 1 **Power Circuit Breaker Theory and Design** C.H. Flurscheim (Editor)
Volume 4 **Industrial Microwave Heating** A.C. Metaxas and R.J. Meredith
Volume 7 **Insulators for High Voltages** J.S.T. Looms
Volume 8 **Variable Frequency AC Motor Drive Systems** D. Finney
Volume 10 **SF_6 Switchgear** H.M. Ryan and G.R. Jones
Volume 11 **Conduction and Induction Heating** E.J. Davies
Volume 13 **Statistical Techniques for High Voltage Engineering** W. Hauschild and W. Mosch
Volume 14 **Uninterruptible Power Supplies** J. Platts and J.D. St. Aubyn (Editors)
Volume 15 **Digital Protection for Power Systems** A.T. Johns and S.K. Salman
Volume 16 **Electricity Economics and Planning** T.W. Berrie
Volume 18 **Vacuum Switchgear** A. Greenwood
Volume 19 **Electrical Safety: A guide to causes and prevention of hazards** J. Maxwell Adams
Volume 21 **Electricity Distribution Network Design, 2nd Edition** E. Lakervi and E.J. Holmes
Volume 22 **Artificial Intelligence Techniques in Power Systems** K. Warwick, A.O. Ekwue and R. Aggarwal (Editors)
Volume 24 **Power System Commissioning and Maintenance Practice** K. Harker
Volume 25 **Engineers' Handbook of Industrial Microwave Heating** R.J. Meredith
Volume 26 **Small Electric Motors** H. Moczala *et al.*
Volume 27 **AC–DC Power System Analysis** J. Arrillaga and B.C. Smith
Volume 29 **High Voltage Direct Current Transmission, 2nd Edition** J. Arrillaga
Volume 30 **Flexible AC Transmission Systems (FACTS)** Y.-H. Song (Editor)
Volume 31 **Embedded Generation** N. Jenkins *et al.*
Volume 32 **High Voltage Engineering and Testing, 2nd Edition** H.M. Ryan (Editor)
Volume 33 **Overvoltage Protection of Low-Voltage Systems, Revised Edition** P. Hasse
Volume 36 **Voltage Quality in Electrical Power Systems** J. Schlabbach *et al.*
Volume 37 **Electrical Steels for Rotating Machines** P. Beckley
Volume 38 **The Electric Car: Development and future of battery, hybrid and fuel-cell cars** M. Westbrook
Volume 39 **Power Systems Electromagnetic Transients Simulation** J. Arrillaga and N. Watson
Volume 40 **Advances in High Voltage Engineering** M. Haddad and D. Warne
Volume 41 **Electrical Operation of Electrostatic Precipitators** K. Parker
Volume 43 **Thermal Power Plant Simulation and Control** D. Flynn
Volume 44 **Economic Evaluation of Projects in the Electricity Supply Industry** H. Khatib
Volume 45 **Propulsion Systems for Hybrid Vehicles** J. Miller
Volume 46 **Distribution Switchgear** S. Stewart
Volume 47 **Protection of Electricity Distribution Networks, 2nd Edition** J. Gers and E. Holmes
Volume 48 **Wood Pole Overhead Lines** B. Wareing
Volume 49 **Electric Fuses, 3rd Edition** A. Wright and G. Newbery
Volume 50 **Wind Power Integration: Connection and system operational aspects** B. Fox *et al.*
Volume 51 **Short Circuit Currents** J. Schlabbach
Volume 52 **Nuclear Power** J. Wood
Volume 53 **Condition Assessment of High Voltage Insulation in Power System Equipment** R.E. James and Q. Su
Volume 55 **Local Energy: Distributed generation of heat and power** J. Wood
Volume 56 **Condition Monitoring of Rotating Electrical Machines** P. Tavner, L. Ran, J. Penman and H. Sedding
Volume 57 **The Control Techniques Drives and Controls Handbook, 2nd Edition** B. Drury
Volume 58 **Lightning Protection** V. Cooray (Editor)
Volume 59 **Ultracapacitor Applications** J.M. Miller
Volume 62 **Lightning Electromagnetics** V. Cooray
Volume 63 **Energy Storage for Power Systems, 2nd Edition** A. Ter-Gazarian
Volume 65 **Protection of Electricity Distribution Networks, 3rd Edition** J. Gers
Volume 66 **High Voltage Engineering Testing, 3rd Edition** H. Ryan (Editor)
Volume 67 **Multicore Simulation of Power System Transients** F.M. Uriate
Volume 68 **Distribution System Analysis and Automation** J. Gers
Volume 69 **The Lightening Flash, 2nd Edition** V. Cooray (Editor)
Volume 70 **Economic Evaluation of Projects in the Electricity Supply Industry, 3rd Edition** H. Khatib
Volume 72 **Control Circuits in Power Electronics: Practical issues in design and implementation** M. Castilla (Editor)
Volume 73 **Wide Area Monitoring, Protection and Control Systems: The enabler for smarter grids** A. Vaccaro and A. Zobaa (Editors)
Volume 74 **Power Electronic Converters and Systems: Frontiers and applications** A.M. Trzynadlowski (Editor)
Volume 75 **Power Distribution Automation** B. Das (Editor)
Volume 76 **Power System Stability: Modelling, analysis and control** B. Om P. Malik
Volume 78 **Numerical Analysis of Power System Transients and Dynamics** A. Ametani (Editor)
Volume 79 **Vehicle-to-Grid: Linking electric vehicles to the smart grid** J. Lu and J. Hossain (Editors)
Volume 81 **Cyber–Physical–Social Systems and Constructs in Electric Power Engineering** Siddharth Suryanarayanan, Robin Roche and Timothy M. Hansen (Editors)
Volume 82 **Periodic Control of Power Electronic Converters** F. Blaabjerg, K. Zhou, D. Wang and Y. Yang
Volume 86 **Advances in Power System Modelling, Control and Stability Analysis** F. Milano (Editor)
Volume 87 **Cogeneration: Technologies, optimisation and implementation** C.A. Frangopoulos (Editor)
Volume 88 **Smarter Energy: From smart metering to the smart grid** H. Sun, N. Hatziargyriou, H.V. Poor, L. Carpanini and M.A. Sánchez Fornié (Editors)
Volume 89 **Hydrogen Production, Separation and Purification for Energy** A. Basile, F. Dalena, J. Tong, T.N. Veziroğlu (Editors)
Volume 90 **Clean Energy Microgrids** S. Obara, J. Morel (Editors)
Volume 91 **Fuzzy Logic Control in Energy Systems with Design Applications in Matlab/Simulink®** İ.H. Altaş
Volume 92 **Power Quality in Future Electrical Power Systems** A.F. Zobaa and S.H.E.A. Aleem (Editors)
Volume 93 **Cogeneration and District Energy Systems: Modelling, analysis and optimization** M.A. Rosen and S. Koohi-Fayegh
Volume 94 **Introduction to the Smart Grid: Concepts, technologies and evolution** Salman K. Salman
Volume 95 **Communication, Control and Security Challenges for the Smart Grid** S.M. Muyeen and S. Rahman (Editors)
Volume 97 **Synchronized Phasor Measurements for Smart Grids** M.J.B. Reddy and D.K. Mohanta (Editors)
Volume 98 **Large Scale Grid Integration of Renewable Energy Sources** Antonio Moreno-Munoz (Editor)
Volume 100 **Modeling and Dynamic Behaviour of Hydropower Plants** N. Kishor and J. Fraile-Ardanuy (Editors)
Volume 101 **Methane and Hydrogen for Energy Storage** R. Carriveau and David S.-K. Ting
Volume 104 **Power Transformer Condition Monitoring and Diagnosis** Ahmed Abu-Siada (Editor)
Volume 108 **Fault Diagnosis of Induction Motors** Jawad Faiz, Vahid Ghorbanian and Gojko Joksimović
Volume 110 **High Voltage Power Network Construction** K. Harker
Volume 124 **Power Market Transformation** B. Murray
Volume 130 **Wind and Solar Based Energy Systems for Communities** Rupp Carriveau and David S.-K. Ting (Editors)
Volume 131 **Metaheuristic Optimization in Power Engineering** J. Radosavljević
Volume 905 **Power System Protection, 4 volumes**

Power Transformer Condition Monitoring and Diagnosis

Edited by
Ahmed Abu-Siada

The Institution of Engineering and Technology

Published by The Institution of Engineering and Technology, London, United Kingdom

The Institution of Engineering and Technology is registered as a Charity in England & Wales (no. 211014) and Scotland (no. SC038698).

First published 2018

The Institution of Engineering and Technology
Michael Faraday House
Six Hills Way, Stevenage
Herts, SG1 2AY, United Kingdom

www.theiet.org

British Library Cataloguing in Publication Data
A catalogue record for this product is available from the British Library

ISBN 978-1-78561-254-1 (hardback)
ISBN 978-1-78561-255-8 (PDF)

Typeset in India by MPS Limited

Contents

About the editor

A. Abu-Siada received Ph.D. degree in Electrical Engineering from Curtin University, Perth, Australia, in 2004. He is currently an Associate Professor with the Electrical and Computer Engineering Department at Curtin University. His research interests include condition monitoring, asset management, power system stability and power electronics. He has published more than 200 research papers in his area of expertise. In 2018, he was named top 10 Science and Technology Innovators in Australia.

Preface

Asset efficiency and optimisation has been prioritised on the agenda among utilities around the globe. A large proportion of the budget invested in the electrical energy sector in all countries every year goes to network upgrade and asset replacement. The mindset within the global electricity utility industry is centred on getting the most of existing equipment rather than replacements. According to recent surveys, the average age of worldwide in-service power transformers has reached or already exceeded the proposed transformer design life. As transformer aging increases the likelihood of catastrophic failure, reliable diagnostic and condition monitoring tools should be adopted to continuously monitor the health condition of the transformer, allow taking a timely maintenance action and avoid any potential catastrophic failures. Within the last few decades, several monitoring systems and diagnostic techniques have been developed with a common aim of extending the power transformer operational life and minimise the possibility of catastrophic failure. This book presents the concepts and current industry practice of various popular condition monitoring and fault diagnosis techniques such as dissolved gas analysis, partial discharge, moisture analysis, frequency response analysis and vibration analysis. The book also presents some practical techniques for transformer lifecycle management and remaining operational life estimation. The book will be a very useful material for transmission and primary distribution networks technicians in deeply understanding strategy concepts to maintain asset, optimize planned replacements, and minimize the possibility of catastrophic failures including accidental deaths associated with sudden explosions, fire and environmental hazards due to oil spillage. The book will also benefit postgraduate research students, upper-division electrical engineering students and practicing engineers.

List of acronyms

DGA	dissolved gas analysis
CO	carbon monoxide
CO_2	carbon dioxide
H_2O	water
H_2	hydrogen
CH_4	methane
ASTM D3612-02	standard test method for analysis of gases dissolved in electrical insulating oil by gas chromatography
IEC 60567 ED. 4.0	oil-filled electrical equipment – sampling of gases and analysis of free and dissolved gases – guidance
GC	gas chromatograph
ppm	parts per million
ASTM D923-15	standard practices for sampling electrical insulating liquids
IEEE C57.104	guide for the interpretation of gases generated in oil-immersed transformers
IEC 60475	method of sampling insulating liquids
IEC 60567	oil-filled electrical equipment – sampling of gases and analysis of free and dissolved gases – guidance
OLTC	on-load tap changer
C_2H_6	ethane
C_2H_4	ethylene
C_2H_2	acetylene
O_2	oxygen
N_2	nitrogen
PD	partial discharges
D1	discharges of low energy
D2	discharges of high energy
T1	thermal faults below 300 °C
T2	thermal faults between 300 °C and 700 °C
T3	thermal faults above 700 °C
S	stray gassing below 200 °C
C	carbonisation of paper

DETCs	de-energised tap changers
O	overheating with temperature less than 250 °C
SF_6	hexafluoride
HV	high-voltage
LV	low-voltage
TEAM	combined effect of thermal, electrical, ambient and mechanical stresses
CBM	condition-based maintenance
DP	degree of polymerization
RH	Rydberg constant
VHF	very high frequency
UHF	ultra high frequency
qapp	apparent charge
CA	capacitance of the healthy part of the insulation in parallel with the defect
CB	capacitance of the healthy part of the insulation in series with the defect
CC	capacitance of the defect
tPD	the time in which the PD take place
EUT	equipment under test
DC	direct current
f	frequency
Z	impedance
WT	wavelet transform
HFCT	high-frequency current transformers
RC	Rogowski coil
ILS	inductive loop sensor
M	mutual inductance
fc	cut-off frequency
PRPD	phase-resolved PD
GAs	genetic algorithms
PSO	particle swarm optimization
TDOA	time difference of arrival
SVMs	support vector machines
ANN	artificial neural networks
PDIV	PD inception voltage
XLPE	cross-linked polyethylene
PRL	power ratio for low-frequencies

PRH	power ratio for high-frequencies
teff	effective pulse time-width
EMN	electromagnetic noise
Sk	skewness
μ	average or mean:
σ^2	variance
Ku	kurtosis
cc	cross-correlation factor
ϕm	mean phase angle
P(q)	probability that a discharge occurs
$K(X_i,X)$	kernel functions
SNR	signal-to-noise ratio
APC	affinity propagation clustering
RBF	radial basis function
RS	relative saturation
Ws_oil	oil saturation limit
T_K	oil temperature
Ar	total aromatics content
Ce	concentration of water in the insulation under equilibrium
KF	Karl Fischer
Aw	water activity
ε_0	vacuum permittivity
χ	susceptibility
$D(t)$	electric displacement
$f(t)$	dielectric response function
RVM	recovery voltage method
PDC	polarisation and depolarisation currents
FDS	frequency dielectric spectroscopy
k	Boltzmann's constant
Ea	activation energy
td	discharging time
RS	relative saturation
Psat	saturation pressure of water vapour
Wc	amount of water in the cellulosic insulation
Ws_c	amount of water in saturation conditions in cellulosic insulation
Ws_oil	amount of water in saturation conditions in oil

PT	power transformer
ABM	arithmetic-Brownian motion
TUP	thermal upgraded papers
θHS	hot-spot temperature
SCI	short-circuit impedance
FRA	frequency response analysis
OFRA	online FRA
Ish	short-circuit current
LVI	low-voltage impulse
FFT	fast Fourier transform
$fsweep$	variable frequency
LFs	low frequencies
HFs	high frequencies
$\Re$	core reluctance
SD	standard deviation
RXY	relative factor
MF	medium frequency
B	magnetic flux density
J	current density vector
DDF	dielectric dissipation factor
WCO	water content in oil
WCP	water content in paper
$MAMD$	mean absolute magnitude distance
$MAPD$	mean absolute phase distance
CDC	current deformation coefficient
HF CT	high-frequency current transformer
H	magnetic field intensity
SCADA	supervisory control and data acquisition
RMS	root mean square
AC	alternating current
TCP	tap-changer position
ODWF	oil directed water forced
GIC	geomagnetically induced currents
HVDC	high-voltage direct current
ΦDC	magnetic flux offset

OECD	Organisation for Economic Co-operation and Development
IRENA	International Renewable Energy Agency
PV	photovoltaic
RCM	reliability centered maintenance
SME	subject matter expert
IEEE Std. C57.91-2011	Guide for Loading Mineral-Oil-Immersed Transformers and Step-Voltage Regulators
2-FAL	2-furfural
AI	artificial intelligence
HI	health index
IFT	interfacial tension
MF	membership function
M/DW	moisture by dry weight

Chapter 1
Dissolved gas analysis, measurements and interpretations

Carlos Gamez[1]

1.1 Introduction

Oil-filled power transformers were first conceived and manufactured in the mid-to-late-1800s [1,2]. They were (and to this day are) an essential component of the electrical power transmission and distribution infrastructure.

Power transformers perform a power conversion function. A transformer takes electrical power of certain voltage and current characteristics through its primary terminals and delivers this power on its secondary terminals with changed voltage and current levels. By increasing voltage, and thus reducing the current level, power transformers minimise losses and enable the economical transmission of electrical power over long distances.

While performing this primary function, the core and coils (or active part) of a power transformer produce a certain amount of losses which generate heat. The function of the cooling system is to remove this heat at a steady and controlled rate in order to maintain an acceptable temperature differential between the active part and its surrounding medium.

Additionally, the active part and its components are subject to a wide range of voltage stresses. The function of the insulation system is to keep these stresses below the desired maximum stresses, both under normal and abnormal operating conditions. This is accomplished by the careful selection and arrangement of the materials that comprise all active components of the transformer.

The engineers that first designed and manufactured transformers in the latter part of the nineteenth century realised that they could improve the cooling and electrical insulating performance of those early power transformers by immersing their active parts in mineral oil. This gave rise to the liquid-filled transformers of our days.

Mineral oil, as well as other types of liquids, have excellent dielectric insulating properties. Mineral oil provides a great insulation medium when used to

[1]Engineers Tools, Australia

impregnate the Kraft paper typically used to wrap winding conductors and other elements of the windings.

The capacity to remove heat from the active part is also improved by using a liquid instead of a gas (such as air) as the medium that surrounds the active part.

A secondary benefit of using a liquid insulating medium can be realised by the fact that this liquid is in direct contact with all active elements inside the transformer. A great deal of knowledge about the condition of the transformer can be inferred by analysing certain properties of the insulating liquid.

In this chapter, we will discuss why this is the case and how we can use this behaviour to the advantage of the person or team in charge of operating and maintaining liquid-filled power transformers.

Assessing the condition of power transformers has the main objectives of determining whether a particular transformer displays an abnormal behaviour and, if so, identifying what type of failure mode is the cause of this behaviour.

For the asset owner of a power transformer fleet, the main value of having a dissolved gas analysis (DGA) program rests in gaining the capability to detect failure modes which, if gone undetected, would otherwise increase the operational risk of that fleet to unacceptable levels. By detecting issues and opportunely reacting to them, an organisation can mitigate the probability of facing the consequences (financial, safety, environmental, reputational, etc.) of major transformer failures.

The field of study created by power transformers is vast and rich. There is an abundance of generalised behaviours as well as individual cases and experiences. In writing this chapter, I have tried to substantiate all statements with adequate references. However, it is inevitable that some of my own opinions, shaped by 20 years working in this field, have made it through the various sections in this chapter. I hope the reader does not take this as a lack of rigour, but rather as a humble attempt to provide him or her with a valuable collection of knowledge in this field, both past and present.

1.2 Insulating liquids

The relevant laws of physics have not changed since those early days in the nascent electric power industry. For transformer manufacturers, the primary way to gain a competitive edge in the market has been through advancements in materials engineering and improved accuracy in the calculations used to predict their expected behaviour. Over the years, manufacturers and academic researchers have developed and commercialised new materials for each major transformer component and insulating liquids are no exception.

Mineral oil was the first liquid to be used for this purpose and it remains as the most popular option to fill liquid immersed transformers.

Progressively, other types of liquids have been developed and commercialised. Some were introduced and then removed from the market due to health, safety and environmental concerns, like the liquids containing an organic compound known as polychlorinated biphenyl.

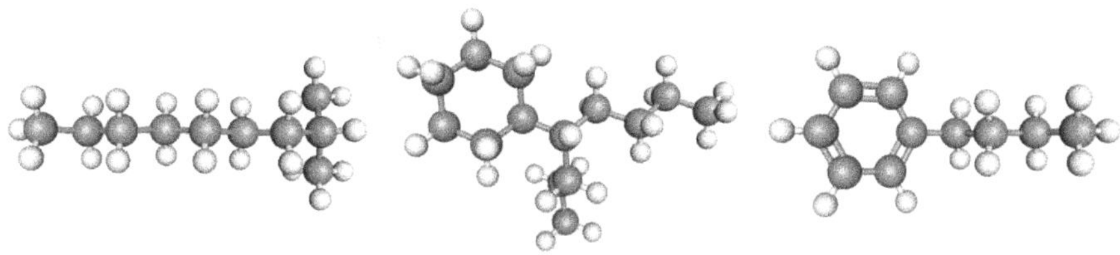

Figure 1.1 Schematic representation of the three main types of structures in transformer oil

Others, like the liquids based on oils extracted from plants and vegetables (esters), are currently trying to find their way into mainstream use by highlighting their biodegradability and higher flash point.

Since mineral oil is the most commonly used fluid in the industry, we will focus on this type of fluid for our discussion in this chapter.

1.2.1 Mineral oil

Crude oil undergoes various refining processes to produce transformer insulating oil. Crude oils can be classified in two main groups, paraffinic and naphthenic, depending on the compounds that constitute them. Transformer oil is typically produced by the following processes: distillation, dewaxing, extraction and hydrogenation. A detailed description of these processes is beyond the scope of this chapter. Interested readers can consult the additional sources listed in the "References" section to gain a deeper understanding of oil refining processes [3].

While the actual composition of the oil molecules is relatively complex, three main types of structures influence the properties of transformer oils. These three structures are shown in Figure 1.1.

Transformer oil is considered naphthenic or paraffinic depending on which structure is more prevalent for a particular oil. In general, if the oil contains between 56% and 65% of carbon-bonded paraffin, the oil is considered paraffinic. If it contains between 42% and 50% of carbon-bonded paraffin, the oil is considered naphthenic. Oils in between these percentages are considered intermediate [3]. The content of these compounds influences the various properties of the oil including, oxidation stability, viscosity, temperature stability, gas absorption, dielectric properties, etc.

The properties that new oils have to meet are established in either national or international standards [4].

1.3 The transformer as a chemical reactor

In the context of DGA, the transformer can be thought of as a chemical reactor.

Inside the tank there are numerous organic and inorganic compounds that can engage in chemical reactions – copper from the windings and steel from the core

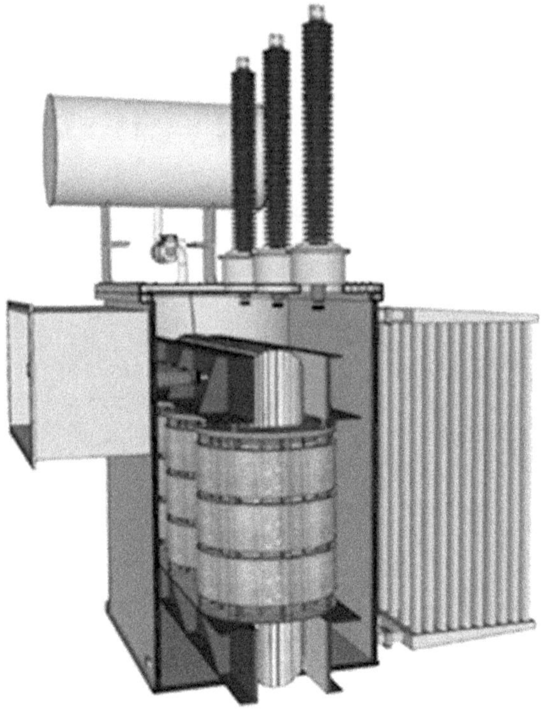

Figure 1.2 Internal view of typical power transformer

and clamping structures; hydrocarbons from the insulating oil; and cellulose from the paper, insulating boards and blocks (Figure 1.2).

Also, there are sources of energy in the form of the losses generated during normal operation as well as the energy contributed by abnormal conditions and faults.

This mix of energy and reactive compounds can be visualised as a chemical reactor where the energy source drives the various reactive materials into decomposing and recombining to form new chemical by-products. Since these by-products are measurable and quantifiable, we can aim to find useful correlations between the energy injected into the system and the chemical behaviour of the components that make it up.

These correlations represent a powerful tool in the assessment of the condition of the transformer. For the purpose of diagnosing the condition of the transformer we are interested in two main components of the insulation system, the liquid and the solid insulation. The liquid insulation comprises oil, and the solid comprises paper and pressboard.

Each of these elements contributes in very specific ways towards the formation of gases. We will see how in the next section.

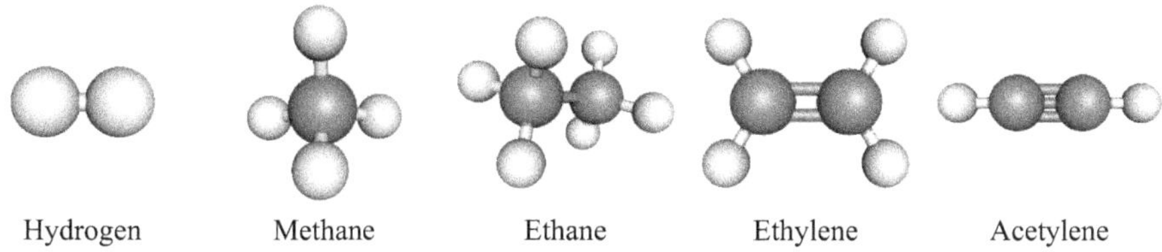

Figure 1.3 Main hydrocarbon gases produced by the oil

1.3.1 Gas production mechanisms

As mentioned above, the oil comprises hydrocarbon molecules. As early as 1919 [5], researchers noticed that abnormal conditions, such as disruptive discharges under the oil, produced certain amounts of hydrocarbon by-products.

The heat produced by the losses and certain types of failure modes as well as the energy dissipated by electrical discharges contribute to the breakdown of the oil molecules. The main by-products of these processes are shown in Figure 1.3.

Each type of chemical bond requires a different level of energy to break or join [6,7]. The energy required to break or make single bonds of hydrogen–hydrogen atoms present in the hydrogen molecule is about half of the energy required to form the triple carbon–carbon bond present on the acetylene molecule. As these by-products are produced, they remain dissolved in the oil up to their solubility limits, after which they would evolve as free gases (bubbles).

The fact that these gases are produced at distinct energy levels has been crucial to our ability to infer the type and intensity of the mechanisms that produce them. Efforts to use this behaviour for diagnosis purposes were underway in the early 1970s [8].

By studying the amount, proportion and evolution rates of these various gases dissolved in the transformer oil, engineers have been able to correlate them to known failure modes and their severity levels.

These are not the only gases or by-products produced by the chemical reactions inside the transformer. As we pointed out before, other than the hydrocarbons present in the oil, there are additional materials present in the transformer that also interact during these chemical reactions.

One of the most important components of the insulation system, aside from the oil, is the solid insulation system. This insulation, comprising paper, pressboard and other solid materials, is made up of cellulose fibres. Cellulose is an organic compound made up of long chains of the same repeating building block as depicted in Figure 1.4.

The black coloured atoms in Figure 1.4 are of oxygen. The same energy that causes the oil to breakdown will have a similar effect on the paper. At a molecular level, all cellulosic materials are made up of long chains called polymers.

Another contributor of oxygen is water. A brand new transformer that has been properly manufactured and assembled will tend to have less than 0.5% of residual moisture per dry-weight of insulation throughout the insulation system after it is

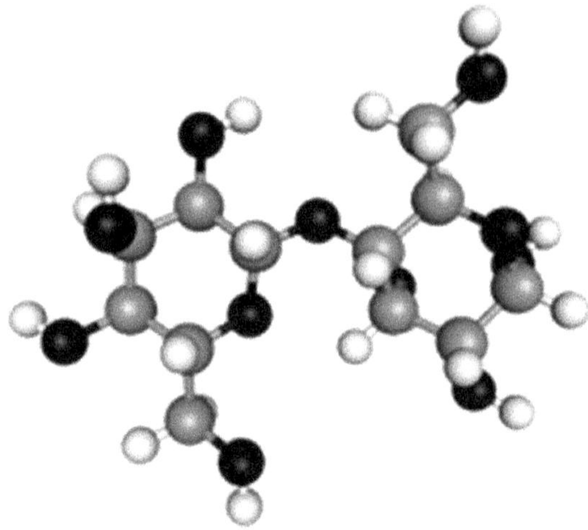

Figure 1.4 Cellulose molecule

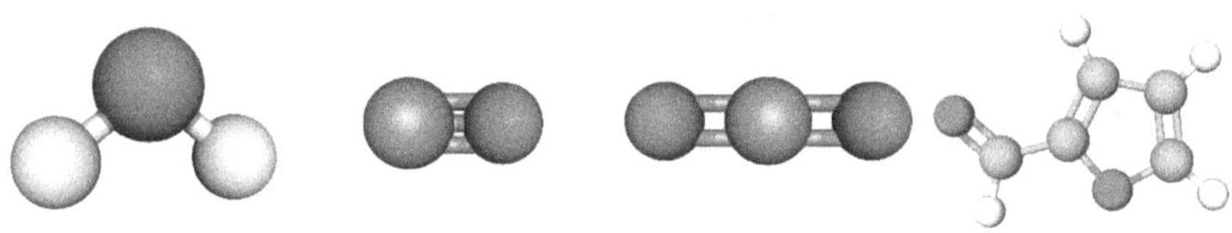

Figure 1.5 By-products of cellulosic and hydrocarbon decomposition in the presence of oxygen

commissioned. However, throughout the life of a transformer it would not be uncommon that additional water enters the system via improperly maintained breathing systems or oil leaks.

Finally, in some cases, such as in free-breathing transformers, oxygen enters via the normal "breathing" process of the oil expanding and contracting due to temperature changes.

For all these oxygen-containing compounds, increases in temperature, either through normal losses or during abnormal events or faults, cause the C–O molecular bonds to break up which releases oxygen, carbon and hydrogen into the oil. These elements recombine to form carbon monoxide (CO), carbon dioxide (CO_2), water (H_2O) and a range of compounds collectively known as furans. It is worth pointing out that neither water nor furans are part of a gas analysis and they are measured using other methodologies beyond the scope of this chapter (Figure 1.5).

In more recent years [9], a gas production mechanism known as "Stray Gassing" has been characterised. Stray gassing is recognised as the production of gases within a temperature range of 90 and 200 °C. Experiments [9] have shown that this mechanism mostly produces hydrogen (H_2) and methane (CH_4). The amount of these gases produced has a correlation with the specific brand and type of oil used in each particular transformer.

While more specific mechanism of gas production is being researched [10], the above mechanisms provide a good overall picture of how oil, paper, water, oxygen and energy interact to produce them.

In summary, we can see that these gas production mechanisms hold a correlation to the chemical elements present inside the transformer and the levels of energy available to enable these chemical reactions to occur. As we will see in the next sections, the transformer engineering community has developed several interpretation techniques that take advantage of this behaviour.

1.4 Oil analysis

In order to extract useful information from the gas production mechanisms explained in the previous section, the dissolved gases in the oil have to be extracted and quantified. As it might be expected, the results of the oil analysis are critical to the correct interpretation and diagnosis of the condition of a particular transformer. It is important that the results available for analysis are representative of the oil inside the transformer and as such, it is expected that they are repeatable and reproducible.

These terms, repeatable and reproducible, have specific meanings in the context of oil analysis. When results are repeatable, it means that a particular laboratory or analyst will produce the same results if the same sample is repeatedly analysed. On the other hand, results are reproducible when two different laboratories or analyst are capable of producing the same results on a given sample.

In practice, achieving perfect reproducibility and repeatability, in other words, zero variation between tests and testing entities, is not practical. Understanding the capacity of the oil laboratory to produce repeatable and reproducible results is important for the end user and ultimately the interpreter of these results.

While each laboratory has to be evaluated by its potential clients, in general, laboratories ensure good quality repeatable and reproducible results by

- Adhering to well-known and reliable test methodologies,
- Regularly verifying the correct calibration of instruments,
- Obtaining and maintaining accreditation by a recognised assessment authority and
- Participating in international round robin tests.

The specific test methodology, used by a particular laboratory, typically depends on the country where the laboratory is located. The majority of modern laboratories around the world analysing transformer oils use the gas chromatography technique. Two of the most well-known standards that prescribe how this technique should be applied are as follows:

- ASTM D3612-02(2009) – Standard test method for analysis of gases dissolved in electrical insulating oil by gas chromatography [11].
- IEC 60567 ED. 4.0 – Oil-filled electrical equipment – sampling of gases and analysis of free and dissolved gases – guidance [12].

1.4.1 Gas chromatography

In general, chromatography is a technique used to separate a mixture of compounds into its constituent parts. It was originally developed in Russia by the botanist Mikhail S. Tsvet in 1900. While researching plant pigments, Tsvet filled a liquid-adsorption column filled with a calcium carbonate as adsorbent. Then he used a mix of ethanol as eluent. As the plant pigments, such as chlorophylls and carotenoids, separated in the column, they formed bands of various colours in the column. This separation of the original mix into coloured bands gave the technique its name: chromatography from the Greek *chroma* (colour) and *graphein* (to write).

More than a 100 years later, the technique we use to analyse dissolved gases in transformer oil is, in principle, based on Tsvet's original work. The technique is called gas chromatography and it is performed with an instrument conveniently named gas chromatograph (GC).

As shown in Figure 1.6, in the case of gas chromatography, a gas sample of unknown composition is injected into the instrument which contains a capillary column filled with a liquid or a polymer which acts as the stationary phase. A carrier gas, typically an inert gas such as helium or nitrogen, is used as the mobile phase. As the mobile phase moves through the stationary phase in the column, the various compounds within the sample interact with the stationary phase causing them to separate and move at different speeds through the column. Each compound would therefore take a different amount of time to travel through the column which is known as the retention time. As the various compounds arrive to the end of the column, a detector or detectors will generate a signal that is recorded on a computer.

The retention time then relates to each individual compound while the area under the curve of each signal is proportional to the amount of that compound in the original mixture.

In order to be able to identify each of the compounds, the laboratory calibrates the GC by running a known mixture of the gases. This provides the laboratory with the necessary data (i.e. retention times and areas) that enable the identification of

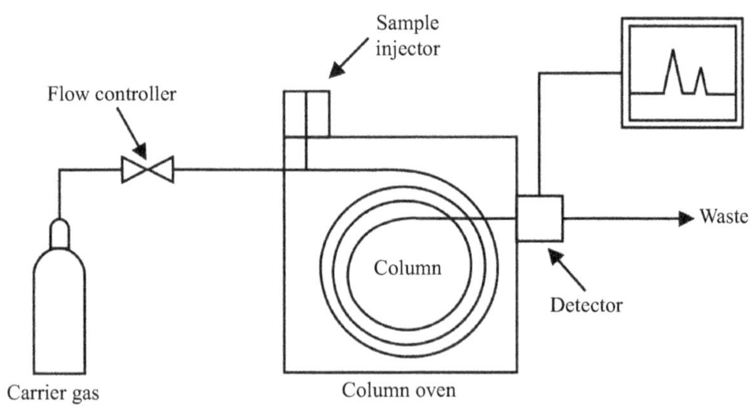

Figure 1.6 Simplified representation of how a gas chromatograph works [13]

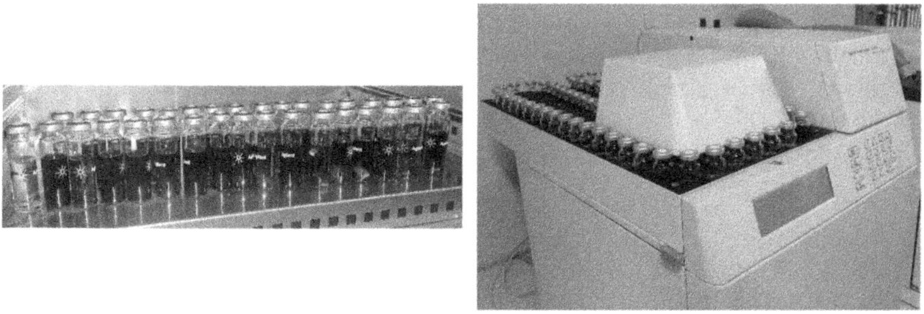

Figure 1.7 Examples of Method C vials and auto-sampler

compounds and calculation of their quantity in a sample with a high level of accuracy.

In order to apply the methodology described above, the first step is to separate the dissolved gases from the oil sample. There are various techniques available to the laboratory to perform this task. In general, these techniques involve increasing the temperature of the sample, mechanically agitating the sample and in some cases apply vacuum in order to extract the gases.

For example, ASTM-3612 [11] lists three different methods, namely A, B and C, that the laboratory can use to extract the gases from the oil.

Method A uses a set of glassware and vacuum to extract the gases. For a schematic representation of the setup used in this method, the interested reader can refer to Figures 1.1 and 1.2 of the aforementioned standard [11].

Method B uses a stripping column that contains a high surface area bead. The oil is run through this column which strips the gases from it and directs the gases to the analysis instrument for analysis.

One of the most common methods currently in use is Method C, which is called Headspace Sampling. This method involves placing the sample in a sealed vial in which the sample is brought in contact with a gas phase (headspace). A portion of the dissolved gases in the oil migrate into the headspace. Using mass equivalence equations, the analyst determines the proportion of gases in the oil and headspace. This method lends itself to automation of the gas extraction and analysis processes, which makes the laboratory more efficient and able to reduce lead times for analysis. Figure 1.7 shows an example of the vials typically used to contain the oil in this method as well as the auto-sampler used to extract the gases.

The results of a gas chromatography are the concentration of each of the gases of interest and reported in either μl/l or ppm.

1.5 Oil sampling

The analysis performed at the laboratory relies on obtaining a representative sample of the transformer's oil. In this sense, the correct execution of the oil sampling process is an essential first step in a successful analysis and therefore interpretation of the condition of a particular transformer. Even the most careful laboratory analysis cannot fix the negative impact of a poor quality sample.

In regards to sampling, there are also useful guidelines and industry best practices published by the main technical bodies around the world. A few examples are as follows:

- ASTM D923-15 – Standard practices for sampling electrical insulating liquids [14]
- IEEE C57.104-2008 – Guide for the interpretation of gases generated in oil-immersed transformers [15]
- IEC 60475:2011 – Method of sampling insulating liquids [16]
- IEC 60567:2011 – Oil-filled electrical equipment – sampling of gases and analysis of free and dissolved gases – guidance [12]

There are also good references [17] available on the internet which present practical demonstrations of how to appropriately sample transformer oil.

In order to obtain a representative sample every time one is taken, it is preferable to select a sampling location that can be consistently accessed over the expected life of the transformer. This provides a good level of consistency and comparability of samples of the same transformer over time.

The most common location for routine oil sampling for condition monitoring purposes is the bottom drain or sampling valve (see Figure 1.8).

Another typical location used to regularly sample oil is the sampling valve of the Buchholz relay. When the transformer is equipped with one of these valves at ground level, they are useful for sampling oil, but particularly used for sampling gas in the event of a Buchholz gas accumulation alarm or trip.

Other valves around the transformer can also be used, such as top filling valves, but not in every transformer this location is easily accessible during routine inspections and sampling while the transformer is energised.

Figure 1.8 Example of bottom drain and sampling valves

Figure 1.9 Examples of Buchholz sampling devices

Figure 1.10 Preparation of oil sampling area

Figures 1.8 and 1.9 provide some examples of these typical sampling locations.

The two most popular sampling containers are syringes and bottles. A summary of what each method entails is given as follows.

Before any sampling takes place, it is important to prepare the area where the sample is to be taken in order to avoid any oil spillage as shown in Figure 1.10.

It is also important to keep in mind that the sampling process relies on the positive pressure of the oil at the point of sampling and as such it is generally not recommended to attempt to take an oil sample while the transformer tank is under vacuum (or negative pressure). Some countries and some companies have developed specialised tools and procedures to enable sampling from transformers under vacuum but this is beyond the scope of this chapter.

Before taking a sample, it is important to properly identify the sample. Most laboratories will supply the correct sampling labels and paperwork. Common data elements required by the laboratory are the sample id, transformer serial number, date, time, temperatures and other factors relevant to the sample identification. The oil laboratory relies on this identification in order to associate the results to the correct transformer. An improperly identified sample makes this task difficult and sometimes impossible.

From the point of view of interpretation of the results, incorrect sample identification is sometimes the root cause of atypical behaviours such as sudden increases or decreases of the otherwise expected gas evolution in the sampling history of a particular transformer.

1.5.1 Bottle sampling

A well-known and traditional sampling method still in use in many countries is to use a bottle or can as a container. The bottle can be a glass bottle or a stainless steel can. This method is easy to execute.

Various guidelines [12,16] describe this procedure with minor variations. In order to obtain a good sample it is important to

- Ensure the cleanliness of the sampling port. In some companies the sampling valves of the transformer are kept capped or blanked-off and a sampling port is attached to them during the sampling process. In others, a sampling port is permanently attached to the valve. In either case, it is important that the sampling port has been properly cleaned and flushed in order to avoid cross-contamination or undue influence from environmental factors. Examples of various sampling ports are shown in Figure 1.11.
- Ensure the sampling bottle is also clean and adequately flushed before taking the sample.
- When filling the bottle with the sample, regulate the flow of oil to a smooth stream without turbulence and foaming as this promotes capturing atmospheric gases and moisture. Figure 1.12 depicts a sample being taken in an ambler glass bottle.

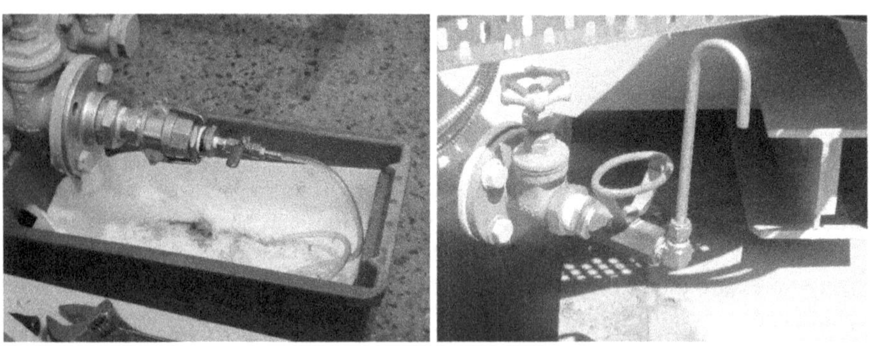

Figure 1.11 Examples of temporarily (left) and permanently fitted (right) bottle sampling ports

Figure 1.12 Example of sampling oil in a bottle

- After the sample has been taken, ensure the bottle cap is correctly tightened and clean the bottle from any oil residue. The sample can now be dispatched to the laboratory for analysis.

While properly taken bottle samples will provide acceptable results, this method has a drawback that could affect the repeatability of the results.

In using this method the oil sample has to be exposed to the atmosphere at least in two instances. Once when the sample is being taken and the oil being introduced to the bottle and a second time when the oil is being prepared for analysis at the laboratory and the chemist has to open the bottle to extract the sample, typically using a syringe, to introduce it to the testing instrument.

In general [12], it is recognised that sampling using a syringe avoids this problem and provides a higher quality sample and more consistent results. This method is described below.

1.5.2 Syringe sampling

In this method, a sampling port with an oil-proof tubing has to be attached to the transformer valve as shown in the left image of Figure 1.11.

A gas-tight syringe is used as the sampling container and it's attached to the sampling tubing through a 3-way stopcock valve. This valve helps one to direct the oil during the flushing and filling of the syringe. Figure 1.13 depicts a sample being taken with a syringe.

A 3-way valve blocks one port and communicates the remaining two. In this technique, the 3-way valve is used as follows:

- The valve is initially placed on a position that allows the flow of oil from the transformer valve into the open port, which is discarded on the drip tray or waste container.
- The transformer is then slowly opened until a smooth and steady stream of oil is observed. The flow of oil should not be turbulent, foam or form bubbles.

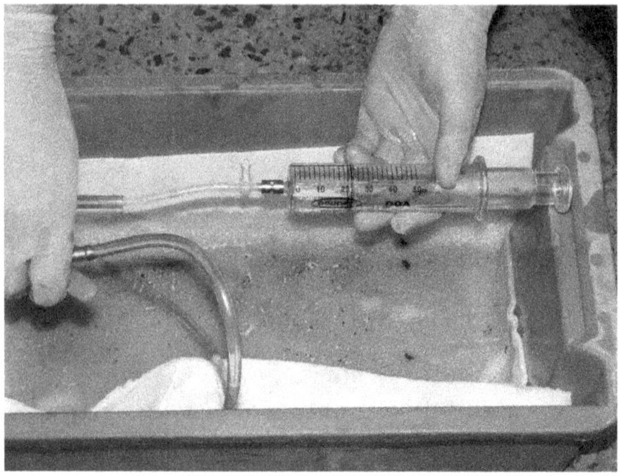

Figure 1.13 Example of sampling with a syringe

- The 3-way valve is now moved to a position that allows the flow of oil from the transformer valve into the syringe. The oil will slowly fill the syringe and push the plunger back.
- When the syringe is full, the 3-way valve is then moved to a position that allows the oil contained in the syringe to flow through the open port and therefore the waste container.
- The plunger is then gently pressed to expel the oil from the syringe and discard it.
- The three previous steps are repeated until a clean, bubble-free sample is obtained in the syringe.
- At this point, the 3-way valve is the moved to its final position in which it communicates the transformer valve to the open port, thereby closing the syringe.
- The transformer valve is then closed and the syringe removed from the sampling tubing.

As can be seen from the steps described above, the oil being sampled is never exposed to the atmosphere. It goes from the transformer directly into the syringe. Similarly, when the syringe reaches the laboratory, the sample is injected into the analysis instrument or vial directly from the syringe.

With an adequate and representative sample in hand and a good quality analysis at a reputable laboratory, the final task is to analyse the results and use them to understand the condition of the transformer. We will delve into the various techniques available in the following section.

1.6 Interpretation techniques

As mentioned in Section 1.3.1, the fundamental principle used to infer the condition of the transformer relies on the correlation between gas generation rates and proportions and the energy that produces them.

Since DGA started being used as a diagnosis technique, a number of patterns and associations have been found, documented and used as diagnosis tools and guidelines.

While there are a variety of interpretation techniques and guidelines, in general they rely on one or more of the following factors:

- The gas profile of a particular sample, typically the latest one.
- The relationship (ratios) between concentrations of different gases in a given sample.
- The rate of change of these concentrations between samples taken at different times.

In almost all cases, the methods attempt to find and establish statistically significant correlations between the observed gas behaviour and particular modes of operation or failure of the transformers in that population. A sense of what is the "normal" or expected gas behaviour of a certain population is achieved by analysing the patterns of transformers without any known active failure modes [18,19]. On the other hand, by selecting and categorising known defects and observing their historical trends, it is possible to establish some general guidelines about gassing behaviours that correspond to particular failure modes [9,20].

Transformers are complex pieces of equipment and their construction varies from unit to unit. Factors such as design and manufacturing practices can have an impact on the behaviour of any particular unit.

While the correlations found for a certain population of transformers are useful to gain generic insights and derive broad interpretation guidelines, an in-depth knowledge of the particular characteristics of the transformer population being analysed will result in more accurate and confident interpretations and condition assessments.

The reader might have heard before that the interpretation of gases dissolved in transformer oil is both an "art" and a "science". As the person using DGA to monitor the condition of power transformers develops expertise and familiarity with a particular fleet or model of transformers, his or her ability to further analyse the fleet's data, correlate with specific failure modes and gain additional insights becomes more granular and specific.

Furthermore, DGA is only one of the tools in the toolbox necessary to produce a meaningful assessment of a transformer's condition. In order to arrive at an actionable conclusion, the diagnosis has to be looked at from multiple angles and using all available data including electrical tests [21], oil quality tests [22], external and internal inspections, paper condition [23], components condition [i.e. on-load tap changer (OLTC), bushings, etc.], known design or manufacturing issues, loading, ambient conditions, undue electrical stresses (i.e. transients and through-faults), overloading, etc.

Analysing all this information in a comprehensive manner is the "art" component of dissolved gas interpretation that engineers in this industry often refer to. Given the variability in the available data as well as the transformers' design and construction, the publicly available guidelines tend to document more universal and broader behaviours. While the overall physics and statistical behaviours, in other words the "science", of gas production is well documented and understood, the more precise and reliable assessments are produced by years of experience in this field and careful consideration of all condition indication factors, the "art".

In the following sections, we will discuss each of the techniques available in the current literature in more detail. In this context we are using the term "gas profile" to refer to all the gases analysed in any given sample. The gases reported, and therefore a profile, consist of

- Hydrogen (H_2)
- Methane (CH_4)
- Ethane (C_2H_6)
- Ethylene (C_2H_4)
- Acetylene (C_2H_2)
- Carbon monoxide (CO)
- Carbon dioxide (CO_2)
- Oxygen (O_2)
- Nitrogen (N_2)

Although not strictly part of the DGA, the oil analysis reports often include results for other oil quality tests such as moisture, interfacial tension, acidity and dielectric breakdown. In this case we'll focus on the interpretation of the DGA component. The "References" section of this chapter points to additional references [22] in the interpretation of this additional oil analysis information for the interested reader.

1.6.1 Fault types

Regardless of the interpretation method being utilised, all techniques are aimed at determining what, if any, failure mode is present in a transformer and its severity.

Liquid-immersed transformers can experience a range of failure modes. Based on their root origin, these failure modes can be categorised into the following broad groups:

- Electrical faults
- Thermal faults

As discussed, these faults will generate characteristic gas profiles depending on their intensity and their location. There is also documentation in the available literature [7,24] of unusual or non-fault causes of gassing, such as catalytic chemical reactions amongst the materials that comprise the internal components of the transformer. However, the two classes of faults mentioned above comprise a large proportion of the cases observed in transformers in service.

The current consensus [7,9,20] is that the common types of faults detectable by DGA are as follows:

- Partial discharges (PD)
- Discharges of low energy (D1)
- Discharges of high energy (D2)

- Thermal faults (T1) below 300 °C
- Thermal faults (T2) between 300 and 700 °C
- Thermal faults (T3) above 700 °C
- Stray gassing (S) below 200 °C
- Carbonisation of paper (C)
- Combinations of the above

Depending on the actual location of the fault within the transformer, it might or might not involve the solid insulation (i.e. paper, pressboard, etc.).

In some cases, there might be faults of more than one type in the same transformer, which should be taken into account when performing the diagnosis on these units. In some other cases, as the faults evolve and become increasingly severe, they might evolve from one category to another.

1.6.1.1 Fault evolution

An important characteristic of using DGA to monitor the condition of transformers is that in a good portion of the cases, the technique is sensitive enough to provide advanced warning (in many cases years) of incipient failures developing in transformers. This affords the organisation that owns this transformer to investigate and correct the issue in a planned and efficient manner rather than reacting to an after-the-fact fault which would typically cause major disruptions in the electrical network.

It is also worth noting, however, that there are cases where certain faults develop in a matter of days and sometimes even hours [25]. Albeit a rather small percentage of all the cases observed, in transformers that are sufficiently critical, this might prompt shorter sampling intervals, with online monitors providing the highest sampling and analysis frequencies, and therefore the earliest warnings, currently available.

Further to the information provided by a single analysis, the capacity of DGA to track the rate of growth of gases over time is a very valuable contribution to the condition monitoring and assessment of power transformers. Further to the awareness, afforded by one particular sample, of the presence of a failure mode, the DGA interpreter can gain additional insights about the intensity and expected development of that failure mode by understanding the evolution of gases in that unit.

This makes DGA an ideal tool in the condition-based power transformer management toolbox of an organisation.

As an example, Figures 1.14 and 1.15 provide a view of the evolution of a common failure mode in de-energised tap changers (DETCs). Note how the intensity of the fault increases over time. These have a correlation with the observed gassing pattern detected in the DGA. These types of faults evolve in a predictable manner to the extent that they can be identified at very early stages of development, providing the asset owner with enough time to address the issue in a controlled way.

1.6.2 Techniques that rely on the gas profile

The first group of techniques uses the concentration of each individual gas present in the sample. The principle of these techniques is that a transformer operating under a particular condition will produce a characteristic profile of gases.

Figure 1.14 Stages of worsening DETC contact failure

Figure 1.15 Failure at crossover inside a coil

As mentioned before, throughout the history of DGA, researchers [8–10,26,27] have found that the profile of each sample has a correlation to the type of condition that generated those gases. In this group, the most commonly known interpretation methods are as follows:

- Key gas analysis [15,28]
- Condition classification [15]
- Typical concentrations [7]

The commonality shared by these techniques is that they aim at using the composition of a particular sample to try to determine the failure mode that this profile is related to.

1.6.2.1 Key gas analysis

The key gas analysis technique relies on the emergence of a particular gas as the main indicator of certain failure modes. The predominance of a specific gas over the others is correlated to a mechanism that produces this particular gas in larger quantities than others.

Key gas: acetylene (C_2H_2)

The presence of acetylene, particularly in a large proportion relative to other gases, correlates with the high energies that can be generated by electrical arcs.

Acetylene is often accompanied by hydrogen and lower amounts of other gases. If solid insulation (i.e. paper and pressboard) is involved in the failure, it will typically also generate a considerable amount of carbon monoxide and carbon dioxide.

An example of a failure mode that can generate acetylene is shown on the right-most image in Figure 1.14. At the latter stages of contact degradation there is typically a considerable growth in the concentration of acetylene. This intense heat also leads to the carbonisation (coking) of oil surrounding the failure area.

Another example is the failure shown in Figure 1.15. In this case, the failure occurred between two strands of the same turn in this coil. As an interesting note, this transformer passed the turn ratio test without problems. The issue was detected purely due to its gassing behaviour, which was rich in acetylene.

As discussed before, this is where experience with a specific populations, models and issues comes in handy when making interpretations. One caveat with the interpretation of acetylene is that in some instances, the presence of acetylene does not always mean a failure is present in the transformer. Some transformers with OLTCs, either by design or due to sealing problems, are exposed to oil cross-contamination between the OLTC and main tank compartments. When this happens, acetylene that is produced during the normal operation of the OLTC finds its way into the main transformer tank and being picked up during normal oil analysis.

Although it might be difficult to discriminate whether the observed acetylene is due to an actual fault or due to OLTC cross-contamination, as described in some of the guidelines [7], the gas ratios produced by normal operation of OLTCs are characteristic enough that can be used to discriminate between D1 faults in the transformer and OLTC operation ratios. As per [7], if the ratio of acetylene to hydrogen is larger than 2–3, it can be indicative of OLTC cross-contamination.

Key gas: ethylene (C_2H_4)

Large quantities of ethylene are typically produced when the oil is exposed to medium-to-high temperatures. This is often accompanied by smaller quantities of hydrogen (H_2) and methane (CH_4).

When there is no evidence of paper decomposition, this type of fault might be occurring due to the oil exposure to a high-temperature component.

This situation can be found on lead structures, connections or joints that are directly exposed to the oil. An example of such type of connection that presented this type of failure mode is shown in Figure 1.16.

Key gas: carbon monoxide (CO)

Thermal failure modes that directly affect the solid insulation, i.e. paper and pressboard, will produce large amounts of carbon monoxide (CO). Since most of

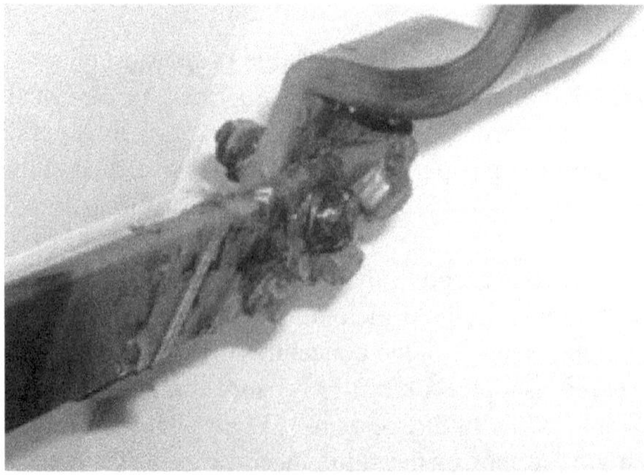

Figure 1.16 Example of an overheated connection

Figure 1.17 Example of coils at their end of life

the time these insulating structures are also impregnated with oil, these failure modes are also typically accompanied by methane (CH_4) and ethylene (C_2H_4).

Generalised overheating of the coils would generate this gas in significant quantities. Situations like overloading, through faults or winding connections that are in direct contact with insulating paper can produce CO.

Figure 1.17 shows an example of coils at the end of their life. In this case, sludge built up over the years and as a result the coils were subject to higher than normal temperatures and produce CO.

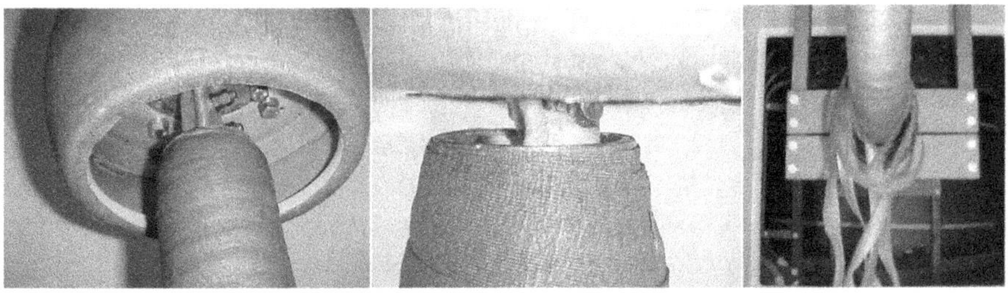

Figure 1.18 Examples of incorrectly installed stress controlling components that might give rise to partial discharges

Key gas: hydrogen (H_2)

The PD phenomena are related with localised electric field stresses. This failure mode does not typically carry the energy that thermal failure modes do and as such is identified as a low-energy failure mode.

It does however generate large quantities of hydrogen (H_2) often accompanied by smaller amounts of methane (CH_4).

As it might be expected, PD tend to occur in areas where the electric field stress is high. Electric field stress is produced by both the voltage and the geometry of the component subject to that voltage. As such, these failures tend to appear in regions where the stress has not been properly shaped or controlled.

The reader that has had the opportunity to work with the internal structures of transformers might be familiar with the stress control components often seen in high voltage structures. These structures might include electrostatic shields and tapes placed around high stress areas. The appearance of PD might be instigated by the incorrect placement or installation of these field-shaping structures. Figure 1.18 shows a couple of examples of this situation.

In recent years, some other sources of a similar gassing behaviour have been identified such as the catalytic reaction of the oil with other substances used in the manufacturing of transformers [25], the production of hydrogen at low temperatures by certain types of oil [24] or by the thin film of oil formed between core laminations [10].

1.6.2.2 Condition classifications and typical concentrations

Also within the group of techniques that rely on the gas profile of any given sample, we can count the condition classification and typical concentrations techniques.

Both these techniques make use of statistical behaviour to attempt to determine whether any given sample can be considered within a "normal" range or is indicative of any issue.

In the ideal case, each utility or manager of a transformer fleet would calculate typical concentrations based on their own results and failure rates. However, when no other reference is available, the available guidelines provide some values that can be referred to during a given interpretation.

The IEC guidelines [7] propose the use of what it calls the 90% typical values. Quoting from this guideline, the concentrations are calculated as follows.

> *The simplest method of calculation consists in gathering all the DGA results concerning a specific type of equipment. For each characteristic gas considered, the cumulative number of DGA analyses where the gas concentration is below a given value is calculated, then plotted as a function of gas concentration. Using the plotted curve, the gas concentration corresponding to a given percentage of the total cumulative number of analyses (for instance 90%) is the 90% typical concentration value for that gas and type of equipment.* [7]

The aim of this calculation method is to allow the asset manager to concentrate on the top 10% of transformers that are at higher risk of failure or outside the "norm".

It is important to keep in mind that when using the values provided by these guidelines, the analyst has to take into account the specifics of the transformer being analysed, such as age, oil volume, manufacturer, voltage and loading practices.

The typical concentrations listed by the IEC guidelines can be found on Table A.2 of [7].

On the other hand, the IEEE interpretation guidelines [15] have opted to settle on a categorisation system in which a particular sample is classified within one of four conditions given the concentration of total dissolved combustible gases.

These conditions, numbered 1–4, range from what is considered a transformer operating satisfactorily under Condition 1 to a state which is considered at high risk of failure represented by Condition 4. While the gas levels and full description of each of these categories can be found on the referenced guidelines, in general, these categories can be summarised as follows:

- Condition 1: Transformer considered to be operating satisfactorily.
- Condition 2: The gas concentration is higher than normal and further investigation is recommended.
- Condition 3: This range indicates high level of decomposition and immediate action should be taken to establish a trend.
- Condition 4: Samples in this range are considered to be the result of excessive decomposition and it is deemed that the transformer is at risk of failure.

The gas limits for these categories are presented in Table 1 of [15].

This categorisation system recommends that if any individual gas concentration is within the range specified by the category this particular sample gets classified under that condition.

In the author's experience, both of the classification systems described above should be understood in the context of each transformer being analysed. The volume of oil of a particular transformer will heavily influence the concentration of gases produced for any given failure mode.

In other words, the same failure mode will produce different concentrations in transformers with different oil volumes, i.e. higher concentrations tend to appear on the smaller transformers.

As a point of reference, it is also worth noting that the ranges of 90% typical gas concentration values identified in the IEC guidelines [7] are roughly equivalent to what the IEEE guidelines [15] would call a "Condition 2" sample.

1.6.3 Techniques that rely on ratios

Another group of techniques commonly used to analyse DGA results are those that relay on the relationship between two or more gases.

When ratios rather than absolute concentrations of gases are analysed, these tend to be less sensitive to the volume of oil in the transformer. Ratios also have characteristic patterns that can be correlated with certain classes of faults which make them useful as interpretation techniques.

As mentioned by the IEEE interpretation guidelines [15]:

> *The use of gas ratios to indicate a single possible fault type is an empirical process based upon the experience of each individual investigator in correlating the gas analyses of many units with the fault type subsequently assigned as the cause for disturbance or failure when the unit was examined.*

Of the various techniques that rely on ratios and gas relationships, we will review the following:

- Dörnenburg ratios [7,8,19,29]
- Rogers ratios [8,15,26]
- Duval triangles and pentagons [18,20,23,25,30]

1.6.3.1 Dörnenburg ratios

In a report published in 1970 as part of the CIGRE International Conference on Large High Tension Electric Systems, the member of the 15-01 Working Group included the results of investigations of decomposition gases produced under normal and abnormal operating conditions.

In this report, three main fault types were analysed:

- Oil decomposition due to PD
- Oil decomposition due to arc discharges under oil
- Oil decomposition due to thermal stress

The report highlighted the empirical correlations between these fault types and certain gas ratios. The results were presented in a chart that included the ratios of methane to hydrogen in one axis and the ratios of acetylene to ethylene on the other.

The IEEE guidelines [15] also include the Dörnenburg method and recommend the use of four gas ratios:

- Methane to hydrogen,
- Acetylene to ethylene,
- Acetylene to methane
- Ethane to acetylene

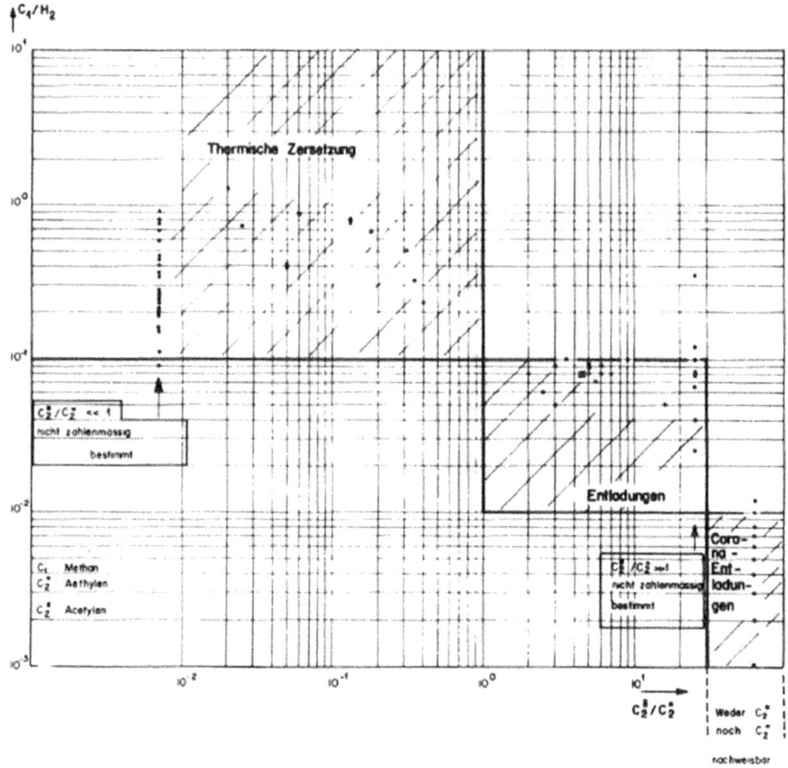

Figure 1.19 Figure 9 from CIGRE 15-07 1970 paper [8,29]

In Table 5, the IEEE C57.104 [15] guidelines correlate the profile presented by each of these ratios to each of the three fault types originally identified in the CIGRE report.

Figure 1.19 shows the graphical representation of these ratios as originally reported in the 1970 paper [8] that Dörnenburg contributed to.

Since the graphical method of representing the gas ratios is very effective and easy to visualise, it has become a very common tool in the DGA diagnosis toolkit. The IEC guidelines [7] further expand this concept in Annex B by representing three gas ratios in a three-dimensional chart used to identify various "fault zones" that include

- PD – partial discharges
- D1 – discharges of low energy
- D2 – discharges of high energy
- T1 – thermal fault, $t < 300$ °C
- T2 – thermal fault, 300 °C $< t <$ 700 °C
- T3 – thermal fault, $t > 700$ °C

1.6.3.2 Rogers ratios

As mentioned by the IEEE C57.104 guidelines [15], the methodology originally proposed by Dörnenburg was subsequently confirmed by Rogers on European systems.

The Rogers methodology comprises three ratios:

- Acetylene to ethylene,
- Methane to hydrogen, and
- Ethylene to ethane

In the case of Rogers, the fault diagnosis was stated as a six different fault cases, numbered from 0 to 5. This number has sometimes been referred to as the "Rogers Code". The cases listed by the IEEE guidelines are as follows:

- 0 – Unit normal
- 1 – Low-energy density arching – PD
- 2 – Arcing – high-energy discharge
- 3 – Low-temperature thermal
- 4 – Thermal <700 °C
- 5 – Thermal >700 °C

Table 6 from the IEEE C57.104 guidelines [15] summarises these codes and their related gas ratios.

1.6.3.3 Duval triangles and pentagons

Perhaps one of the most widely recognised and utilised interpretation methods in the industry is the graphical representations created by Michel Duval.

Michel Duval has been a senior scientist with Hydro Quebec's Institute of Research in Canada since 1970. First developed in 1974 [31], Duval devised a graphical representation format that allows one to correlate the relationship of three gases using a triangular chart. In 2008 [32], this method was extended to include six more triangles capable of identifying new faults types and included non-mineral fluids and on-line tap changers. More recently, this concept has been further extended by creating additional representations using pentagons [33], which allow one to correlate the content of five gases.

Each triangle or pentagon is subdivided into zones which correlate to the various failure modes that these methods are able to detect.

For each chart, the total amount of gases is calculated and then the proportion of each gas in relation to that total is calculated as a percentage. These percentages are then used to identify a point in the "triangular" or "pentagonal" coordinate system.

To date, the available literature identifies seven triangles and two pentagons which are listed in the following table.

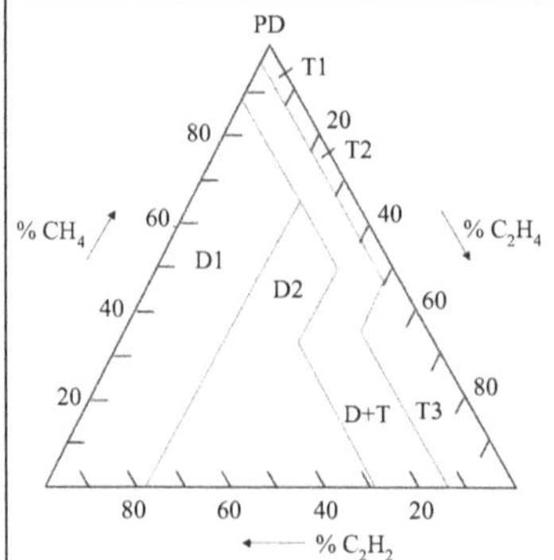 Triangle 1	This was the first and it is considered the "classical" Duval triangle. It uses methane, ethylene and acetylene to diagnose the various failure modes. It can be used for transformers, bushings and cables filled with mineral oil.
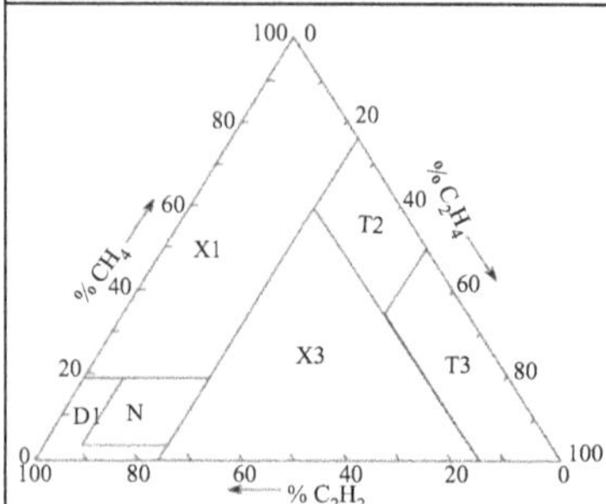 Triangle 2	This triangle applies to conventional, compartment-type on-load tap changers where normal operation involves arc breaking in mineral oil. Below are comments provided by Mr. Duval in his calculation Excel® file [34]: *A few resistive LTCs of this type (e.g. UZBs) may have their normal operation in the X3 zone. For OLTCs of the conventional, vacuum bottle-type with no sparking of the selector in the cooling oil, use Duval Triangle 1. For LTCs of the in-tank type (e.g. Reinhausen) where most or a significant portion of current is dissipated in transition resistors and heats up the resistors, the normal operating zone may be located in a different part of the Triangle (e.g. in T3 or T2). New versions of the Duval Triangle 2 (Triangles 2a, 2b and 2c) are in preparation for such LTCs.*
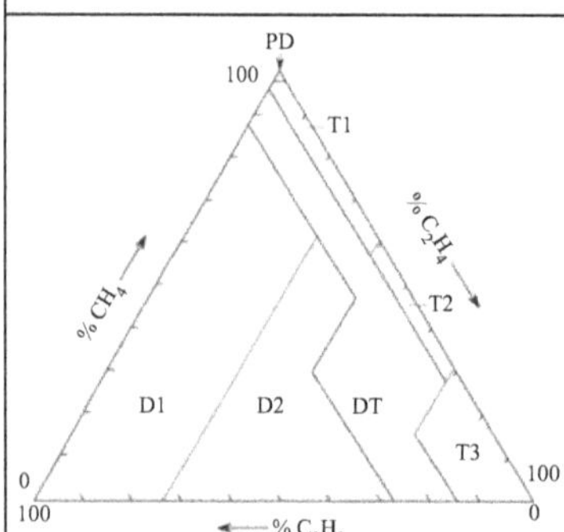 Triangle 3 –Midel®	The Triangle 3 is a family of triangles which deal with non-mineral fluids and includes BioTemp,® Midel, Silicone and FR3® fluids. These triangles use methane, ethylene and acetylene. The example on the left is the triangle for the Midel fluid.

(Continues)

(*Continued*)

<table>
<tr>
<td>
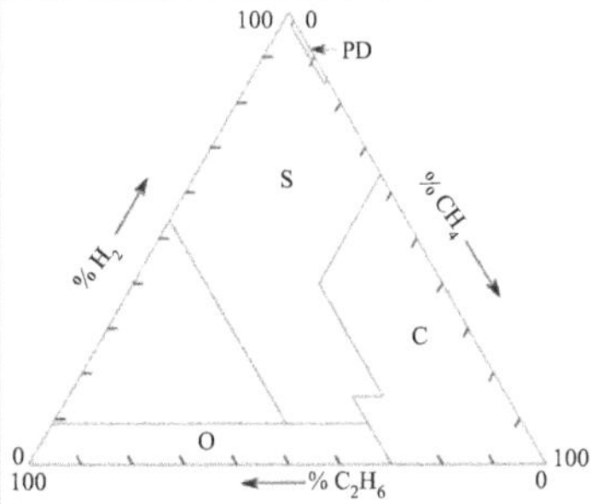

Triangle 4
</td>
<td>
Triangles 4 and 5 are used as a refinement to the diagnosis obtained with Triangle 1.
Below are comments provided by Mr. Duval in his calculation Excel file [34]:
The Duval Triangle 4 for low temperature faults in mineral oils uses Hydrogen, Methane and Ethane. The Duval Triangle 4 is used to get more information about faults identified as low temperature faults (PD, T1 or T2) with Duval Triangle 1 for mineral oils (using Methane, Ethylene and Acetylene).
Do not use the Duval Triangle 4 for faults D1, D2 or T3.
If Triangle 4 and Triangle 5 do not agree, this probably indicates a mixture of faults.
Some new oils on the market may produce stray gassing inside zone PD. Verify with stray gassing tests in the laboratory.
In zone C, the probability of having carbonisation of paper is 80%, not 100%.
</td>
</tr>
<tr>
<td>
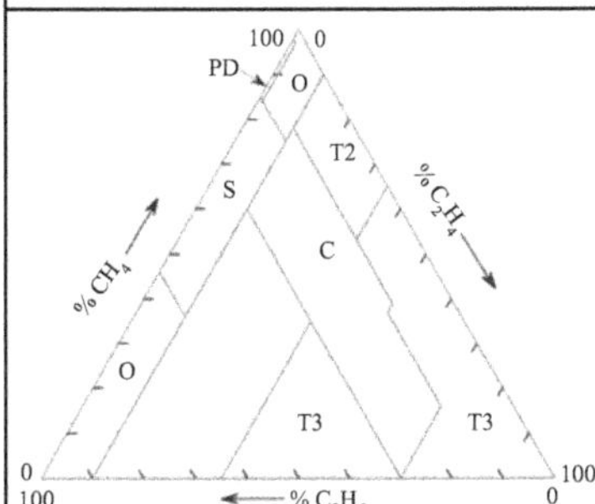

Triangle 5
</td>
<td>
Triangles 4 and 5 are used as a refinement to the diagnosis obtained with Triangle 1.
Below are comments provided by Mr. Duval in his calculation Excel file [34] for this triangle:
The Duval Triangle 5 for thermal faults in mineral oils uses gases Methane, Ethylene and Ethane.
The Duval Triangle 5 is used to have more information about faults identified as thermal faults (T2 or T3) with the Duval Triangle 1 for mineral oils (using Methane, Ethylene and Acetylene).
Do not use the Duval Triangle 5 for faults D1, D2.
If Triangle 5 and Triangle 4 do not agree, this probably indicates a mixture of faults.
In zone C the probability of having carbonisation of paper is ~90% (not 100%).
</td>
</tr>
<tr>
<td>
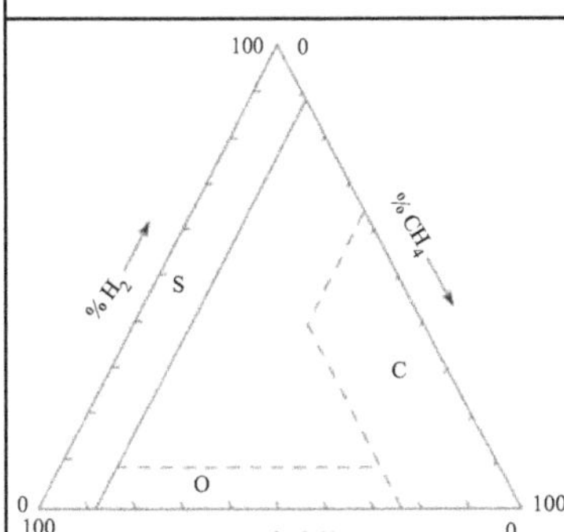

Triangle 6
</td>
<td>
Below are comments provided by Mr. Duval in his calculation Excel file [34] for this triangle:
The Triangle 6 for low temperature faults in FR3 oils uses gases Hydrogen, Methane and Ethane.
The Duval Triangle 6 is used to have more information about faults identified as low temperature faults (PD, T1 or T2) by the Triangle 3 for FR3 oils (using Methane, Ethylene and Acetylene).
Do not use the Triangle 6 for faults D1, D2 and T3 in FR3 oils.
</td>
</tr>
</table>

(*Continues*)

(Continued)

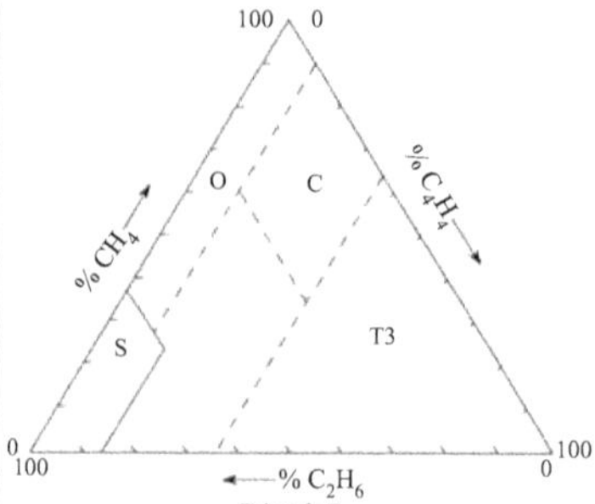 Triangle 7	Below are comments provided by Mr. Duval in his calculation Excel file [34] for this triangle: *The Triangle 7 for thermal faults in FR3 oils uses gases Methane, Ethylene and Ethane. The Duval Triangle 7 is used to have more information about faults identified as thermal faults (T1, T2 or T3) with the Triangle 3 for FR3 oils (using Methane, Ethylene and Acetylene).* *Do not use the Triangle 7 for faults D1, D2 in FR3 oils.*
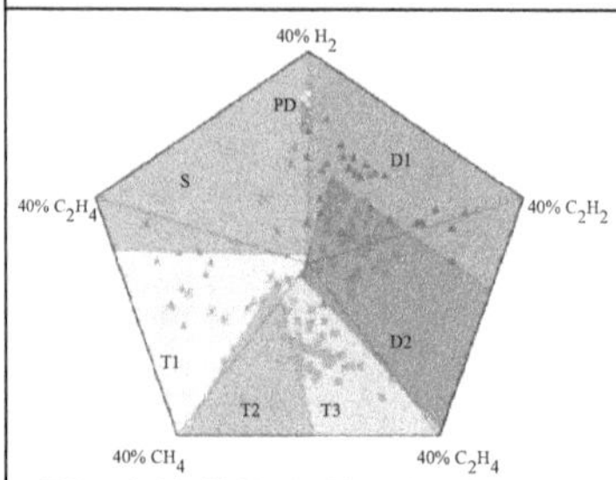 DGA results identified by visual inspection as due to faults: PD D1 D2 T3 T2 S>200C S–120C T1 Pentagon 1 (taken from (33))	The first pentagon can be used to determine Partial Discharges, Low- and High-Energy Discharges as well as various levels of Thermal Faults. It can be applied on transformers, bushings and cables filled with Mineral Oil [25,33,46].
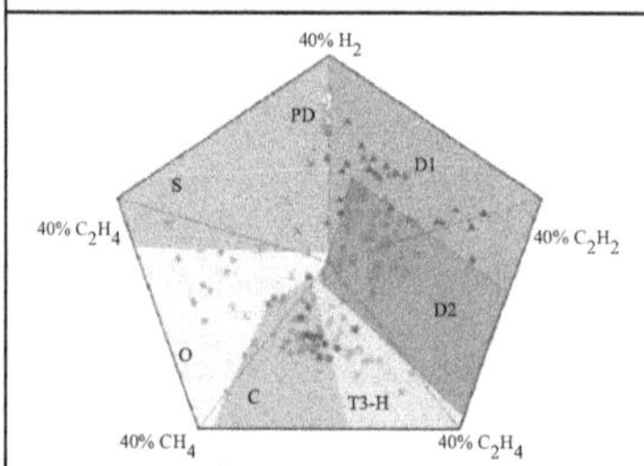 DGA results identified by visual inspection as due to faults: PD D1 D2 S>200C S-120C T3-H T2-H T1-H T3-C T2-C T1-C T1-Q Pentagon 2 (taken from (33))	The second pentagon has a similar function to what Triangles 4 and 5 have to Triangle 1. This pentagon provides further refinement to the diagnosis reached by the first pentagon. In addition to the three basic faults of Partial Discharges and Low- and High-Energy Discharges, provides additional diagnosis for Thermal Faults in oil only (T3-H), Thermal Faults that involve carbonisation of paper (C), generalised Overheating (O) and Stray Gassing (S) [25,33,46].

These triangles and pentagons have been implemented in various monitoring and diagnosis platforms. They have also been included in international guidelines, for example, Triangle 1 is included in the IEC 60599 interpretation guidelines [7].

1.6.4 Techniques that rely on rates of change

While the measurement of gases provides a good idea of the condition of the asset at that particular point in time, additional insight can be obtained by analysing how the gases evolve over time between one sample and the next.

The increase or decrease of specific gases can be indicative of

- Increasing or decreasing intensity of the fault activity
- Changing or overlapping of various failure modes

In general, a decrease in the concentration of gases is considered a sign of diminishing intensity of the fault. However, this must not be taken for granted and an adequate interpretation should be done every time new sample analysis information becomes available.

The analysis of the rate of change of gases can also be utilised to refine priorities for transformers that are part of a larger fleet.

For example, by now it should be clear that the presence of acetylene is almost invariable indicative of arcing faults. However, if two transformers have the same level of acetylene but one is increasing faster relative to the other, it would be prudent to assume that this transformer should be attended to first since the intensity of failure mode is likely to be more severe.

Both of the main guidelines that I have been referring to throughout this chapter include guidance around how to interpret the rate of change of gases in two consecutive samples.

The IEEE C57.104 guidelines [15] correlate each of the condition classifications with a range of rate of increase for the Total Dissolved Gas Content. These in turn are used to provide general recommendations about the course of action to be taken. All this is summarised in Table 3 of these guidelines.

In a similar fashion, the IEC 60599 guideline [7] includes a flowchart in which the verification of the gas increase rate forms part of the determination of whether the transformer's condition is classified as Normal, Alert or Alarm.

The ranges of 90% typical rates based on a survey of 20,000 samples are listed in Table A.3.

This guideline also recommends to convert the values in this table to ml/day when the volume of the oil for a particular transformer is known.

1.6.5 Putting it all together

All the information provided in the preceding sections is aimed at giving the reader a broad understanding of all the tools available in the diagnosis toolkit.

In my experience, DGA has proven to be an invaluable instrument for two main reasons.

1. It provides a reliable and cost-effective method to track and prioritise the condition of large populations of transformers. Although DGA does not cover every possible failure mode that a power transformer can experience, it does provide one of the most extensive coverage of any of the condition monitoring techniques currently available. Coupled with a sensible inspection and electrical testing programme, the fleet manager is equipped with the necessary tools to utilise the maintenance budget in the most efficient manner possible.
2. Once an issue has been detected in any particular transformer, this condition monitoring technique provides the means to more closely watch the evolution of that potential failure mode and make intervention decisions in an opportune and planned manner.

While there are plenty of information and recommendations about how to use the various interpretation guidelines available, I have most successfully used them as follows.

1.6.5.1 As a fleet condition monitoring tool

The application of both IEC and IEEE recommended calculations can be used to devise an overall ranking system that attempts to provide a sense of priority and identify whether there are any transformers in the population that might be showing early signs of fault development.

In addition to the interpretation of DGA results, this ranking process can also include considerations about the physical and chemical properties of the oil. Both IEC [22] and IEEE have their own recommended interpretation guidelines for these types of properties.

The analysis of Furan compounds can be also included in order to gain a rough understanding of paper degradation on each transformer. Please note that Furan content in a particular sample can be affected by a various factors not related to the actual ageing state of the transformer. These results should be taken as indicative and any suspect outliers should be analysed and understood in more depth.

While there are many possible ways in which transformers could be ranked on the basis of the tests outlined above, a methodology that generally renders practical results of sufficient quality is to use the "final" outcomes of each of the guidelines and then establish an overall assessment. Each of the guidelines uses some form of algorithm or classification system to provide an overall assessment of the transformer. Using each of these outcomes, a ranking table can be constructed that simplifies the high-level overview and management of these data.

This criterion is easy to tabulate and understand for all parties involved in this task. A conceptual example of how a ranking system like this could look like is given in Table 1.1.

Rather than being taken as an automated "set and forget" system, this method of ranking transformers is useful as a "first pass" of the data, particularly if the asset manager is in charge of monitoring a high volume of transformers.

Table 1.1 Example of possible ranking criteria using oil analysis information

Transformer ID	IEC 60599	IEC 60422	IEEE C57.104	Calculated DP (based on furans)	Final assessment	Active failure mode?
TX 1234	Normal	Good	Condition 1	800	Good	Unlikely
TX 5678	Alert	Fair	Condition 2	500	Fair	Uncertain
TX 9123	Alarm	Poor	Condition 3	300	Bad	Likely
Etc.	Etc.	Etc.	Etc.	Etc.	Poor	Very likely

By creating and documenting this ranking process, the following benefits are realised.

- The inherent knowledge, risk preferences and prioritisation rules are well-documented in a manner that it allows the organisation owner of the transformer fleet to pass this knowledge onto new technicians and engineers tasked with managing these fleets.
- This also provides a good level of traceability in the decision-making process should the need to audit this process arise in the future.
- By applying a consistent ranking process for the overall fleet, the relative distance within the ranking table between units is maintained for a given set of ranking rules.
- As the asset management team learns from events that occur in the fleet, this practice provides a centralised repository where this new knowledge can be integrated.
- Information presented in this simplified way, rather than large tables with raw oil properties data, is easier to understand to the non-specialist stakeholders around the organisation.

Generally speaking, the method described above is useful in

- Identifying transformers that are likely to require more individualised attention,
- Prioritising the expenditure in preventive or corrective maintenance, and
- Identifying transformers that show signs of active failure modes.

This method can be applied each time a fresh round of sampling is performed and analysis data are available. Once the fleet has been ranked, a review of each of the results should be performed, looking for any anomalies or specific cases that the process has flagged as high priority.

1.6.5.2 As a fault-monitoring mechanism

Once a possible active failure mode has been identified on a particular transformer, further attention and a more in-depth assessment is required. The same set of tests and analyses described in the previous section can be used to track the condition of that specific unit.

Any of the fault identification techniques explained in this chapter as well as some additional alternatives [10,27,28,35] can be used to attempt to determine the most likely failure mode present in that unit.

It is not uncommon that additional electrical tests or inspections are requested in order to confirm or disprove a particular hypothesis.

While Table 1.2 is not exhaustive, it provides a general overview of what types of electrical tests or additional inspections might be used to further enhance a particular diagnosis.

In addition to recommended additional tests and inspections, those familiar with the routine management of large fleets will be aware of the concept of re-sampling at a reduced time interval (i.e. increased sampling frequency).

Table 1.2 Example (non-exhaustive) list of typical fault types and investigative tests and inspections

Fault type	Common failure modes	Investigative tests or inspections
Thermal (might involve paper carbonisation)	Generalised overheating (due to overloads or through faults)	• Check for correlation with load • Analyse frequency and magnitude of through faults
	Generalised overheating (due to cooling)	• Verify the proper condition and operation of the external cooling system (i.e. fans, pumps, radiators, etc.) • Verify the condition of the internal cooling systems (i.e. winding ducts, directed cooling components, deflectors, etc.) • Check for excessive sludge build-up on the windings' cooling ducts and passages (old transformers with high acidity)
	Tap changer (de-energised or selector contacts) or winding/ bushing connections	• Winding resistance • Internal visual inspection • Correlation to load
	Circulating currents	• Excitation current test • Insulation resistance test • Inspection of core and clamping structures • Inspection of internal and external earthing bonds
Partial discharges	Improperly installed or damaged electric field stress management components	• Acoustic PD localisation • Internal visual Inspection • Review of baseline records (i.e. factory tests, commissioning tests, previous maintenance events and inspections, etc.)
	Overvoltage, switching surges	• Check correlation to transient events
High energy discharges	Tap changer (de-energised or selector contacts) or winding/ bushing connections	• Winding resistance • Internal visual inspection • Correlation to load
	Improperly grounded components (i.e. core, clamps, etc.)	• Insulation power factor (or dielectric dissipation factor) • Insulation resistance test • Excitation current test • Internal visual inspection

As discussed before, the intensity of a particular fault is correlated with the energy that it dissipates. The higher the energy levels, the higher the gas volumes produced by that fault over time. In many cases, faults have a quasi-exponential behaviour in which early warnings of the fault produce a steady increase in the gas levels. However, when the fault evolves into a more critical point, the gases are produced at an accelerating rate, which would be a good sign that is time to intervene.

Increasing the sampling frequency on a unit with a suspected (or confirmed) active failure mode allows the asset manager to more closely monitor the evolution of that fault and take the necessary action before the failure mode poses an unacceptable safety or operational risk.

The sequence of images shown in Figure 1.14 show the evolving nature of this particular failure mode. These images show the surface of a moving DETC contact fault that has progressed in this manner.

1.7 Future of oil analysis

As the pace of technological advancements continues, new opportunities for their application to the condition assessment of power transformers and other assets in the electrical power networks will continue to arise.

On one hand, the cost of monitoring technologies and sensors continues to decrease while their capabilities grow. It is possible now to instrument a transformer with a variety of devices which can monitor a wide range of properties, from dissolved gases to PD.

On the other hand, consumer-driven internet technologies have rapidly advanced the state of the art in the capabilities afforded by data analysis techniques. While the media publicity around these mathematical techniques makes them seem more capable than what they really are, their value, at least for now, rests more in their capability to assist the expert doing the condition assessment rather than replacing him or her.

Predicting the future is impossible, but there is no harm in allowing ourselves to imagine how these new technologies could be integrated to more effectively and efficiently manage power transformers, particularly large fleets.

1.7.1 Online monitors

The advent of online monitors has brought both new capabilities and new challenges to the condition monitoring field.

The main advantage that online monitoring brings to the condition assessment activities is their higher rate of data acquisition when compared with traditional methods such as manual sampling and subsequent laboratory analysis. With sampling, analysis and reporting rates in the range of hours, these monitors afford the asset manager the capability of detecting incipient failures almost in real-time.

While a large proportion of failure modes take months or years to develop, there are documented cases of failure modes developing in a matter of weeks, days and even hours [36]. These failures are rare but on sufficiently critical transformers the use of online monitors to detect these failures is typically well justified.

Once a decision has been made to make use of online monitors, the management of the monitors themselves should be considered. These devices need to be specified, selected, purchased, installed, commissioned and maintained as any other asset in the fleet. The value that can be extracted from having them in the asset base is directly correlated with the ability to integrate the data coming from these monitors into the condition monitoring processes of the organisation.

Authors have proposed holistic schemes [37–39] that look at the system-wide deployment, collection and analysis of data from these monitors. However, the current reality is that there is still work to be done to be able to easily and coherently integrate technologies from a diversity of vendors into a unified Enterprise Asset Management system. International Standards bodies have made great strides in defining vendor-agnostic configuration and communication protocols such as the IEC 61850 series.

1.7.2 Larger datasets

It does not take a great number of these monitors to start producing large amounts of data that, if not properly managed, can become a challenge for the asset engineer.

While data can be downloaded and analysed utilising Microsoft Excel workbooks, this method quickly becomes impractical when the number of monitors and data sources grows beyond non-trivial numbers.

Most of the vendors of online monitors offer software suites that allow the collection and analysis of data in an automated or semi-automated manner. These software packages help one to alleviate the burden of managing and analysing these amounts of data.

On the other hand, there is a certain amount of commitment required when a certain vendor is selected and their software packages become the central means to collect, analyse and distribute these data.

Furthermore, as any other asset management industry, power transformer management has lacked standardise mechanisms for the modelling, collation and analysis of data from disparate sources.

It would not be unreasonable to expect that the standardisation of the online monitoring data through international standards like IEC 61850 might enable in the near to mid-term the improved integration of data in larger and common datasets.

1.7.3 Analysis automation

If the above is true, we will need to develop and deploy systems that take advantage of these data in a way that has not been needed up until now.

These rich datasets will contain behaviours and nuances that a traditional yearly sampling programme could not have provided. Much better correlation to loading cycles, environmental conditions and specific failure modes will be afforded.

In order to extract knowledge from these datasets in a timely fashion, the asset manager will need the assistance of some form of automated or semi-automated analysis system. These systems would monitor the incoming information and perform a series of algorithmic checks in order to detect outliners and anomalies.

While it is unlikely, at least in the near future, that these systems make decisions on the best course of action by themselves, they will prove to be a valuable tool in assisting the asset manager with the identification of emerging patterns and trends which can then be reasoned about and turned into new knowledge. The value of the asset engineer will be in this reasoning and knowledge extraction process rather than on its ability to wrangle data between spreadsheets.

Various authors [40–42] have proposed techniques and software systems that allow the automated analysis of these data. However, a large commercial system that makes use of these techniques is yet to be developed and popularised [43–47].

Machine Learning, a branch of Artificial Intelligent, is being continuously reported about in the modern popular media. These techniques have made great strides in the development of algorithm capable of classifying, trending and detecting pattern in data. However, they require vast amounts of data to train and tune these algorithms to a reasonable accuracy level. To my knowledge, this amount of data is not currently available in one central database that could be used to develop similar algorithms for transformer DGA data.

Maybe as online monitors, electronic databases and analysis techniques develop hand in hand in the future; a better and more efficient future awaits in the transformer DGA space.

References

[1] Coltman, J. W. The transformer [historical overview]. *IEEE Industry Applications Magazine*. Jan–Feb, 2002, Vol. 8, 1, pp. 8–15.

[2] Schossig, W. Protection History, Transformer Protection, Next Steps. *PAC Magazine*. Winter, 2009, pp. 1–7.

[3] Nynas Naphtenics, A. B. *Transformer Oil Handbook*. Stockholm, Sweden: s.n., 2001.

[4] International Electrotechnical Commission (IEC). Fluids for electrotechnical applications – Unused mineral insulating oils for transformers and switchgear. s.l.: TC 10 – Fluids for electrotechnical applications, 2012. IEC 60296:2012.

[5] Eschhoz, O. H. Some Characteristics of Transformer Oils. *The Electrical Journal*. 1919, Vol. XVI, 2, p. 74.

[6] National Science Foundation. Bond Energies. *Chemistry – LibreTexts*. [Online] 20 01 2017. [Cited: 02 04 2017.] https://chem.libretexts.org/Core/Physical_and_Theoretical_Chemistry/Chemical_Bonding/General_Principles_of_Chemical_Bonding/Bond_Energies.

[7] International Electrotechnical Commission (IEC). Mineral oil-filled electrical equipment in service – Guidance on the interpretation of dissolved and free gases analysis. 2015. IEC 60599:2015.

[8] Fallou, B., Davies, I., Rogers, R.R., *et al. Application of Physico-chemical Methods of Analysis to the Study of Deterioration in the Insulation of Electrical Apparatus*. Paris: CIGRE, International Conference on Large High Tension Electric Systems, 1970. 15-07.

[9] CIGRE. Recent developments in DGA interpretation. s.l.: Joint task force D1.01/A2.11, 2006. 296.

[10] Oommen, T. V., Ronnau, R. A. and Girgis, R. S. *New Mechanism of Moderate Hydrogen Gas Generation in Oil-Filled Transformers*. Paris: CIGRE, 1998. 12-206.

[11] American Society for Testing and Materials (ASTM). Standard test method for analysis of gases dissolved in electrical insulating oil by gas chromatography. 2009. ASTM D3612-02(2009).

[12] International Electrotechnical Commission (IEC). Oil-filled electrical equipment – Sampling of gases and of oil for analysis of free and dissolved gases – Guidance. 2011. IEC 60567:2011.

[13] Offnfopt. Gas_chromatograph-vector. *Wikimedia Commons*. 2015. [Online] [Cited: 23 04 2017.] https://commons.wikimedia.org/wiki/File%3AGas_chromatograph-vector.svg

[14] American Society for Testing and Materials (ASTM). Standard practices for sampling electrical insulating liquids. 2015. ASTM D923-15.

[15] Institute of Electrical and Electronics Engineers (IEEE). IEEE guide for the interpretation of gases generated in oil-immersed transformers. 2008. C57.104-2008.

[16] International Electrotechnical Commission (IEC). Method of sampling insulating liquids. 2011. IEC 60475:2011.

[17] TxMonitor Pty Ltd. Transformer Oil Sampling Training by TxMonitor. *YouTube*. [Online] 18 02 2014. [Cited: 23 04 2017.] https://www.youtube.com/playlist?list=PLU3qAQVb80h4akXLu9L1VwpKtyRUsFoxH.

[18] Duval, M. Calculation of DGA Limit Values and Sampling Intervals in Transformers in Service. *IEEE Electrical Insulation Magazine*. Sep–Oct, 2008, Vol. 24, 5, pp. 7–13.

[19] Institute of Electrical and Electronics Engineers (IEEE). IEEE guide for dissolved gas analysis in transformer load tap changers. 2015: s.n. C57.139-2015.

[20] Duval, M. A Review of Faults Detectable by Gas-in-Oil Analysis in Transformers.*IEEE Electrical Insulation Magazine*. May–Jun, 2002, Vol. 18, 3, pp. 8–17.

[21] Institute of Electrical and Electronics Engineers (IEEE). IEEE guide for diagnostic field testing of fluid-filled power transformers, regulators, and reactors. 2013. C57.152-2013.

[22] International Electrotechnical Commission (IEC). Mineral insulating oils in electrical equipment – Supervision and maintenance guidance. s.l.: TC 10 – Fluids for electrotechnical applications, 2013. IEC 60422:2013.

[23] Duval, M., Langdeau, F., Gervais, P. and Belanger, G. *Influence of paper insulation on acceptable gas-in-oil levels in transformers*. Conference

on Electrical Insulation and Dielectric Phenomena, Leesburg, VA. 1989, pp. 358–362. doi: 10.1109/CEIDP.1989.69572.

[24] Höhlein, I. Unusual Cases of Gassing in Transformers in Service. *IEEE Electrical Insulation Magazine*. Jan/Feb, 2006, Vol. 22, 1, pp. 24–27.

[25] Duval, M. Michel Duval Presents the Duval Pentagon in a Qualitrol Sponsored Webinar. *YouTube*. [Online] 24 06 2015. [Cited: 02 04 2017.] https://www.youtube.com/watch?v=gPeWIYMe7B0.

[26] Rogers, R. R. IEEE and IEC Codes to Interpret Incipient Faults in Transformers, Using Gas in Oil Analysis. *IEEE Transactions on Electrical Insulation*. 1978, Vol. EI-13, 5, pp. 349–354.

[27] Jakob, F. and Dukarm, J. J. Thermodynamic Estimation of Transformer Fault Severity. *IEEE Transactions on Power Delivery*. 2015, Vol. 30, 4, pp. 1941–1948.

[28] Horning, M., Kelly, J., Myers, S. and Stebbins, R. *Transformer Maintenance Guide – Third Edition*. s.l.: Transformer Maintenance Institute, Division of S.D. Meyers, Inc., 2004. 0-939320-02-9.

[29] Dörnenburg, E. and Strittmatter, W. *Brown Boveri Review*, 1974, Vol. 61, 5, pp. 238–247.

[30] CIGRE. DGA in non-mineral oils and load tap changers and improved DGA diagnosis criteria. s.l.: Working Group D1.32, 2010. 443.

[31] Duval, M. Fault gases formed in oil-filled breathing EHV power transformers – The interpretation of gas analysis data. s.l.: IEEE PAS, 1974. Conf. Paper No C 74 476-8.

[32] Duval, M. The Duval Triangle for Load Tap Changers, Non-Mineral Oils and Low Temperature Faults in Transformers. *IEEE Electrical Insulation Magazine*. Nov–Dec, 2008, Vol. 24, 6, pp. 22–29.

[33] Duval, M. and Lamarre, L. The Duval Pentagon-A New Complementary Tool for the Interpretation of Dissolved Gas Analysis in Transformers. *IEEE Electrical Insulation Magazine*. Nov–Dec, 2014, Vol. 30, 6, pp. 9–12.

[34] Duval Triangle – SPX Transformer Solutions. [Online]. Available: http://www.spxtransformersolutions.com/wp-content/uploads/2018/01/duval-triangle-032016.xls.

[35] Church, J. O., Haupert, T. J. and Jakob, F. *Electrical World*. 1987, Vol. 201, 10, pp. 40–44.

[36] Duval, M. Identifying and Analyzing Quick Developing Transformer Faults with DGA. http://serveron.qualitrolcorp.com. [Online] 6 11 2013. [Cited: 9 7 2017.] http://serveron.qualitrolcorp.com/webinar-on-dga-with-michel-duval-1-google.

[37] Snow, T. *Deployment of Monitors System-Wide for Condition based Maintenance of Substation Equipment*. Sydney: TechCon, 2013.

[38] Roizman, O. *Intelligent Sensor Networks for Transformer Health Monitoring*. Sydney: TechCon, 2015.

[39] McDonald, J. D. *Transformer Monitoring, Communications Network and Data Marts: Extracting Full Value from Monitoring and Automation Schemes to aid Enterprise Challenges*. Sydney: TechCon, 2015.

[40] Mirowski, P. and LeCun, Y. Statistical Machine Learning and Dissolved Gas Analysis: A Review. *IEEE Transactions on Power Delivery*. 2012, Vol. 27, 4, pp. 1791–1799.

[41] Tapson, J. *Predicting Black Swans: New Machine Learning Methods for Very Rare Failure Events*. Manly, NSW: Austorque – Machines, 2014.

[42] Abu-Siada, A. *Software based Technique for the Identification of Transformer Faults in Real Time*. Sydney: TechCon, 2013.

[43] International Electrotechnical Commission (IEC). Mineral oil-filled electrical equipment – Application of dissolved gas analysis (DGA) to factory tests on electrical equipment. 2012. IEC 61181:2007 + AMD1:2012 CSV.

[44] Haupert, T. J. and Jakob, F. *A Review of the Operating Principles and Practice of Dissolved Gas Analysis*. Philadelphia: ASTM Electrical Insulating Oils, 1988. STP998.

[45] Duval, M. and dePablo, A. Interpretation of Gas-in-Oil Analyses Using New IEC Publication 60599 and IEC TC10 Databases. *IEEE Electrical Insulation Magazine*, 2001, Vol. 17, 2, p. 31.

[46] Duval, M. *Use of Duval Triangles and Pentagons for DGA in Transformers, LTC's and Non-mineral Oil*. Sydney: TechCon, 2015.

[47] Bergeld, L. *Effects of Transformer Materials on Stray Gas Generation*. Sydney: TechCon, 2015.

Chapter 2

Partial discharges: keys for condition monitoring and diagnosis of power transformers

Ricardo Albarracín[1], Guillermo Robles[2], Jorge Alfredo Ardila-Rey[3], Andrea Cavallini[4] and Renzo Passaglia[5]

Abstract

The combination of thermal and electrical stress, in conjunction with the synergic effects of moisture in the insulation, might dramatically reduce the lifetime of a power transformer, which is roughly estimated to be 40 years if properly maintained, meaning that many transformers do not experience any failures before being dismantled. Mechanical stresses during short-circuits can also play a role leading, in the worst case, to the radial buckling of the winding and damage of the connection cables at the bushing terminations. A number of these factors can, in the course of time, give rise to partial discharge (PD) phenomena which can be a cause of failure or a symptom of degradation. PD measurements can be carried out conventionally, following the standards IEC 60270 and IEC 60076. Alternative methods, such as those presented in the IEC TS 62478 standard based on acoustic or ultra-high frequency sensors, have been proposed and might offer better sensitivity, particularly for monitoring purposes, and the capability to locate the PD sources. For all these methods, identification of the PD source is a key to achieve a correct course of action (run, repair and replace).

2.1 Introduction

Partial discharge (PD) activity is one of the key factors for aging of insulation systems within electrical assets, which result in physical–chemical degradation in

[1]Department of Electrical, Electronic, Automatic Engineering and Applied Physics, Escuela Técnica Superior de Ingeniería y Diseño Industrial (ETSIDI), Universidad Politécnica de Madrid, Spain
[2]Department of Electrical Engineering, Universidad Carlos III de Madrid, Spain
[3]Department of Electrical Engineering, Universidad Técnica Federico Santa María, Santiago de Chile, Chile
[4]Department of Electrical, Electronic and Information Engineering "Guglielmo Marconi", University of Bologna, Italy
[5]Consultant at CESI S.p.A. of Milan, Italy

the area of its occurrence [1]. The monitoring of this phenomenon in power transformers plays an important role when assessing the degradation of its insulation system, especially when the transformer is in service. Factors such as mechanical, thermal, electrical and environmental effects must be considered as possible sources of degradation that can lead to the apparition of PDs activity in the insulation system of these electrical equipment [2,3]. In addition, PDs measurement can be used as an important indicator in quality control tests, in order to assess possible damage to the insulation system during transport and deployment in the field before its commissioning [4–6].

Currently, different techniques of PDs condition monitoring in transformers have been implemented based on the measurement of macroscopic effects occurring as a result of their activity (optical, mechanical, chemical and electrical, among others) [3,6–18]. Many of these techniques have some advantages over others in addressing the challenge of separation, identification and location of PD sources that are active during measurements. For example, great strides have been obtained in separating and identifying sources of PD by electrical techniques even when multiple sources are acting at the same time or in the presence of high-noise levels. However, the location of sources in transformers is a difficult task and requires complementary acquisition methods such as acoustic and ultra-high frequency (UHF) techniques. These techniques could allow the location of a PD activity to be identified by utilizing an appropriate deployment of multiple sensors, such as in dielectric windows of the transformer enclosure tank. Several studies have been carried out to address the advantages and limitations of each of these techniques [19].

2.2 Dielectric materials used in power transformers

Transformers can be classified as follows:

- *Air-cooled transformers*, which are dry-type transformers with air-cooling or cast resin.
- *Oil-immersed transformers*, whose dielectric cooling fluid is mineral oil or (synthetic) silicon-based liquid.
- *Hexafluoride (SF_6) gas-insulated transformers.*

Dry-cooling transformers presents lower risk of fire, because the materials used in their construction such as epoxy resin, quartz powder and alumina are self-extinguishing and do not produce toxic or poisonous gases [20]. Air-cooling present a clear advantage compared with oil-cooling transformers, the materials used in this kind of transformers have a higher temperature at which they start to decompose, from 300 °C, and the smoke they produce is tenuous and non-corrosive. Another advantage is their lower installation cost. However, their main disadvantage is their higher cost per MVA. Moreover, dry transformers present higher electrical noise level, lower resistance to overvoltages, higher vacuum loses and are not suitable for outdoor installation or for contaminated environments compared with oil-immersed transformers. For these reasons, transformers rated in the 10 kVA–1,500 MVA range and with line voltages up to 1,000 kV are commonly manufactured with oil–paper insulation systems [21].

The relative cost per unit according to the cooling type in a power transformer is mineral oil 1, silicon 1.25−1.5, air-cooled 1.8 and cast resin 2.0 [22].

In oil-filled transformers, the fluid is the main insulation and one of the main advantages for oil-cooled transformers is that this equipment can withstand prolonged overloads. However, the relatively low inflammation temperature of the oil, about 140 °C for mineral oil, increases the risk of fire with high fuming.

In addition to natural oil, there are also insulation liquids with combustion points above 300 °C [23], such as *biodegradable natural ester* for application in electrical transformers [24], *dielectric liquid silicone* class K [25] and *biodegradable synthetic ester* [26]. These insulation media have excellent dielectric properties maintaining high-dielectric strength values in presence of high humidity with a high saturation point of water.

Since the late 1990s, new research is being carried out using conductive nanoparticles to be dissolved in oil-based nanofluids to improve its thermal and dielectric attributes [27,28]. Nanofluids have slower positive streamer velocities and higher positive voltage breakdown than that of pure conventional oil [29]. A slower propagation velocity allows more time for impulse voltage to be extinguished, so the positive voltage breakdown is higher than that obtained in conventional transformer oil. This good performance, thermal and electric, suggests that nanofluids can be used in near future for power transformers.

In SF_6 gas-insulated transformers, polymer film is used as insulation for turn-to-turn and SF_6 gas offers insulation for the majority of the interstices in the transformer. The cooling system is composed by cooling ducts through which the gas circulates for cooling the winding. This type of transformer can be used where oil-filled transformer is not appropriate, such as those deployed in some high-rise buildings [30].

In addition to the main insulating medium, other insulation materials can be found in a transformer [31].

- *Conductor insulation*. In oil-filled transformers, the windings (conductors) are coated with cellulose layers, Kraft or thermally enhanced paper and epoxy resin and/or varnish. In addition, pressboard is deployed between layers. In transformers working at high temperatures Nomex fibre is often used.
- *Coated steel tank*. The surface of steel plates/walls of the enclosure are treated and painted to prevent impurities in which weaknesses of insulation can be produced.

Other insulating materials that can be found, mainly, in transformers for transmission grids [31–33] are

- Aramid paper and board. The same application as cellulose-based paper and board for oil-filled transformers.
- Rigid laminates such as glass fibre reinforced plastic. Screws, nuts, core supports and so on.

- Flexible laminates (Nomex, Mylar, etc.).
- Wood (laminated, dried, steamed) to be used instead of cellulose-based board. It is exclusively applied in the fastening rings of the coils [they are wedges that introduced between the high-voltage (HV) and low-voltage (LV) windings, tertiary and the core]. In armoured transformers, wood also is used to separate the core from the phase and the core from the tank.
- Gases such as SF_6 and fluoroketones used in switchgear connection to generator transformers.
- Synthetic insulation such as polymer composite is also used in bushings.
- On-load tap changers used in some transformers, also mica is deployed.

The electrical insulation used in transformer, i.e. paper for the windings, can be subjected to different types of failure. The four main failure locations of substation transformers (>100 kV) are windings (37.69%), tap changer (31.16%), bushings (17.16%) and lead exit insulation (8.96%) [34]. Figure 2.1 shows the failure mode in this type of transformers, according to [35].

Once the most common failures in transformers have been identified and considering that most of them are due to PDs activity is key to identify the different types of stress to which insulation system is subjected. These ageing effects are presented in the next section.

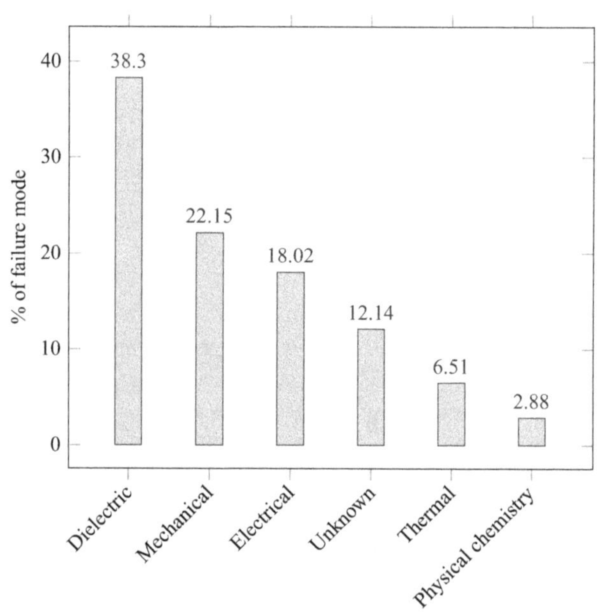

Figure 2.1 Failure modes in substation transformers

2.3 Effects of ageing in insulation systems of power transformers

Ageing of electrical insulation in transformers is due to a combined effect of thermal, electrical, ambient and mechanical stresses, the acronym TEAM is often used to evoke these stresses factors. In transformers, the most important is the thermal stress, followed by mechanical stress resulted from over-current during short-circuit events. Ambient stress is less likely, since, in principle, the transformer is a sealed device. However, water and oxygen can leak into the insulation system if the bushings or the cooling pipe flanges are not properly sealed. Electrical stress is, probably, the less important stress factor. However, the effect of the former stresses can lead to the formation of PDs. Therefore, electrical stress can be both a cause of ageing (or, more properly, degradation, when PDs are incepted) or just a symptom. There are cases where electrical ageing can be both a cause and a symptom of localized degradation. As for any other electrical equipment, failure is only rarely due to a single cause. Most often, failure is due to a complex cause–effect chain and the combined effect of several mechanisms that act in a synergistic way, so monitoring the PD effect allow to carry out an active electric condition-based maintenance (CBM) of electrical equipment [36].

2.3.1 Thermal stress

Temperature activates or accelerates chemical processes that degrade the dielectric characteristics of the insulation. Temperature in the transformer is raised upon ambient temperature by all losses found in electrical machines (Joule losses, hysteresis and stray currents). Generally, the temperature of the cellulosic insulation should not exceed 105 °C and, since cellulose is in direct contact with conductors, oil should be at lower temperatures. Hot spots, where higher temperatures can be found, observed if the core lamination is shorted or improperly cut or if the conductor connections are not properly realized. The authors shall review here the main degradation mechanisms for cellulose and oil separately.

2.3.1.1 Cellulose

Cellulose is a natural polymer, where alpha and beta glucose monomers aggregate to create chains. In the new Kraft paper, cellulose chains include roughly 1,200 monomers (thus, its degree of polymerization, reported below as DP, is around 1,200). The chains are held together by oxygen bridges between the monomers in the same chain, and by hydrogen bonds that form between the hydroxyl groups that protrude laterally from the chains. These bonds are weak and can be broken by polar species, in particular by water, which is the most important threat to cellulose stability over time. As a matter of fact, under temperature and in the course of time, the intra- and inter-chain bonds usually break down leading to a brittle paper having a lower DP. When DP is around 250 monomers, the tensile strength is almost halved. This is generally considered the end point of the insulation.

As a consequence of depolymerization, both the paper mechanical resistance and its elastic module decrease, exposing the winding to mechanical faults (as shown in the following). Sometimes, its weak mechanical shock is sufficient to collapse the cellulose insulation. Thermal ageing does not have noticeably effect on the dielectric performance of the paper (which can work reasonably well unless a short-circuit or a mechanical shock does not destroy it), although some detrimental phenomena may ensue. As an example, all thermally activated degradation mechanisms lead to the formation of water and other polar species (weak acids). Water and polar species in general increase the dissipation factor of the insulation, reinforcing thermal stress. This might become a self-breeding process leading to thermal runaway. Also, thermal decomposition products (including gases as carbon monoxide and dioxide) dissolve in the oil, reducing its dielectric performance with time.

Thermal ageing proceeds due to three different mechanisms, which are activated or not depending on the temperature of the cellulose:

- Hydrolysis: water molecules attack the covalent bonds between the oxygen atoms in different glucose monomers. The main effect is the separation of a chain into several shorter chains. A possible by-product of hydrolysis is carboxylic acid which is a catalyst of the reaction and which can attack interchain hydrogen bonds.
- Oxidation: oxygen attacks the carbon atoms in the cellulose molecule to form aldehydes ($C_nH_{2n}O$, e.g. formaldehyde) and acids (formic acids), releasing water, carbon monoxide and carbon dioxide. The bonds between monomers are weakened, leading to lower DP. Water released by this mechanism can also contribute to the hydrolysis effect previously mentioned.
- Pyrolysis (>140 °C): heat in the extreme will result in charring (carbonization) of the fibres, but at lesser levels, it contributes to the breakdown of individual monomers in the cellulose chain. A solid residue (sludge) is formed and gases are liberated, namely, water vapour, carbon monoxide, carbon dioxide and hydrogen.

These phenomena can enhance/activate PDs via different routes. Gases due to thermal decomposition could give rise to bubbles within the interturn insulation or in the oil if carbon monoxide (low solubility in the oil: 9% by volume) is generated. Bubbles might become site of PD activity since the field within the bubble is larger than that in the surrounding medium and since the dielectric strength of the gas within the medium is lower than the oil. Conductive by-products deposited on the barriers can give rise to creeping discharges.

2.3.1.2 Oil

The ageing of mineral oil is associated with oxidation. As per any chemical reaction, high temperatures favour chemical reaction, with reaction rates following the Arrhenius law. Thus, oil degradation becomes more important in reactors or heavily loaded transformers, particularly in hot climates. Unstable hydrocarbons in the oil react with oxygen and moisture. Oxygen attacks initially hydrocarbons forming peroxides. These dissociate to form free radicals which can act as initiators for chain reactions involving free radicals and hydrocarbons. The propagation reactions are further capable of repetition over a number of cycles for each hydrocarbon free radical supplied by the initial reaction. These reaction occur in any transformer, since it is very difficult to prevent that oxygen and moisture leak into power transformers. In fact, oxygen can leak in from deteriorated gaskets or from the conservator for free breathing transformers. Water can also leak in but, as discussed above, can also be produced by the deterioration of the cellulose.

The final products of oxidation reactions are sludge, in sufficient quantity to impair the heat transfer of the windings, and soluble acids (polar species) that reduce the dielectric properties of the oil and damage the solid insulation as well as other elements of the transformer. The acids attack on the cellulose fibres and metals forming metallic soaps, lacquers, aldehydes, alcohols and ketones which precipitate as heavy tarry acidic sludge on the insulation. The insulation shrinks leaching out varnishes and cellulose materials. Oil oxidation can be retarded by oxidation inhibitor that interrupt and terminate the free radical process of oxidation. If a compound (Rydberg constant (RH)) with labile hydrogen atom is added to the system to compete with the hydrocarbon, it will produce a relatively inactive radical (A') which will break the chain sequence and inhibit the overall reaction.

Chemical reactions are remarkably influenced by temperature. Indeed, several other physical parameters of the oil change as well. For example, if temperature increases, there is an increasing number of water molecules that can dissolve in the oil before saturation occurs, the Oommen curve represents the equilibrium between moisture in the oil and the moisture in the paper [37,38]. Besides that, oil viscosity decreases, and the conductivity of transformer oil increases.

When cooling the oil from an high temperature, the water migrates from the oil to the cellulose. Since penetration in the cellulose is time-consuming, the layers of oil adjacent to the pressboard might have high moisture and, therefore, low capability to withstand the electric field. In the regions were the field is sufficiently large and tangential to the pressboard surface, PD can develop along the pressboard surface itself, leading to tracking.

PD activity can degrade the oil leading to, mostly, hydrogen formation. Since hydrogen solubility in mineral oil is low (7% by volume), this could lead to the formation of bubbles in the oil and to inception of PD in the bubbles, as explained in the next section.

2.3.2 Mechanical stress

Short-circuit events in the lines connected to the transformer cause large currents in the winding and in the cables connecting the windings to the bushing. Often, the movement of overhead lines connected to a transformer tend to pull out the cables from the transformer. This pull force is normally counteracted by the interface between the bushing and the cable and, as the fault is cleared, operation can start again. However, under a number of conditions (the transformer is poorly manufactured/designed/matched to the short-circuit power of the network, or the paper is brittle due to thermal degradation, the copper of the conductor is oxidized and some strand breaks down) the friction resulting from this pull force can degrade the cable insulation. PDs can thus be incepted in the degraded region, and the region will further degrade till breakdown (most likely, a short-circuit between the cable and the tank in the bushing area).

In the windings, the short-circuit currents give rise to axial and radial forces, which must be withstood by the winding. Clamping rings on the top and bottom of the winding exert strong compressive forces and preload the paper in the insulation, which can be regarded as a spring. If the paper is in good condition, the mechanical resonance frequency of the winding is, by design, well above the maximum frequency of the forces induced by short-circuit. However, if the paper is in bad condition, the winding becomes loose and the forces might result in plastic deformations of the winding of two types: (a) radial buckling (the winding section is not anymore circular) and (b) tilting of conductors. The movement associated with these deformations might abrade the turn insulation leading to a turn-to-turn short-circuit.

2.3.3 Electrical stress

Electrical stress does not degrade cellulose or oil in a significant way, unless PDs are incepted. There are a number of reasons for PD inception within the transformer. Most common causes for PD are elaborated below.

2.3.3.1 Tracking on barriers

Below 60 kV, the barriers extend a bit above (and below) the winding ends to create a barrier between HV and LV windings. A sufficiently long clearance is left between the energized conductors and the yoke (at ground potential) to prevent discharges. Above 60 kV, the former design would lead to an inefficient design (the resulting transformer size would be very large). Thus, above 60 kV, at the winding ends, the barriers bend to protect the HV conductors from the yoke. The design of the barriers in this region is done in a way that barriers are at right angles with the electric field. This ensures that the electric field tangential to the barrier surface is zero or very low. However, due to bad design, improper manufacturing of barriers (e.g. barriers that undergo deformations during drying operations) or contamination, the tangential component of the electric field might increase up to a point where PDs start to occur. PDs, such as surface PD, activity leads to the formation of carbonized tracks (tracking) having a high electrical field at their tips and, thus, leading to self-propagation and, eventually, to flashover.

2.3.3.2 Bubbling

Gases can form in the oil due to a number of reasons (overheating of metals or cellulose, PDs, arcs on the core, improper sealing). If the concentration of some gases exceeds the saturation level of this specific gas in the oil, bubbles might appear. Since the dielectric strength of the gas is lower than that of the oil and the electric field in the gas bubble is larger than that in the oil, PDs can be incepted inside the bubble. PD activity due to gas bubbles is referred to as 'bubbling'.

Bubbling is generally unstable with time. In fact, bubbles originate where the gas concentration is above the saturation level, but generally a concentration gradient exists and bubbles can move away into regions where the gas in the bubble dissolves into the oil. Also, PD within the bubble may break it into tinier bubbles, each having higher PD inception field. Eventually, if the electric field is divergent, dielectrophoretic forces will tend to move the bubble (a 'particle' having lower permittivity than the surrounding medium) towards low electric field regions. The combined action of these phenomena tends to quench PD activity. However, if gas generation continues, bubbling can restart. In general, bubbling appears as random bursts of PD activity such as internal discharges.

2.3.3.3 Incomplete filling of turrets

The 'turrets' where bushing are lodged have a ceiling which is higher than the tank. If air has not been extracted during filling operations, arcs might be incepted in the air pocket between the cable and the turret ceiling. These arcs may differ from PDs as their frequency content might be around a few hundreds of kHz. This type of problem is easily recognizable (in general the three turrets are interested) and solved by opening the turret valves.

2.3.3.4 Metallic objects

Metallic parts are normally firmly grounded, at a safe distance from energized conductors and covered by cellulose (paper or pressboard) to prevent injection of charge carriers in the liquid. Yet, there conditions are not met, sometimes. For instance, after maintenance operations, some tools or ropes can be left inside the transformer (on the bottom of the tank, on the yokes or on the clamping rings). Metal object can give rise to arcs in the oil (often indicated as corona) which have large energy and peculiar waveshapes with large energy content in the frequency range of the hundreds of kHz.

2.3.3.5 Improper spacing of tap changer conductors

The potential voltage between tap changer cables is not very high, in general. However, these cables are often grouped in narrow regions, and discharges might occur at insulation interfaces existing between cables of different phases.

2.3.3.6 Static electrification

At the interface between a liquid and a solid, an electric double layer is commonly observed. In an electric double layer, electrically charged species of one sign are created at the solid surface, while species with opposite-sign species are dispersed within the liquid. If the solid is a dielectric, charges may accumulate due to trapping. If the liquid is in motion, the species in the liquid are transported away from the interface, and new charges can accumulate at the dielectric surface. In transformers, static electrification consists of a preferential adsorption of negative ions from the oil into the pressboard. This yields on the one hand a space charge in the oil which can relax in contact with grounded metallic walls and on the other hand a space charge in the pressboard which can accumulate depending on the leakage paths.

If the electric field due to charge displacement exceeds the breakdown strength of the oil, discharges of large magnitude might take place. Traces of carbonization and 'wormholes' have been found in the pressboard of a number of power transformers. Static electrification is enhanced by oil speed. Thus, it becomes a threat in forced cooled transformers where the oil circulates at speed larger than 500 mm/s.

2.3.4 Ambient stress

2.3.4.1 Water leaks

Water can leak into the tank from bushings where o-rings are cracked. Water has a number of detrimental effects. In the paper, it causes additional dielectric losses and depolymerization due to hydrolysis. An excessive content of water in the oil (>30 ppm) leads to a drastic reduction of the oil breakdown voltage. As a result, faults are more likely to occur. In extremely cold climates, when the transformer is not in service, the capability of oil to solvate water can be very low; thus, water can deposit on the bottom of the tank. The low temperatures turn water into ice which, eventually, floats to the top of the transformer (the region where, during service, the highest electrical fields exist). Therefore, during cold start of transformers, ice can trigger flashover between the phases and the ground or between different phases. Water leaks are a threat especially for bushings and instrument transformers, which are directly exposed to atmosphere and rain. Instrument transformers, with a rubber membrane on the top to allow oil expansion, are particularly prone to water leaks as the membrane can crack due to corona discharges and ultraviolet radiation.

2.3.4.2 Moisture transients

Cellulose can store a large amount of water (moisture accounts for about 5% of the weight of undried cellulose), while the saturation level of oil is a few tens of ppm (above that level, moisture starts to aggregate in micro bubbles). The transformer is a closed system where the partition of moisture between cellulose and oil is strongly dependent on temperature. At low temperatures, the saturation level of oil is extremely low, and moisture tends to stay in the cellulose. As temperature increases, the saturation level of the oil increases too, and moisture migrates from the cellulose to the oil. The opposite occurs when the temperature of the oil decreases. The migration of moisture from the cellulose to the oil and vice versa gives rise to moisture transients, where the concentration of moisture is inhomogeneous in the oil, and tends to be larger in proximity of the solid insulation. Moisture transients are slower when moisture flows from the oil to the cellulose (i.e. when the transformer cools down). As mentioned before, the dielectric strength of oil drops dramatically above 30 ppm. Thus, during moisture transients of poorly dried or contaminated transformers, it is possible that some regions of the oil have extremely low dielectric strength. If these regions are in high electric field areas, flashover might occur.

Moisture transient are particularly dangerous in instrument transformer since, by design, the mass of the solid insulation is predominant compared with that of the oil.

2.3.4.3 Insulating particles

The analysis of transformer oil often reveals that insulating particles are dispersed in the oil. These particles usually consist of cellulose released by the solid insulation into the oil. Since these particles have higher permittivity than the oil, under the effect of divergent electric fields (sometimes it is sufficient some micrometric protrusion on a bare electrode, i.e. an electrode not wrapped with dielectric such as a bolt in the core), particles might start to align forming bridges in the oil. These bridges reduce the dielectric strength of the oil gap and might lead to flashover or breakdown.

2.3.4.4 Metal particles

Metal particles can be formed by corrosion or mechanical wear of the gears in the pumping system. These particles are removed by filters and are not a threat for the transformer windings. However, they might cause a problem for the tap changer.

2.3.4.5 Oxygen and moisture leaks

As explained above, oxygen assists the formation of acids and sludge. Thus, keeping oxygen levels as low as possible is of prime importance. Indeed, in open-breather transformers, the oxygen supply is almost unlimited, and oxidative deterioration is much faster than in sealed transformers. Atmospheric oxygen is not the only source of oxygen available for the oxidation of insulating oil; water also serves as a carrier of oxygen and leaky gaskets constitute a real hazard, causing both oxidation and moisture contamination. The rate of oxidation also depends on the temperature of the oil; the higher the temperature, the faster the oxidative breakdown. This points to the importance of avoiding overloading of transformers, especially in summer time.

2.3.4.6 Sludge

Sludge originates from oxidation under temperature of oil and paper. Copper acts as a catalyst in the process. In the early stage of sludge formation, precursors as peroxides, organic acids, alcohols, aldehydes, ketones, lacquers and other aromatic compounds, particularly those have polar functional groups are formed. These materials attack paper, iron and copper, forming intermediate by-products in the oil. These by-products polymerize together to form a solid type material (sludge). This is the terminal stage of this degradation process. Sludge tends to precipitate out in the coldest and hottest regions of the transformer. Sludge sticks to the solid insulation, reducing the heat transfer capability of the windings.

Since the most common stresses cause PDs, PD monitoring is one of the key tools to determine whether degradation processes have not exceeded some limit threshold, thus ensuring that proper maintenance actions can be decided.

2.4 Condition monitoring techniques in power transformers

2.4.1 Electrical measurements

PDs give rise to a variety of physical phenomena within a transformer. The most obvious one is the circulation of conducted currents. However, apart from circulation of conducted currents, PD trigger also other phenomena as ultrasonic pressure waves in the oil, electromagnetic waves in the very high frequency (VHF) and UHF range and the production of gases (mostly H_2) that can be used to detect PD within the transformer. In this section, the aim is to recall the basis of conventional PD detection, i.e. that based on the conducted currents.

2.4.1.1 ABC circuit

The ABC circuit is generally used to explain the behaviour of conducted currents associated with PD events. The circuit is made of capacitors (*A*, *B* and *C*, also named as C_A, C_B and C_C) that are defined below:

- C_A: capacitance of the healthy part of the insulation in parallel with the defect (generally schematized as a cavity);
- C_B: capacitance of the healthy part of the insulation in series with the defect;
- C_C: capacitance of the defect.

In general, considering the geometry of the insulation system (see sketch in Figure 2.2), and considering that $D \gg d$ and $S_A \gg S_B$, it is possible to observe that

$$\varepsilon_D \frac{S_A}{D} \gg \varepsilon_D \frac{S_B}{D-d} \rightarrow C_A \gg C_B \tag{2.1}$$

$$\varepsilon_0 \frac{S_B}{d} \gg \varepsilon_D \frac{S_B}{D-d} \rightarrow C_C \gg C_B \tag{2.2}$$

To investigate what happens when a PD takes place, it is assumed that according to the model, the capacitance C_C gets shorted. Thus, the voltage across C_C before the

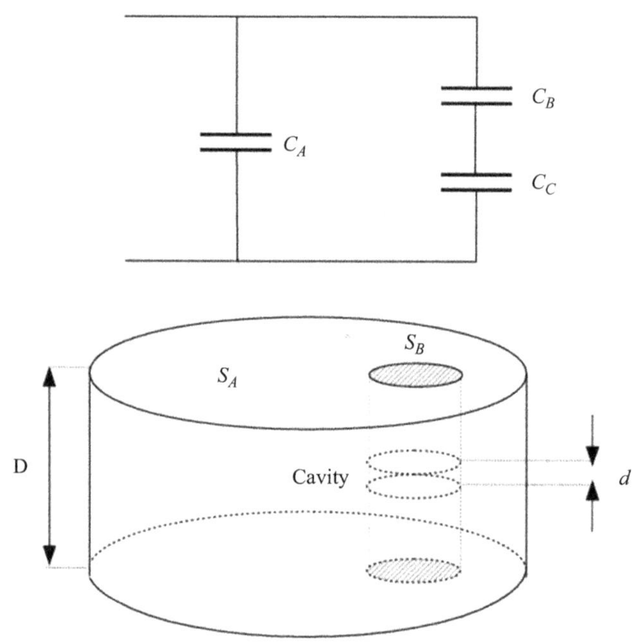

Figure 2.2 ABC circuit

discharge, V_C, is (approximately) transferred to C_B. This causes a charge transfer from C_A to C_B (it is considered that the generator is inductive and will not be able to provide any significant contribution during the time the discharge takes place, i.e. hundreds of ns) defined by

$$q_{app} \approx C_B \times V_C \tag{2.3}$$

The charge q_{app} is known as 'apparent charge'. The term stems from the fact that the charge associated with PD event is $q_{PD} = C_C \times V_C$, while the only electrical quantity which is able to be detected is q_{app}. According to the relationships existing between C_A, C_B and C_C [Eqs.(2.1) and (2.2)], the PD charge is much larger than the apparent charge:

$$q_{PD} \gg q_{app} \tag{2.4}$$

In particular, considering the geometry of the system, the thicker the insulation, the lower will be the apparent charge compared with the PD charge.

Considering the above discussion, it comes out that

1. To model the circuit electrically, the defect can be replaced by an ideal current generator injecting an ideal (Dirac) pulse $q_{app}\delta(t - t_{PD})$ (where t_{PD} is the time in which the PD take place).
2. In the absence of any external circuit, the charge transfer will occur integrally within the equipment under test (EUT). Thus, it is necessary to place a capacitor (termed coupling capacitor) in parallel to the EUT to be able to measure any current associated with the PD (note that the PD will slightly modify the voltage at the EUT terminal, but the voltage drop will be too low to ensure that PD can be measured indirectly through this voltage drop).

The fraction of the total PD current can be calculated using the capacitive divider between C_A and C_K as (I_K in Figure 2.3)

$$I_K/I_{PD} = \frac{1}{1 + (1/(C_K/C_A))} \tag{2.5}$$

The results of (2.5) are reported in Figure 2.4, which shows that only 50% of the PD current can be detected when $C_K = C_A$ (note that C_A for HV transformers is from tens to hundreds of pF). The situation is further aggravated by the stray capacitances existing between the HV conductors and the earth.

2.4.2 Apparent charge estimation: quasi-integration and calibration

2.4.2.1 Principle of quasi-integration and calibration

The PD apparent charge should be estimated by integrating the PD current pulse or by evaluating its DC component, considering that T should be large enough so that the pulse is extinguished:

$$q = \int_0^T i(t)\,dt = I_{DC}T \tag{2.6}$$

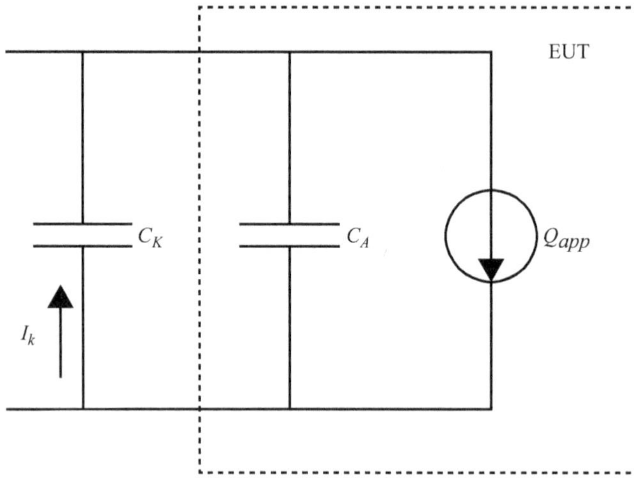

Figure 2.3 ABC circuit with coupling capacitor C_K

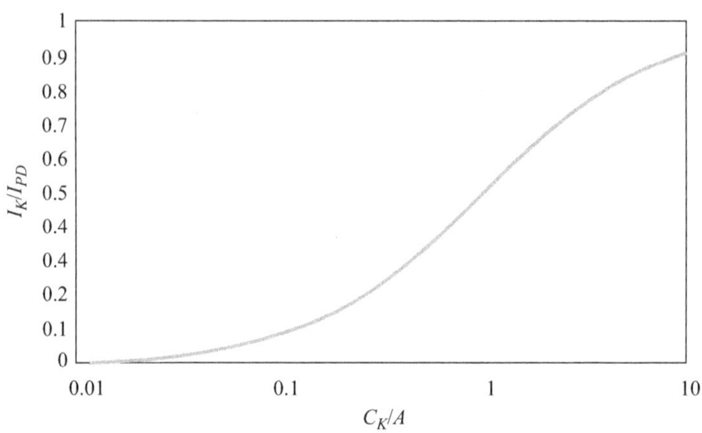

Figure 2.4 Transfer ratio I_K/I_{PD}

This operation can be done by coupling the PD pulse current and processing it via electronic (analog or digital) circuitry. However, practical reasons prevent to do that. In practice, in the frequency domain, the DC residual component is generally larger than the PD signal and should be filtered out. However, filtering out this residual component keeping the DC component is not realistic, since it would require filters with extremely large components. Therefore, all PD detection systems have high-pass filters that remove all harmonics starting from a cut-off frequency of 20–30 kHz (f_1 in Figure 2.5).

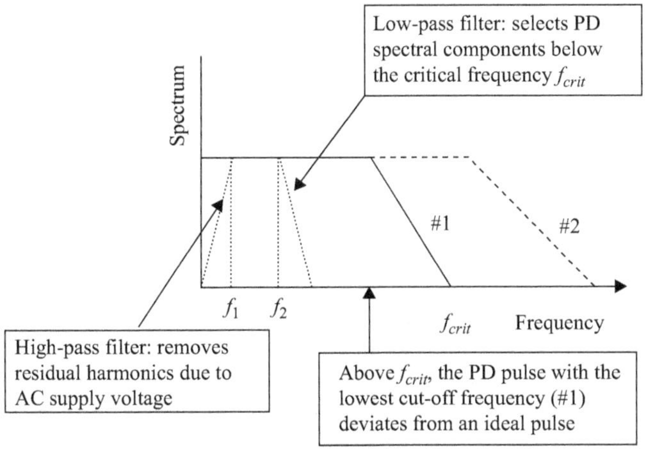

Figure 2.5 PD pulse spectrum

Since measuring the DC component of the current is infeasible, the Std. IEC 60270 suggests an indirect estimation procedure. This procedure is based on the hypothesis that the PD pulse resembles, up to some cut-off frequency (see f_{crit} in Figure 2.5) an ideal pulse, i.e. all spectral components have exactly the same magnitude, equal to the DC component. Thus, if the PD pulse is processed by a cascade filter consisting of a high-pass filter [transfer function $HP(f)$] with cut-off frequency f_1 and a low-pass filter [transfer function $LP(f)$] with a cut-off frequency f_2 satisfying

$$f_2 \ll f_{crit} \tag{2.7}$$

The output of the cascaded filter will be a signal whose spectrum is

$$X(f) = HP(f)LP(f)I_{DC} \tag{2.8}$$

Being I_{DC} the average current of the PD pulse.

In other words, if $h(t)$ is the impulse response function of the cascaded filter [i.e. the inverse Laplace transform of $HP(s)LP(s)$], the output of the processing unit, in the time domain, is

$$x(t) = h(t)I_{DC} \tag{2.9}$$

Hence, any feature of $x(t)$ that scales linearly with a constant (e.g. peak value, root mean square (RMS) value) provides information about the average current of the pulse, since I_{DC} times the duration of the pulse provides the apparent charge of the PD. Thus, the determination of I_{DC} is the key to get the apparent charge. Since the maximum signal-to-noise ratio is obtained measuring the peak value of $x(t)$ (and is also easy to implement using analog circuitry), it is customary to measure the peak voltage of the cascaded filter output (known as quasi-integrator filter) as a mean to determine, indirectly, the apparent charge

$$\max(x(t)) = \max(h(t))I_{DC} = kI_{DC} \tag{2.10}$$

In order to get the actual DC component of the pulse, a calibration procedure must be used. Calibration involves injecting a pulse of known amplitude using a calibrator device, as the peak voltage is measured in mV, while the apparent charge is expressed in pC, so that the calibration constant k can be derived as

$$k = \frac{\max(x(t))}{I_{DC}} \tag{2.11}$$

2.4.2.2 Limits

It is observed that the Std. IEC 60270 provides guidelines for selecting the cut-off frequencies, f_1 and f_2, of the quasi-integration filter. From the above discussion, it comes out that in $[f_1, f_2]$,[1] the spectral components of the PD pulse should be equal to the DC component of the pulse itself. However, there exist (at least) a couple of potential problems that might hamper remarkably a correct estimation:

- The frequency f_2 is too high. As a result, (2.8) does not hold anymore. In transformers, this happens when corona in oil is incepted. Corona in oil is due to conductive or semi-conductive protrusions (e.g. metal tools or wet ropes forgotten after maintenance) inducing arcs in the oil, as it was mentioned in Section 2.3.3.4. These arcs have generally low-frequency content.
- In inductive systems, where inductances and turn/turn or turn/ground capacitances might resonate at frequencies that are within the frequency range of the PD detector, if the transfer function from the injection point to the PD detector is the same for the calibrator and the PD source, the calibration procedure deals with the departure from the ideal condition depicted in Figure 2.5. However, the injection point of the calibrator and that of the PD source are different, and so are, in most cases, their transfer functions. In the presence of resonances, this might lead to gross estimation errors.

2.4.3 PD detection in transformers

2.4.3.1 Circuits for PD detection

There exist different scenarios to be considered in practice. The first consideration is whether the PD testing is to be performed in the factory or in the field. In the first case, external capacitors can be used to couple the PD signal. An indirect detection circuit as that sketched in Figure 2.6 will thus be assembled. The PD pulse will pass through the measurement impedance Z_m to be acquired for the acquisition system that will be adjusted and calibrated with the calibrator. Compared to other detection techniques, the indirect circuit is more sensitive and easier to implement. Furthermore, the apparent charge can be evaluated. The evaluation of apparent charge is extremely important for tests made in the factory, since standards have established apparent charge limits (e.g. apparent charge

[1] In reality, the range is a bit larger, since filters are not ideal.

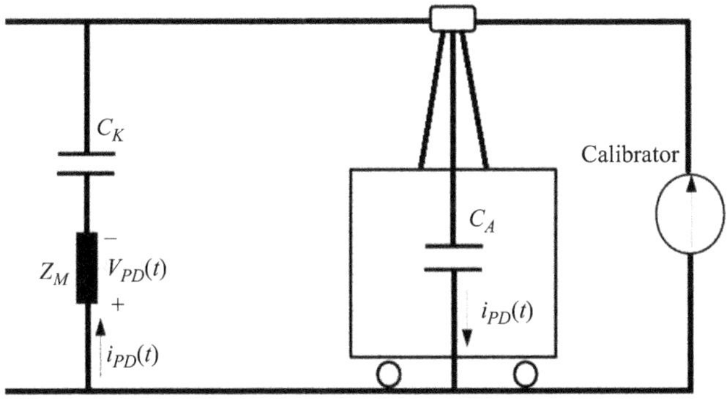

Figure 2.6 Indirect circuit using an external coupling capacitor

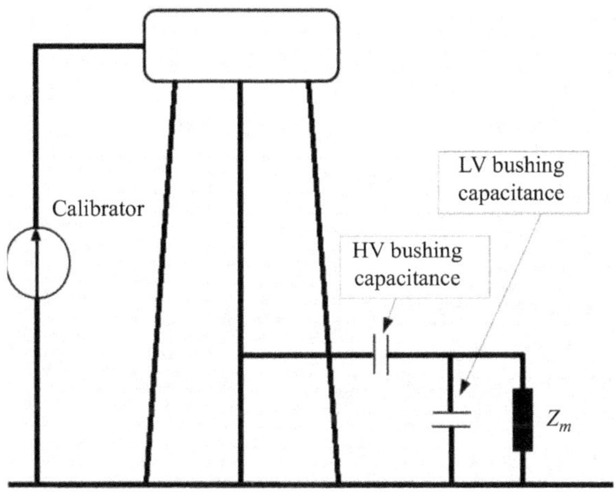

Figure 2.7 Coupling PD from bushing taps

should be lower than 300 pC) that are used to identify whether a new transformer is in good or bad conditions.[2]

In the field, dealing with HV transformers, this type of configuration is hardly usable due to the weight and size of HV coupling capacitors. If the bushing is installed with a bushing tap, detection can be performed via the bushing tap using the schematic shown in Figure 2.7.

2.4.3.2 Interference and noise

Most transformers are installed in air-insulated substations. Besides, noise due to radio stations, a number of PD events that are external to the transformer can be

[2] The authors do not support the use of apparent charge in transformers, since resonances can lead to very large errors in apparent charge estimation.

coupled during on-line monitoring. The most important disturbances arc corona and dry band arcs.

Corona, in air, are discharges that originate from bare conductors in the overhead lines connected the transformer. Corona phenomena are generally not very intense, but their repetition rate can be extremely high. This type of PD has a large number of discharge points, especially with old conductors, and low memory effect of space charge in the gas, that allows several PD events to take place in the same site during the same half cycle. Recognition of corona PD is generally easy (their PD pattern is easily recognizable), but their high repetition rate might prevent detection of PD from the transformer.

Dry band arcs are pulses that develop on the surface of the bushing (except when hydrophobic surfaces are used). Dry band arcs originate from hydrophilic conductive or semi-conductive pollution deposited on the bushing surface. When moisture is deposited on the surface, leakage current starts to flow on the surface itself. If the leakage current dries out, some parts of the surface create dry bands, and an arc develops on the surface of the band (dry band arcs). In general, dry band arcs are highly energetic and can be detected by the PD measuring system. However, their repetition rate is not very high, and their spectral characteristics might help to reject them from the measurements. One tool used to reject all kinds of noise for PD measurements with very good results is the wavelet transform (WT) [39].

2.4.3.3 Inductive sensors

To carry out electrical measurements according to the IEC 60270 standard, the inductive sensors commonly used for PD detection are high-frequency current transformers (HFCT) [40–42] and Rogowski coil (RC) [43–46]. Additionally, there exists an alternative inductive sensor with a marginal use for PD measurements, the inductive loop sensor (ILS) [47,48]. All inductive sensors operate on the basic principle of the Faraday's law. Accordingly, the core, generally toroidal to facilitate the construction and assembly of the sensor [40,49,50], is placed around the conductors through which the current pulses $i(t)$ associated to PD can be propagated. The circulating current produces a magnetic field which links the secondary of the coil and induces a voltage $e(t)$ proportional to the rate of change of the current in the conductor and the mutual inductance M between the coil and the conductor, as per the following equation:

$$e = M\frac{di}{dt} \tag{2.12}$$

The magnetic coupling allows the two circuits involved, the primary through which current flows and the secondary formed by the sensor coil, to be electrically isolated. Thus, allowing to obtain a non-intrusive measurement system, which requires not to alter the circuit under test. After analysing the operating principle of inductive sensors, it is possible to analyse the terminals output according to the simplified electrical model in Figure 2.8.

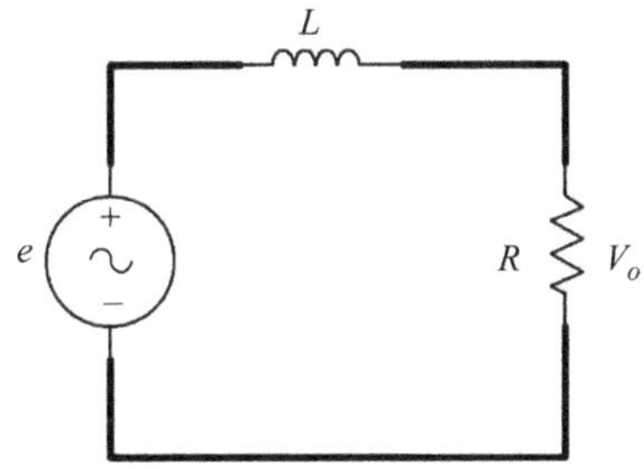

Figure 2.8 Electrical circuit of an inductive sensor

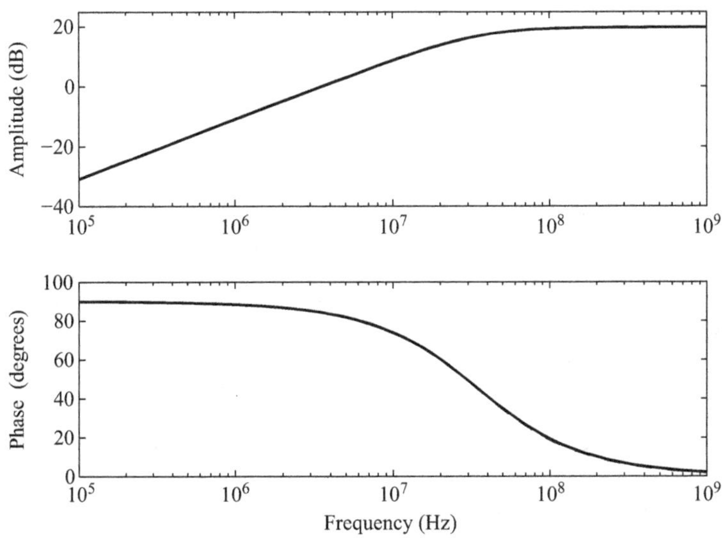

Figure 2.9 Transfer function of an inductive sensor

According to Figure 2.8, when the inductive sensor is in open-circuit mode, its output voltage $v_o(t)$ is equal to the induced voltage $e(t)$. However, when it connects to measuring systems with a finite input impedance *R* (acting as a load), the sensor output is modified, and its response together with the measuring system is determined from dividing the induced voltage, by the impedance (inductance *L* in series with resistance *R*). Therefore, a transfer function defined by circuit impedance is expressed as follows:

$$Z = \frac{V_o(s)}{I(s)} = \frac{RM_s}{Ls + R} \tag{2.13}$$

Z is depicted in Figure 2.9.[3]

[3] Any capacitive effect is neglected, or its effect is far above the frequencies of interest.

According to the above, the frequency response of any inductive sensor has two clearly distinct regions: a low-frequency region where the output signal is the derivative of the current and a flat region where the output signal is proportional to the current (high-frequency range). In Figure 2.9, these two regions are divided by f_c which represents the cut-off frequency or the derivative limit of the sensor response and can be calculated as

$$f_c = \frac{R}{2\pi L} \tag{2.14}$$

- HFCT: These are sensors with ferromagnetic cores with high-sensitivity and high self-inductance. As a result of this last characteristic, the cut-off frequency f_c and bandwidth are at lower frequencies, so the derivative behaviour is limited. This allows to obtain output signals proportional to the current for a high bandwidth [50].
- RCs: These toroidal inductive sensors differ from HFCT mainly because they have an air core or non-magnetic material, which provides linearity and low self-inductance [46,51]. This core allows to design sensors with lower weights, cheaper and more flexible, enabling wider application and ease of use. However, their main disadvantage when compared to HFCT is their sensitivity to resonance phenomena, due to parasitic capacitances at certain frequencies that vary with the geometry of the sensor [45]. RCs are, in general, less sensitive than HFCT for PD measurements [43].

Other type of inductive sensor is the ILS that is based on RC configurations, since the sensor is manufactured with a single turn or loop that can be of a square or a circular shape. This kind of sensors is marginally used for PD measurements due to its poor sensitivity. It differs from HFCT and RC in having an electric model without capacitive parameters, which makes it free of resonance or oscillation in the high-frequency range. This sensor has a lower value of mutual inductance M, which reduces its sensitivity. However, its low self-inductance L makes it to have a derivative behaviour for a wide range of frequencies and is virtually unaffected by low-frequency signals. More advantages include simple of design and low cost [47,48].

2.4.4 Unconventional methods of partial discharge measurements in power transformers

Other methods to measure PDs in transformers are based on acoustic and UHF emissions. Though they are considered unconventional methods, their use is now widely extended and applied as a complement to the measurements with conventional methods. The IEC 62478 standard [7] involves all these methods as a new framework using more sophisticated measuring and signal processing techniques to classify and identify PDs. The two methods presented below are best applied in oil-insulated power transformers since the propagation of acoustic waves and the insertion of UHF sensors is favoured by the liquid insulation.

2.4.4.1 Acoustic

As mentioned before, a PD is generated by a sudden electron avalanche resulting in a release of pulsed energy, and some of the released energy heats up the surrounding material adjoining the discharge site that results in liberating gases and creating a small explosion [52,53]. In an open volume full of transformer oil the acoustic pulse emission would propagate spherically at a speed in a range from 1,240 to 1,440 m/s depending on the temperature and the type of oil [11,54]. This emission is read with ultrasonic sensors based on piezoelectrics that are preferably placed inside the tank. In most cases, the discharge is located inside cavities formed by different types of insulating materials. The emission is then modulated by the cavity favouring the creation of standing waves and resulting in a damped resonant pulse. The wave may be subjected to some obstacles which attenuate the amplitude and can change the direction of the initial path. In general, acoustic waves have many propagation modes, but the most common are longitudinal and transverse (shear) modes. When the propagation is in oil, the main mode is only longitudinal. However, if the front wave hits the metallic tank, the propagation changes to longitudinal and transverse modes along the wall [11,52,53], and the propagation speed also changes to 5,900 and 3,200 m/s, respectively. This means that there are several types of signals arriving to an acoustic sensor mounted on the surface of the transformer tank depending on the path they follow. These paths can be classified as

- Direct path from the source to the sensor is travelling only through oil. Understanding as *direct* the path from the external surface of the area where the PD is located to the sensor.
- Alternative path through the steel tank for the structure-borne longitudinal mode signals which are the wavefront with highest speed.
- The same path through the tank for the structure borne transverse mode signals which are the second with the highest velocity.

Figure 2.10 shows an acoustic signal generated by a PD. The part of the signal inside the box arrives before the direct front wave and can induce errors in the detection and localization of PDs. Depending on the incidence angle and the distance to the sensor, wavefronts travelling through the direct path can be the first or the last to arrive [54]. Figure 2.11 shows different paths followed by the acoustic emission. If $\alpha = \beta = 90°$, the signal travels the shortest distance 1 through oil and the longest distance d_1 through the steel wall. On contrary, path 4 is the direct emission and the overall shortest distance though not necessarily the fastest path.

The acoustic methods have some advantages when compared to electrical methods to detect PDs [54,55] as they are easy to manipulate, install and replace because most of the sensors are placed on the external side of the transformer tank which is grounded. Acoustic methods are immune to electromagnetic noise induced by corona and other sources, and placing several sensors can help to locate PDs. Unfortunately, there are also some disadvantages. The location of the source is a difficult task considering all the problems derived from attenuation and structure-borne signals.

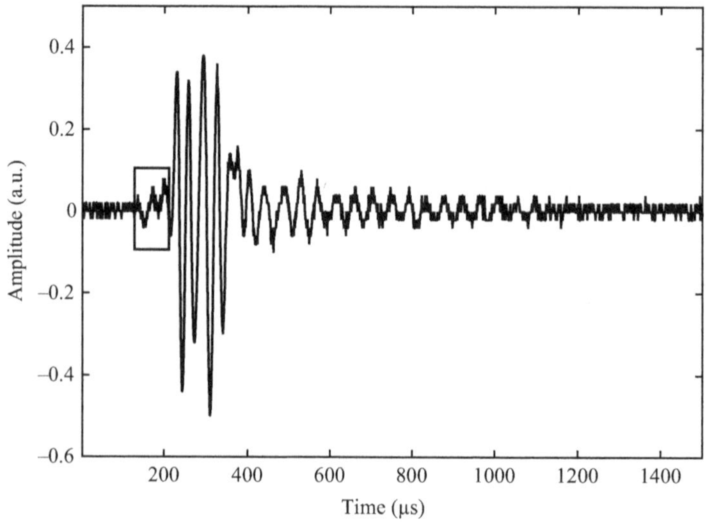

Figure 2.10 Acoustic signal created by a PD. The boxed portion is due to signals travelling faster than the direct front wave

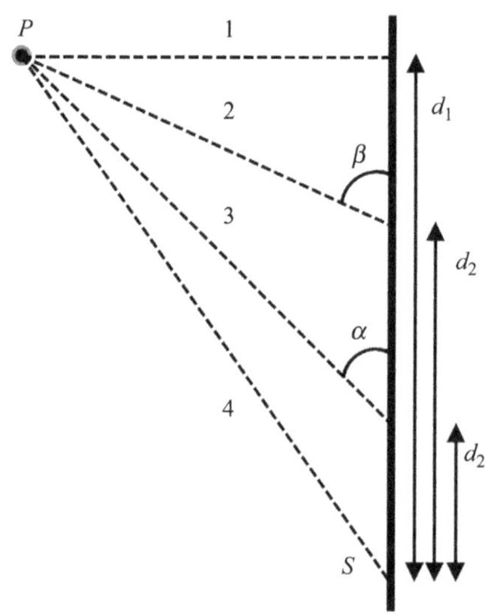

Figure 2.11 Different paths connecting the PD source (P) to the sensor (S)

This makes it very difficult to detect more than one source simultaneously discharging, and in most cases, it is necessary to combine several detection methods that include electrical detection [19] or UHF detection [11]. Additionally, to represent the data in a phase-resolved PD (PRPD) pattern, Section 2.4.5.1, it is necessary to have the phase voltage reference available which, in most cases, requires a capacitive divider that makes the measuring set-up more invasive.

There are two types of acoustic sensors, piezoelectric and fibre optic. Though they measure the same effect, their characteristics are completely different. The piezoelectric sensors are narrowband detectors with a window around 150 kHz and are installed in the external surface of the tank, so they have to cope with all the problems presented above. The acoustic sensors can be based on fibre optic intrinsic interferometers such as all-fibre Michelson interferometers and Mach–Zehnder interferometers, Fabry–Perot cavities and fibre Bragg gratings [56,57]. They can be installed inside the transformer close to the expected PDs location. This would reduce the multipath emission issues and would permit an easier detection and localization of the source.

2.4.4.2 Ultra-high frequency

The derivative of the current pulse due to the electron avalanche involved in the PD creates an electromagnetic emission proportional to that derivative. UHF sensors based on antennas or capacitive coupling can detect these emissions when placed inside the transformer. There are several types of sensors: monopoles, printed antennas, aperture antennas and disc couplers, but all of them have some basic characteristics in common. They all are classified as sensors because the frequency emission of the discharge can be very different depending on the type and the site of inception; and all sensors are omnidirectional to cover the widest area as possible. The range of frequencies can start as low as 100 MHz invading the VHF zone, whereas the top frequency does not usually surpasses 2,000 MHz because at these frequencies, the emission is very low.

A common classification of these sensors is based on the type of mounting on the transformer [58]. Thus, it is possible to find window sensors, probe sensors and internal sensors [59].

The window sensors have the advantage of being able to be installed or replaced with the transformer in use. Tests have been carried out using dielectric windows to attach UHF capacitive sensors in [60–62] for the detection of PDs.

Probe sensors are introduced inside the transformer tank through oil valves such as in [63,64].

Internal sensors are similar to window sensors with the advantage that the former have better sensitivity being effectively inside the transformer. This kind of sensors have to be included by the manufacturer in their designs.

In any case, it is very uncommon to count with more than one dielectric window or an available oil valve unless the manufacturer is committed to include them in their transformers. This makes it very difficult to take advantage of one of the main characteristics of UHF detection of PDs, namely the possibility of locating the source of the discharge. Most of the work done in this direction is based on experiments in transformer models or adapted transformers [62,65].

Recently, the use of an external coil capable of measuring the magnetic field of the radio-frequency emission of PDs inside the transformer has been reported in [66]. The sensor is placed on the external plate of the tank and is based on the fact that the electric field of the electromagnetic emission is attenuated in the metallic interface while the magnetic field can pass through it.

Localization

Locating the position of the source of emission in three dimensions is straightforward in open, noise-free and homogeneous environments with an array of four sensors. In such ideal conditions, and considering that the source is located at $\mathbf{P_s} = (x_s, y_s, z_s)$, the distance to sensor i at $\mathbf{P_i} = (x_i, y_i, z_i)$ is

$$\|\mathbf{P_i} - \mathbf{P_s}\| = \sqrt{(x_i - x_s)^2 + (y_i - y_s)^2 + (z_i - z_s)^2} \tag{2.15}$$

Therefore, the difference of distances between the source and the sensors i and j would be $D_{ij} = \|\mathbf{P_i} - \mathbf{P_s}\| - \|\mathbf{P_j} - \mathbf{P_s}\|$. Since the electromagnetic or acoustic wave propagates at a constant speed v, the difference of distances can be written as $D_{ij} = v \cdot (t_i - t_j)$ with $t_i - t_j$ the time difference of arrival (TDOA) of the front wave to sensors i and j. Then

$$v \cdot (t_i - t_j) - \|\mathbf{P_i} - \mathbf{P_s}\| + \|\mathbf{P_j} - \mathbf{P_s}\| = 0. \tag{2.16}$$

With four sensors, the total number of equations in (2.16) is six though only three are needed to solve the position of the source. The TDOA and the positions of the sensors are known, so the unknown position of the source $\mathbf{P_s}$ can be obtained through solving the non-linear set of equations in (2.16) using an iterative method such as Newton–Raphson.

Another approach to the solution is the use of optimization methods such as genetic algorithms (GAs), ant colony or particle swarm optimization (PSO) to minimize an objective function $f(\cdot)$. Let $\hat{\mathbf{P}}_\mathbf{s} = (\hat{x}_s, \hat{y}_s, \hat{z}_s)$ be the estimation of the source position. Extending (2.16) to all pairs of sensors and summing all equations together gives

$$f(\hat{x}_s, \hat{y}_s, \hat{z}_s) = \sum_{i=1}^{L-1} \sum_{j=i+1}^{L} \left(v \cdot (t_i - t_j) - \|\mathbf{P_i} - \hat{\mathbf{P}}_\mathbf{s}\| + \|\mathbf{P_j} - \hat{\mathbf{P}}_\mathbf{s}\|\right)^2 \tag{2.17}$$

where L is the number of sensor. Equation (2.17) would be 0 when $\hat{\mathbf{P}}_\mathbf{s}$ is located in the correct position of the source so the objective of the optimization method is to minimize $f(\hat{x}_s, \hat{y}_s, \hat{z}_s)$.

Considering that the TDOA are measured with a certain degree of uncertainty, it is found that the most accurate results for the position of the source are obtained when the sensors are placed surrounding it; otherwise, the solutions to (2.16) or (2.17) are dispersed along the direction of the emission.

In the particular case of a transformer, the deployment of sensors should be placed on the vertices of the tank to ensure that the source is completely surrounded. However, the discharge site would usually be located in the HV coil so the electromagnetic and acoustic front waves will be transmitted through different media, paper and oil or other dielectrics that will change the speed of propagation and reducing its amplitude. Other important effects, mostly affecting the radio frequency front wave, are multipath reflections in conductive surfaces such as copper, aluminium, iron or steel. Consequently, the signal received by the sensors has been modified by the windings and core and the multiple reflections due to the resonant cavity formed by the transformer tank hindering the finding of a correct solution to the position of the source. There are some works proposing techniques to take these effects into account, but the models are not exhaustive having to assume simplifications and lesser constraints to be able to formulate a solution.

2.4.5 Methods of partial discharge analysis

During the past decades, there has been an extensive research for automated identification of the origin of PDs to evaluate the condition of the equipment. With the development of new techniques in machine learning during the decades of 1960s–1990s in the last century such as artificial neural networks (ANN) and support vector machines (SVMs) or optimization techniques such as PSO, GAs or ant colony, many different approaches have been applied in the study of the signals derived from PDs. All processes can be more or less complicated; however, they all have some common characteristics and follow the same procedure below:

- Selection of the features that better describe PDs against noise or other types of PDs. These features can be statistical parameters extracted from the histograms of the pulses in phase-resolved patterns or time–frequency (TF) characteristics of the pulses.
- Preprocessing of the pulse in the time, frequency or TF domains to extract those features more accurately.
- Use of SVMs or neural networks or other techniques to reduce data dimension such as principal component analysis. These methods can classify events in different classes with a previous training with individual types of PDs. This means that the intelligence of the method relies on an appropriate data base of PDs belonging to the types that are going to be identified.
- Use of a clustering technique to separate different types of pulses. The parameters of the clustering or the identification techniques can be selected using optimization techniques such as PSO.
- Finally, the use of the identification technique (SVM, ANN, ...) to label the clusters and identify the type of discharge.

All approaches, shown in next sections, pursue three main goals sorted by their degree of difficulty:

- Identification of PDs and noise rejection. In this step, it is not necessary to classify the type of PD source, it is sufficient to know if the pulse is effectively from a PD source.
- Classification of PDs based on the interpretation of $\phi - q - n$ and PRPD patterns, Section 2.4.5.1. These patterns are independent from the set-up so the features can be extrapolated to any type of HV asset.
- Classification of PDs based on time, frequency or statistical characteristics of the pulses. In most cases, these features are strongly dependent on the electrical set-up, so the pulse acquired by the acquisition system is a convolution

of the actual pulse and the transfer function of the physic and electrical set-up that transports the signal. TF maps and power-ratio representations are examples used in identifying PD events and are explained in Sections 2.4.5.2 and 2.4.5.3, respectively, but there are many more tools based on other parameters.

2.4.5.1 Phase-resolved partial discharge patterns

Detecting PDs is the first stage in a condition-based monitoring system. The truly valid information should determine the nature and origin of the PD in order to discriminate the insulation problem and set up an efficient maintenance. Many efforts have been directed to obtain fingerprints that are common to PD from similar sources. One of the most popular and trusted methods among maintenance engineers is based on the collection of signals with the same phase referred to the grid frequency, and, hence, PRPD patterns are analysed. The shape and duration of the pulses are common characteristics for pulses of the same nature, so other methods are based on the extraction of the time and spectral characteristics of the signal such as time-resolved tools and TF maps. Power spectral density and spectral power share in different frequency bands have also been proved to be reliable techniques to extract characteristics of the nature of the discharge. PRPD patterns are a conventional electrical classification method used to identify PD sources and noise [67]. This pattern representation is based on the amplitude or peak acquisition from the pulses, thus measurements in a frequency band below 500 kHz are enough according to IEC 60270 standard. The amplitudes of the signals, PD or noise, in mV or pC versus the phase in which they occur referred to the AC voltage reference from the grid are represented. In this plot, the whole PD phases and their amplitudes are represented in one cycle of the ac voltage. Each cluster of points has a typical feature representation (pattern) depending on the type of PD or noise that can be identified by an expert, in some cases after preprocessing. By identifying the patterns, their evolution in time and by a post-processing that allow the location of the source it would be possible to find the origin of the fault. In order to show typical PD features in power transformers, below are shown PRPD from tests made in the laboratory with artificial defects. Internal discharges are the most dangerous PD for the insulation system in a transformer such as for cellulose insulation type and its typical cluster is shown in Figure 2.12(a). The test object is composed by five layers of Kraft paper immersed in oil, and the PD inception voltage (PDIV) in this case is produced at 19 kV.

Surface discharges can be established on pressboard paper surface, and their clusters are typically accumulated around the maximum and minimum of the AC signal. Figure 2.12(b) represents surface PD on pressboard paper surface with a PDIV of 20 kV.

In a transformer, apart from internal and surface PD, it is possible to measure activity from other PD sources such as corona in oil, PD gas bubbles in oil and PD in floating metal particles. These three types of PD are of less dangerous because affect the oil that can be dielectrically regenerated [68].

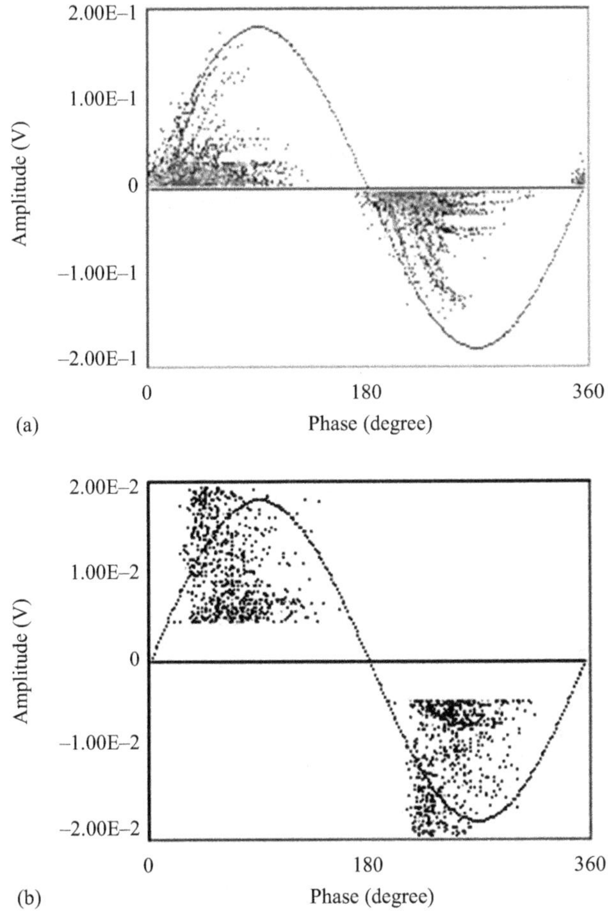

Figure 2.12 Internal and surface PD: (a) internal PD in Kraft paper and (b) surface PD on pressboard paper surface. © 2010 IEEE. Reprinted, with permission, from [68]

Figure 2.13(a) shows corona PD in a point/plane configuration in the presence of moisture: 60 ppm, $V = 27$ kV, with a distance between the electrodes of 20 mm. In order to obtain PD due to bubbles moving freely in oil, a test object with two vertical plane electrodes separated 12 mm was used. To avoid flash-over, à sheet of cellulose paper was deployed on the surface of each electrode.

Air bubbles was produced by a pump and their number and size could be adjusted by controlling pump's valve, the PRPD pattern is plotted in Figure 2.13(b) and the PDIV was 28 kV.

Figure 2.14(a) shows the PRPD representation due to small floating metal particles, e.g. rust from tank walls, in oil with a PDIV of 4 kV.

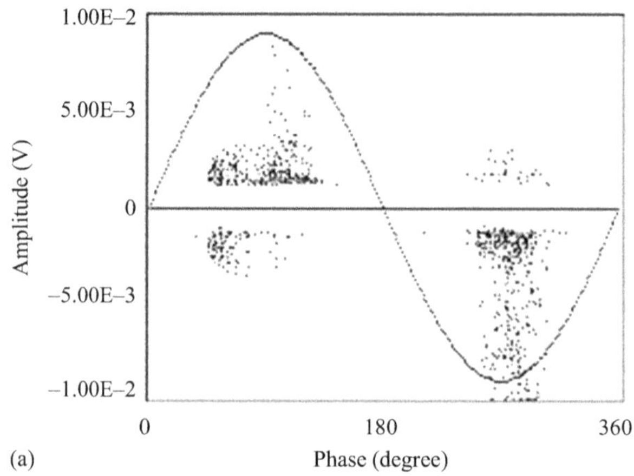

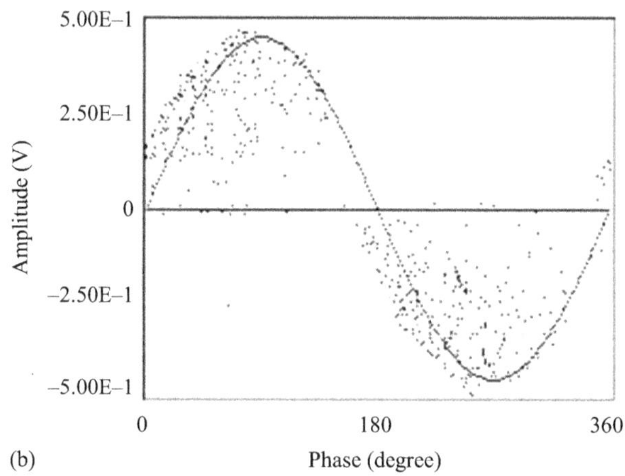

Figure 2.13 Corona PD and PD in bubbles in oil: (a) corona PD in oil and (b) bubbles floating in oil. © 2010 IEEE. Reprinted, with permission, from [68]

The main characteristics of PRPD patterns in transformers are summarized in Table 2.1.

In a transformer, several signal sources can be acting at the same time. Figure 2.14(b) shows a PRPD pattern in a 220-kV distribution transformer with simultaneous pulse sources during the acquisition. To distinguish the type of source it would be required additional methods such as those explained below.

2.4.5.2 Time–frequency maps

TF maps is a classification tool commonly used in the signal theory field and considers the separation of several groups of clusters from signals with similarity in their waveform and duration [69,70]. This technique allows to separate PD and noise pulses from different sources that have different waveforms and pulse time-length.

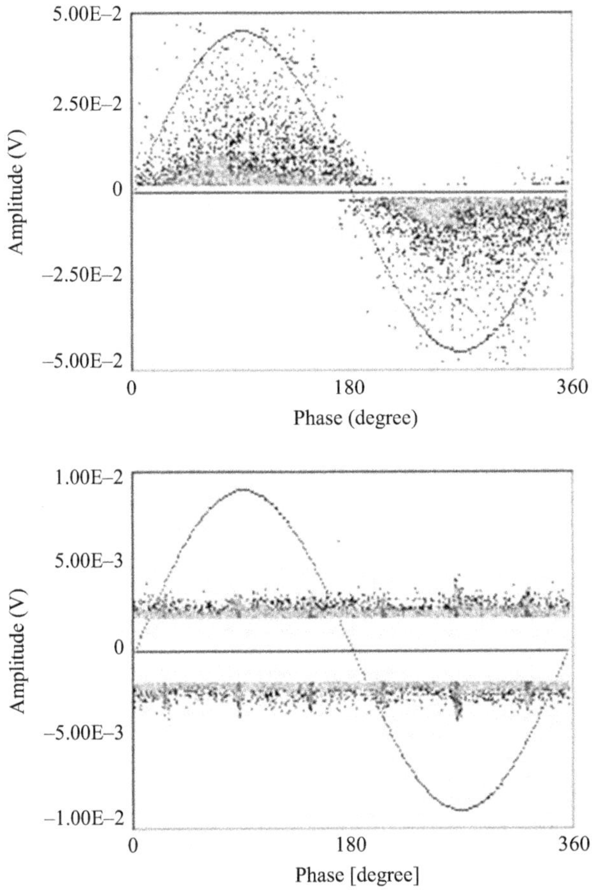

Figure 2.14 PD in floating metal particles in oil and internal PD and noise: (a) PD in floating metal particles in oil and (b) internal PD source 'submerged' in noise in a transformer. © 2010 IEEE. Reprinted, with permission, from [68]

As an example, noise with similar duration and spectrum will be grouped in the same cluster; interference pulses are expected to have different characteristics than PD and noise, so they will be represented in other zone of the TF map; finally, PD with different waveforms and time duration will be plotted in other areas.

To represent these maps, it is necessary to calculate the equivalent time duration and the equivalent bandwidth for all pulses. The signal-processing begins by normalizing all pulses with time, $s(t)$:

$$\tilde{s}(t) = \frac{s(t)}{\sqrt{\int_0^T s(t)^2\,dt}} \tag{2.18}$$

T being the time-length for each pulse.

Table 2.1 PD in transformers and their PRPD

Type of source	Features
Internal PD in paper layers	Marked symmetry between positive and negative pulses Clusters span from zero crossings, 0° or 180°, of the ac voltage
Surface PD on paper surface	Marked symmetry between positive and negative pulses Clusters are typically accumulated around the maximum and minimum of the ac signal
Corona PD in oil	Marked asymmetry between positive and negative pulses PD clusters symmetrically concentrated around 90° and 270° Remarkable difference between positive and negative PD pulses
Gas bubbles in oil	Symmetry between positive and negative pulses Bubble disruption leads to PD activity extinction due to PD activity in small bubbles displays larger inception fields It has stochastic cycles in PD activity
PD in floating metal particles	Symmetric behaviour with most of discharges range between 0° and 90° (180°–270°)
Noise	Noise is not correlated with the phase of the sinusoidal signal, so in a PRPD plot it is represented as a dispersed band of points at all phases

Second, the standard deviation of the normalized signal in the time domain is calculated in (2.19), and in the frequency domain in (2.20).

$$\sigma_T = \sqrt{\int_0^T \left(t - t_g\right)^2 \tilde{s}(t)^2\, dt} \tag{2.19}$$

$$\sigma_F = \sqrt{\int_0^\infty f^2 \left|\tilde{S}(f)\right|^2\, df} \tag{2.20}$$

where f is the frequency, $\tilde{S}(f)$ is the signal's Fourier transform $\tilde{s}(t)$ and t_g is the time gravity centre for the normalized signal, defined by

$$t_g = \int_0^T t\tilde{s}(t)^2\, dt \tag{2.21}$$

In Figure 2.15, a TF map obtained from the PRPD data in Figure 2.14(b) is represented. With this tool, it is possible to obtain two different clusters that allow to separate different sources. With post-processing and selecting only each cluster, it is possible to obtain a typical PRPD pattern from background noise, and internal PD from a XLPE power cable connected to the HV side of the transformer. So, with this kind of techniques, it is possible to identify each kind of source from noise to PD.

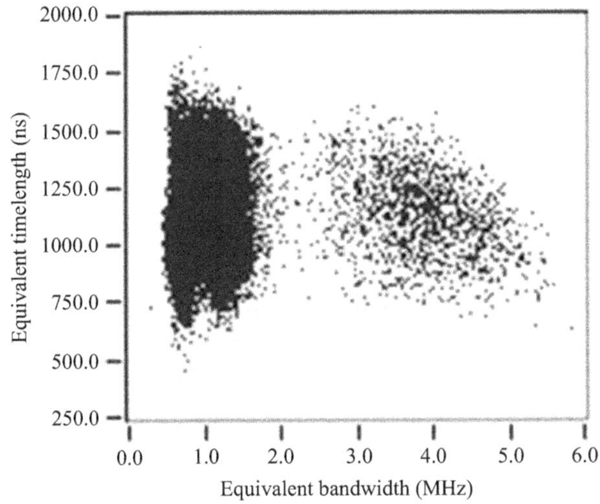

Figure 2.15 TF map with three PD sources in a transformer. © 2010 IEEE. Reprinted, with permission, from [68]

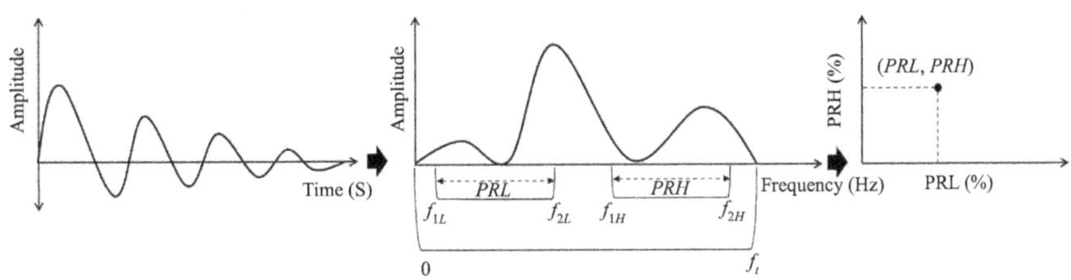

Figure 2.16 Steps to represent a power-ratio map

2.4.5.3 Power-ratio maps

The frequency spectra of the PD pulses change their shape and content depending on the source of PD, even for the same experimental set-up [71]. The power-ratio map technique is based on the analysis of the relative spectral power of each of the signals (PD and noise) captured in the acquisition.

The signal processing starts with the fast Fourier transform applied to each signal $s(t)$ obtaining the spectral magnitude distribution $S(f)$ for each pulse. Next, the spectral power is calculated for two different bands, one in power ratio for low-frequencies (PRL) and another in power ratio for high-frequencies (PRH). The percentages or power ratios are calculated with (2.22) and (2.23) which finally are represented in a two-dimensional map [72], this process is described in Figure 2.16. To prevent the total spectral power of the signal, or amplitude, from having any influence on the characterization of each signal, the spectral power accumulated in the bands *PRL* and *PRH* is divided by the spectral power of the entire signal.

$$\%PRL = \frac{\sum_{f_{1L}}^{f_{2L}} |S(f)|^2}{\sum_0^{f_t} |S(f)|^2} \times 100 \tag{2.22}$$

$$\%PRH = \frac{\sum_{f_{1H}}^{f_{2H}} |S(f)|^2}{\sum_0^{f_t} |S(f)|^2} \times 100 \tag{2.23}$$

where both *PRL* and *PRH* have units in %. f_{1L}, f_{2L}, f_{1H} and f_{2H} are the lower and upper limits of each band. Being f_t the maximum frequency under analysis, i.e. the maximum frequency where spectral content was detected.

Regarding to the frequency bands selected manually or automatically, it is recommended that

- $0 \leq f_{1L} < f_{2L}$, $f_{1H} < f_{2H} \leq f_t$ and $f_{1L} < f_{2H}$. This will avoid losing the perspective of the represented interval.
- Note that as described in the previous point, *PRL* and *PRH* can be complementary or overlapping.

In some cases, there can be sources with very similar spectrum distribution, so the their power ratio (PR) clusters would be overlapped in the map. The effective pulse time-width (t_{eff}) gives the duration of the signal and can be used as third axis to simplify the separation between clusters [73]. Figure 2.17 shows the clusters of several electromagnetic sources measured with an antenna. The t_{eff} corresponds to the width of a rectangular pulse with the same peak as a squared signal whose integral gives the same energy than the original pulse, so both have the same area, and it is calculated as follows:

$$t_{eff} = \frac{\int_0^T \tilde{s}(t)^2\, dt}{\tilde{s}(t)^2_{max}} = \frac{1}{\tilde{s}(t)^2_{max}} \tag{2.24}$$

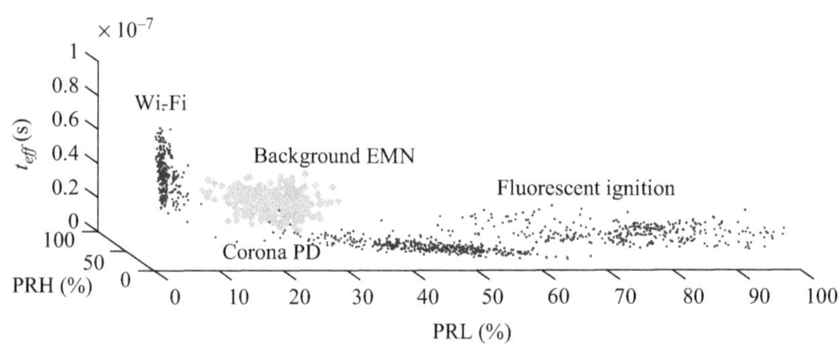

Figure 2.17 Power ratio-time map for several electromagnetic sources and electromagnetic noise (EMN) measured with a Vivaldi antenna. © 2015 Elsevier. Reprinted, with permission, from [73]

where the numerator of the fraction is the integral value of the pulse for the whole acquisition window, being 1 because it is normalized, and $\tilde{s}(t)_{max}$ is the maximum value of the signal.

2.4.5.4 Key characteristics identification

As it is known, the occurrence and magnitude of PD in an insulation system depends on the applied voltage and a series of mechanical–statistical parameters that make it a non-deterministic phenomenon [74]. Therefore, to characterize a single event is not relevant, and the parameters associated with all of the detected pulses should be statistically treated [75–77]. For this reason, the analysis of data obtained during a test of PD requires using a number of statistical parameters that provide relevant information about the type of PD and/or the degree of insulation deterioration [78].

In the case of power transformers, there is no clear consensus on which parameters are most suitable to characterize PD in the insulation system, because many of these parameters may vary depending on the weather, aging and the instrumentation used [79]. However, most of the works focused on identifying PD source are based on the PRPD pattern analysis that allows to identify uniquely parameters associated for each type of PD, without being affected by the experimental set-up used or applied voltage level applied during measurement. This PD statistical parameters are the skewness S_k, kurtosis K_u, mean, variance and the cross-correlation factor, which are obtained for the three different statistical distributions [74]:

- $H_{qmax}(\phi)$: Maximum value from the pulses (mV or pC) produced at each phase angle ϕ.
- $H_{qn}(\phi)$: Mean value of the pulses (mV or pC) produced at each phase angle ϕ.
- $H_n(\phi)$: Number of PD events at each phase angle ϕ.

Considering that $H_{qmax}(\phi)$, $H_{qn}(\phi)$ and $H_n(\phi)$ have different distributions in each half cycle of the voltage waveform reference, each of these distribution is divided into two positive (+) and negative (−) semi cycles, thus six different statistical distributions are obtained: $H^+_{qmax}(\phi)$, $H^-_{qmax}(\phi)$, $H^+_{qn}(\phi)$, $H^-_{qn}(\phi)$, $H^+_n(\phi)$ and $H^-_n(\phi)$. Additionally, there are detailed descriptions of each of the statistical parameters applied to each of the distributions [74,80–83]:

- **Skewness:** Describes the asymmetry of the distribution with respect to a normal distribution. For a normal distribution, $S_k = 0$, if it is skewed to the left, $S_k > 0$, and if it is skewed to the right, $S_k < 0$.

$$S_k = \frac{\sum_{i=1}^{n} (x_i - \mu)^3 f(x_i)}{\sigma^3 \sum_{i=1}^{n} f(x_i)} \tag{2.25}$$

where x_i is the amplitude value for each PD pulse and $f(x_i)$ is the function of interest.

- **Average or mean:** Refers to one value of the central tendency characterized by the distribution.

$$\mu = \frac{\sum_{i=1}^{n} x_i f(x_i)}{\sum_{i=1}^{n} f(x_i)} \tag{2.26}$$

- **Variance:** Is the expectation of the squared deviation of a random variable compared with its mean.

$$\sigma^2 = \frac{\sum_{i=1}^{n} (x_i \mu)^2 f(x_i)}{\sum_{i=1}^{n} f(x_i)} \tag{2.27}$$

- **Kurtosis:** Compares the slope of the distribution with respect to a normal distribution. If the slope of the distribution is equal to that of a normal, then $K_u = 0$, if the distribution is steeper, $K_u > 0$, and if is flattest, $K_u < 0$.

$$K_u = \frac{\sum_{i=1}^{n} (x_i - \mu)^4 f(x_i)}{\sigma^4 \sum_{i=1}^{n} f(x_i)} - 3 \tag{2.28}$$

- **Cross-correlation factor:** It shows the difference in shape between the distributions of positive and negative half cycles. A $cc = 1$ value indicates that the distribution of the two half cycles have the same shape, a $cc = 0$ value indicates that the two distributions are different.

$$cc = \frac{\sum_{i=1}^{n} x_i y_i - \sum_{i=1}^{n} x_i \sum_{i=1}^{n} (y_i/n)}{\sqrt{\left[\sum_{i=1}^{n} x_i^2 - \left(\left(\sum_{i=1}^{n} y_i\right)^2/n\right)\right]\left[\sum_{i=1}^{n} y_i^2 - \left(\left(\sum_{i=1}^{n} y_i\right)^2/n\right)\right]}} \tag{2.29}$$

  where x_i is the amplitude value in the positive half cycle, y_i is the amplitude value in the negative half cycle and n is the number of phase windows per half cycle.

In addition to the parameters described above, also it is used ϕ_i (PD inception phase angle) or ϕ_m (mean phase angle), based on the PRPD, which have been used successfully in PD identifying [84].

Another statistical tool that has been widely implemented in PD characterizing is based on the analysis of the amplitudes of the PD pulses using the Weibull probability distribution (2.30).

$$P(q) = 1 - e^{(q/\alpha)^{\beta}} \tag{2.30}$$

where $P(q)$ is the probability that a discharge occurs with an amplitude equal or lower to x_i (pulse magnitude), α is analogous to the definition of the average for a normal distribution and β is a measure of the variability of the PD magnitudes. A small value of β is associated with high variability, i.e. there will be much difference between the high and low PD magnitude detected in the test. Each of these values can be calculated using the least squares regression [70,84,85].

Then, the use of the approaches that have been explained above, allow to address the problem of identifying PD and noise sources, permitting the classification of PD in function of their dangerousness degree for the insulation system.

2.4.5.5 Machine learning

Two of the most popular algorithms in the machine-learning process are explained in the next sections. These are simply examples that help to understand what is the underlying process of classifying events. Of course, other techniques can be applied as identification algorithms.

Introduction to support vector machines

The aim of the SVM is to find an optimal separation of two classes of features that define the events, namely, PDs [86–88]. This separation is done with a hyperplane that is located where the minimum distance between two points of different classes is maximum, Figure 2.18.

The linear classifier or hyperplane is defined with $f(\mathbf{x}) = \mathbf{w}^T\mathbf{x} + b$ where $\mathbf{x}$ is a set of the space of features, $\mathbf{w}$ determines the orientation of the hyperplane and b is bias parameter or the separation of the hyperplane from the origin. The decision of classifying an event in one class or other is based on the sign of $\mathbf{w}^T\ \mathbf{x}$. As shown in Figure 2.18, if $\mathbf{w}^T\ \mathbf{x} > 1$, positive and greater than 1, the event falls in the class labelled as cross, while if $\mathbf{w}^T\ \mathbf{x} < 1$, negative and lower than -1, the event would fall in the circle class. The dotted lines $\mathbf{w}^T\ \mathbf{x} = \pm 1$ are the margins that separate the classes, and $\mathbf{w}^T\ \mathbf{x} = 0$ is precisely the hyperplane. The training of the SVM with data from two different classes is done to define the hyperplane and the decision of labelling the test data in one or another class is done with the sign of $f(\mathbf{x})$. If the classification is non-linear, the data can be mapped in another

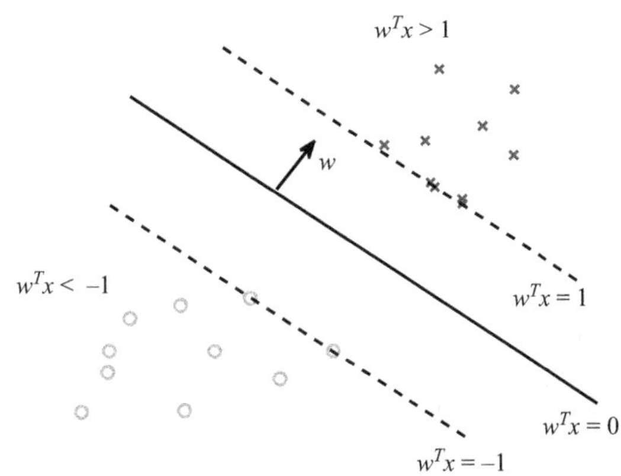

Figure 2.18 Linear discriminant function, solid line, with the classification margin in dotted lines. © 2016 Elsevier. Reprinted, with permission, from [86]

feature space using the kernel functions $\kappa(\mathbf{x}_i, \mathbf{x})$, and the decision is defined by the following equation:

$$f(\mathbf{x}) = \text{sign}\left(\sum_{i=1}^{N} \alpha_i y_i \kappa(\mathbf{x}_i, \mathbf{x}) + b\right). \tag{2.31}$$

The SVM would be now a linear combination of those kernel functions centred on the N training data samples $\mathbf{x}$, with their corresponding labels y. The coefficients of the linear combination, α, and the bias term b are the parameters to be leart in the training stage.

The function in (2.31) is evaluated for all the characteristics extracted from the PD. If the value of $f(x)$ falls clearly on one side of the hyperplane, the event is classified with that the label of all events in the training set that had the same characteristics. Therefore, if the training set was a collection of internal discharges, that event would be labelled as internal PD.

Introduction to particle swarm optimization

This method is mainly used in the geometric localization of PDs finding the solution of an optimizing function that gives the position of the source. Once the position is known, the asset can be further investigated to determine the type of problem that presents. It consists on the random distribution of a flock of particles in the solutions space defined by the parameters that are going to be optimized [89]. In every iteration step, l, all particles are moved around changing their position by the addition of a positive or negative incremental component. The objective function is then evaluated for every entity updating, if necessary, the combination of parameters with the best personal solution for that particle. When all particles have moved, the position of the particle with the overall best is stored as the global best. In the next iterations, the movement of the particles is modified by a weighted component that pulls the particle to its own best and another weighted component that guides the particle to the global best [90]. The following set of equations defines the position $\mathbf{P_n}$ and the speed $\mathbf{v_n}$ of the particle $\mathbf{n}$ in every iteration:

$$\begin{aligned} \mathbf{v_n}(l+1) &= \mathbf{v_n}(l) + \mathbf{U_1}(0,1) \otimes \left[\mathbf{P_{n,b}}(l) - \mathbf{P_n}(l)\right] + \mathbf{U_2}(0,1) \otimes \left[\mathbf{P_b}(l) - \mathbf{P_n}(l)\right], \\ \mathbf{P_n}(l+1) &= \mathbf{P_n}(l) + \mathbf{v_n}(l+1), \end{aligned} \tag{2.32}$$

$\mathbf{U_1}(0,1)$ and $\mathbf{U_2}(0,1)$ are line matrices with as many elements as parameters to optimize uniformly distributed between 0 and 1 that randomizes the movement of the particles towards their own best $\mathbf{P_{n,b}}$ and the swarm best $\mathbf{P_b}$, respectively. The operator $\otimes$ multiplies the random numbers by the coordinates in the n-dimensions space component by component. Equation (2.32) can be complemented with an inertia that modifies the velocity of the particles and two correction factors that multiply the uniform distributions. The process is ended when the changes in the maximum D are negligible or when the number of iterations reaches a defined number.

Then, with the optimum of the function found, the source position is known, and it is possible to analyse the asset and even to know the type of PD. In the case of having more than one source, the function would find as many optima as sources as long as they are sufficiently separated.

2.4.5.6 Comments on different approaches

There exist hundreds of bibliographic references and approaches when dealing with the process of identifying PDs. The following references are examples that have been selected to show how different authors face the process described in the beginning of this section. By no means they represent an exhaustive compilation of the vast literature.

In [91], the authors prepare a test object to measure PDs in transformers and evaluate the possible transmission of the pulses to a remote acquisition system through an electro-optic modulator. The pulses are generated with a calibrator so the charge injected in the circuit is known. They report that the measurement sensitivity may be conditioned to the optical noise introduced by the modulator being the minimum detectable PD level close to 160 pC. The goal is to detect PDs in signals where the signal-to-noise ratio (SNR) is very low and improve the minimum detection level. They use SVMs to classify the events as noise or PDs testing the information contained in the frequency spectrum of the pulses and in their wavelet decomposition. The conclusions shows that, in their case, the spectrum is not an option since the identification of PDs below 160 pC is not consistent. However, they found a level in the wavelet decomposition that is very different from noise and, using that feature, they are able to identify PDs of 30 pC with a hit rate of 99.5%.

Reference [92] also uses a combination of WT and SVM to de-noise PD signals. In this case, the algorithm calculates the modula maxima line of all scales in the WT. The evolution of that line from large to small scales is used as the feature that distinguishes between PDs, where it always has a maximum in the same scale and noise where the shapes of the lines are random. The SVM is trained with characteristic lines for noise and PDs to classify the events.

In [86], the spectra of PDs and noise is used to train a multiclass SVM. The algorithm is capable of separating different events generated in three test objects individually and a setup that generates two types of PDs and noise. Once the classes have been defined, the resulting events are plotted in a PRPD to identify the type of PD.

In [93], the authors define a set of 20 features to describe PDs with the purpose of finding what combination is the best to separate and classify the events with a SVM. These features are related to time: peak amplitude, phase angle, rise time, fall time, definite integral; frequency: mean, standard deviation, skewness, kurtosis, peak frequency; and TF characteristics of the pulses: wavelet energy ratio of nine detail levels plus approximation. Then, four SVM models were trained with these sets of characteristics: time, frequency, TF and all features using four different types of PDs. The results show that the wavelet energy set of features is sufficient to obtain high classification accuracies. The authors also point out an important common fact in all these methods, namely the characteristics of the pulses acquired by the sensor are strongly modulated by the transfer function defined by the path from the source to the sensor or the acquisition system itself. This fact makes difficult an identification of the source with the sole information of the signal and other tools such as PRPD patterns are needed.

In [88], PDs are identified through their phase-resolved patterns using a three-dimensional representation of parameters $\phi - q - n$. The measuring process followed by the authors is aimed at separating the training data from the testing data. In this regard, the training of the SVM is done with a classical PD detector that records the phase, charge and number of PDs, whereas the testing data are measurements taken with a HFCT. With this separation, the training could be done under controlled conditions while the testing would be done using field measurements. During the training, the $\phi - q - n$ features of individual and known types of PDs are fed to the SVM. Then, the same events are recorded with the HFCT and combined manually to have different sources simultaneously. Prior to the identification of multiple signals, it is necessary to gather the events in clusters of PDs of the same type. This is done using the energy of a specific detail of the WT of every type of pulse as pattern and comparing the rest of the data with a correlation technique. Once the clusters are defined and the different classes are separated, the type of discharge is decided by the SVM using the $\phi - q - n$ features of the training set.

The authors in [87] follow a more intricate process including techniques such as S-transform, affinity propagation clustering (APC), SVM and PSO. Their aim behind the process is the same as in the rest of approaches. The first step is the preparation of the signal to facilitate the separation in clusters using the S-transform. This transform can be seen as a time series $h(t)$ windowed with a normalized Gaussian or a continuous WT with Gaussian mother wavelet [94]:

$$S(\tau,f) = \int_{-\infty}^{\infty} h(t)\frac{\|f\|}{\sqrt{2\pi}}e^{-(\tau-t)^2 f^2/2}e^{-i2\pi ft}\,dt \qquad (2.33)$$

where f is the frequency or the inverse of the window width σ and τ is the translation. In a second step, the STA matrix $A(\tau,f) = \|S(\tau,f)\|$ is calculated as for every pulse. With this transformation, similar pulses would have similar STA matrices, and any clustering algorithm could be used to group the events. In the case of this work, the authors use the APC algorithm which is able to select the number of clusters automatically defining a preference parameter adequately. Once the clusters have been separated and the different types of PDs have been individualized, it is necessary to determine to label those types as corona, internal or surface PD. This is done parameterizing the pulses extracting 27 features based on classic statistical histograms of the signals: $H_{qmax}(\phi)$, $H_{qavg}(\phi)$, $H_n(\phi)$ and $H_n(q)$ and statistical operators over those histograms such as kurtosis, skewness, asymmetry and cross-correlation coefficient. An SVM can be trained with these features using single types of PD to label the type of pulse. The authors in [87] use an optimization method to determining two parameters of the SVM, namely the standard deviation of the Gaussian radial basis function (RBF), neural networks kernel and the penalty factor. These factors are included in an iterative optimization applying PSO and testing the classification accuracy of the SVM with a cross-validation on the training samples. The process is tested in four artificial defects models the generate PDs that can happen inside power transformers: cavity discharge in pressboard, surface discharges in oil, corona discharges in oil and discharges in air–oil interface. The results seem to be very promising since the events are separated and then identified correctly. However, there is not an explicit accuracy index of the classification and identification to evaluate the method. Additionally, all the process is very time-consuming specially in the first step when the similarity matrices are calculated and the determination of the preference parameter in the automatic clustering technique has some difficulties. Kuo *et al.* in [95] follow a similar procedure using the same statistical features to identify the discharge type and using a back propagation ANN optimized with PSO to classify the discharges. In this case, there is no preprocessing of the PD pulse in the time or frequency domains.

Other authors such as [96] report a recognition accuracy over 95% and their process is the same as explained at the beginning of the section. The features to characterize the pulse of PD are the peak charge, average charge and number of pulses in a predefined window. Then, the 1-norm, 2-norm and infinite-norm of those features are calculated, and the results are sent to different neural network configurations. These ANN have been previously trained with artificial test objects that generate only one specific type of PD. More examples on the use of ANN can be found in [97–100], for example.

2.5 Conclusions

Power transformers are one of the most important electrical asset in power grids. Besides, transformers are the links between networks, both transmission–distribution, as between consumption–generation and vice versa. Thus, a failure caused by a short-circuit in this type of electrical equipment can represent substantial damage depending on the load connected to its terminals. For this reason, it is crucial to know the keys for condition monitoring and diagnosis of power transformers in order to achieve a correct course of action (Run, Repair or Replace).

Transformers are commonly classified according to their insulation cooling system, such as air-cooled transformers, air cooling or cast resin, oil-immersed transformers, mineral oil, mineral oil with conductive nanoparticles and silicon-based liquid filled. Besides, transformers can be affected by ageing due to thermal, mechanical, electrical and ambient stress. These effects can gradually degraded the insulation system which can lead to weak points in the insulation system where a breakdown due to a short-circuit can be produced.

To prevent a short-circuit in the insulation system, it is key to know which kind of variables it is possible to be monitored. PDs are a clear symptom measurable that a potential weakness in the insulation is present. The aforementioned stresses, can be the source of this type of stochastic and measurable pulse. The use of condition monitoring by measuring PDs through electric, acoustic or electromagnetic techniques allow to know the health status of the insulation and estimate its lifespan if the measurement is carried out in a continuing basis.

The key parameters for electrical condition monitoring are the severity of the PD, mainly the pC measured, the number of pulses per power cycle, the type of PD and the localization of the source. Internal PD are the most dangerous discharges that can lead to an insulation breakdown, so it is crucial to identify them, to know its evolution in time and to localize where are they occurring. One of the main difficulties of PD acquisition is the presence of noise. These can be tackled through the use of digital de-noising such as using WT. Other important issue is to identify the type of PD, this issue can be carried out by using PRPD patterns, TF maps, PR maps, statistical analysis and a mixture of those. In this regard, the PRPD pattern representation is the classical technique and the basis of other techniques for PD classification. The newest trend in electrical condition monitoring is to use the above-mentioned methods of PD analysis and combining them with artificial intelligence and evolved techniques such as machine learning, SVM, PSO, ANN, fuzzy logic and GAs to implement an expert solution that offer a final report with all sources detected with the acquisition system in a transformer to carry out an effective CBM.

The present trend in PD measurement and identification methods for transformers is to have low-cost solutions to carry out condition monitoring of the insulation system when it will be required. This issue can be achieved by measuring PD activity with portable PD diagnosis equipment instead of continuous condition monitoring, or by preprocessing the acquisitions in each transformer point and, then, by sending these preprocessed information to a centralized server for a final deep analysis of a transformers group. In this regard, a combination of several analysis techniques are required, such as

- Noise rejection, i.e. using WT, for a better analysis of the pulses acquired.
- PD source identification and classification, i.e. using PRPD patterns assisted with artificial intelligence tools, i.e. ANN, SVM, Fuzzy logic, among others to improve the accuracy in identification and classification of PD.
- Statistical and tendency analysis of the PD evolution to know the severity of the PD source for the insulation system.
- Localization of PD sources combining different measurements techniques such as electrical, acoustic and UHF techniques.

Finally, in the two last decades, a significant progress is being made in the field of nanomaterials. In this regard, the new synthetic oils, in which conductive nanoparticles are added, can improve the dielectric strength of the oil, increasing the breakdown voltage level, and can enhance the heat dissipation of the transformer outwards of the tank. This technology is currently being investigated in depth and could be widely accepted as insulation system for transformers in the coming years.

Acknowledgements

The authors would like to thank the Spanish Ministry of Economy and Competitiveness for partially funding this work under projects DPI2015-71219-C2-2-R and DPI2015-66478-C2-1-R, and they would like to acknowledge the support of the Chilean Research Council (CONICYT), under the project Fondecyt 11160115.

References

[1] P. Gill. *Electrical Power Equipment Maintenance and Testing, Second Edition*. Power Engineering (Willis). Taylor & Francis, Boca Ratón, FL, 2008.

[2] H. L. Willis and R. R. Schrieber. *Aging Power Delivery Infrastructures, Second Edition*. Power Engineering (Willis). CRC Press, Boca Ratón, FL, 2016.

[3] J. López-Roldán, T. Tang, and M. Gaskin. Optimisation of a sensor for onsite detection of partial discharges in power transformers by the UHF method. *IEEE Transactions on Dielectrics and Electrical Insulation*, 15(6): 1634–1639, December 2008.

[4] Y. P. Nerkar, R. N. Narayanachar, and R. S. Nema. Characterisation of partial discharges in oil impregnated pressboard insulation systems. In *High Voltage Engineering, 1999. Eleventh International Symposium on (Conf. Publ. No. 467)*, volume 5, pages 208–211, 1999.

[5] S. M. Hoek, A. Kraetge, R. Hummel, *et al*. Localizing partial discharge in power transformers by combining acoustic and different electrical methods. In *Electrical Insulation (ISEI), Conference Record of the 2012 IEEE International Symposium on*, pages 237–241, June 2012.

[6] IEC-60270. *High-Voltage Test Techniques – Partial Discharge Measurements, Third Edition*. International Electrotechnical Commission (IEC), Geneva, Switzerland, 2000.

[7] IEC-62478. *High-Voltage Test Techniques – Measurement of Partial Discharges by Electromagnetic and Acoustic Methods. Proposed Horizontal Standard, First Edition.* International Electrotechnical Commission (IEC), Geneva, Switzerland, 2016.

[8] IEC-60840. *Power Cables with Extruded Insulation and Their Accessories for Rated Voltages above 30 kV ($U_m = 36$ kV) Up to 150 kV ($U_m = 170$ kV) – Test Methods and Requirements*. International Electrotechnical Commission (IEC), Geneva, Switzerland, 2011.

[9] R. Schwarz and M. Muhr. Modern technologies in optical partial discharge detection. In *2007 Annual Report – Conference on Electrical Insulation and Dielectric Phenomena,* pages 163–166, October 2007.

[10] X. Wang, B. Li, H. T. Roman, O. L. Russo, K. Chin, and K. R. Farmer. Acousto-optical PD detection for transformers. *IEEE Transactions on Power Delivery*, 21(3):1068–1073, July 2006.

[11] S. M. Markalous, S. Tenbohlen, and K. Feser. Detection and location of partial discharges in power transformers using acoustic and electromagnetic signals. *IEEE Transactions on Dielectrics and Electrical Insulation*, 15(6): 1576–1583, December 2008.

[12] P. Kundu, N. K. Kishore, and A. K. Sinha. Identification and localization of multi-source partial discharges by acoustic measurements in oil-pressboard insulation system. In *2008 Annual Report Conference on Electrical Insulation and Dielectric Phenomena*, pages 383–386, October 2008.

[13] A. B. A. Ghani, Z. A. A. Zarim, J. Abdul Majeed, and H. Osman. Diagnostic criteria based on the correlation of the measurement of DGA, moisture

contents with PD and tan d in MV oil-filled underground cable. In *2011 IEEE International Conference on Dielectric Liquids*, pages 1–4, June 2011.

[14] The Institute of Electrical and Electronics Engineers. IEEE guide for the interpretation of gases generated in oil-immersed transformers. In *IEEE Std C57.104-2008 (Revision of IEEE Std C57.104-1991)*, New York, NY, pages 1–36, February 2009.

[15] S. Coenen, S. Tenbohlen, S. M. Markalous, and T. Strehl. Sensitivity of UHF PD measurements in power transformers. *IEEE Transactions on Dielectrics and Electrical Insulation*, 15(6):1553–1558, December 2008.

[16] M. Muhr and R. Schwarz. Partial discharge measurement as a diagnostic tool for HV-equipments. In *2006 IEEE 8th International Conference on Properties Applications of Dielectric Materials*, pages 195–198, June 2006.

[17] R. Albarracín, J. A. Ardila-Rey, and A. A. Mas'ud. On the use of monopole antennas for determining the effect of the enclosure of a power transformer tank in partial discharges electromagnetic propagation. *Sensors*, 16(2):148, 2016.

[18] E. Lemke, S. Berlijn, E. Gulski, *et al.* Guide for electrical partial discharge measurements in compliance to IEC 60270, No. 241, 2008. Available: https://online.tugraz.at/tug_online/voe_main2.getvolltext?pCurrPk=40331

[19] J. Rubio-Serrano, M. V. Rojas-Moreno, J. Posada, J. M. Martínez-Tarifa, G. Robles, and J. A. García-souto. Electro-acoustic detection, identification and location of partial discharge sources in oil-paper insulation systems. *IEEE Transactions on Dielectrics and Electrical Insulation*, 19(5):1569–1578, October 2012.

[20] Schneider Electric. Technical publication Schneider: PT-004. Technical report, Training Center Schneider, 2000.

[21] T. A. Prevost and T. V. Oommen. Cellulose insulation in oil-filled power transformers: Part I – History and development. *IEEE Electrical Insulation Magazine*, 22(1):28–35, January 2006.

[22] R. K. Mobley. *Plant Engineer's Handbook*. Butterworth-Heinemann, Woburn, MA, 2001.

[23] Ormazabal. Transformer with natural biodegradable dielectric liquid. Technical report, Ormazabal. Velatia. CA-332-ES-1105, 2015.

[24] IEC-61100. *Classification of Insulating Liquids According to Fire Point and Net Calorific Value*. International Electrotechnical Commission (IEC), Geneva, Switzerland, 1992.

[25] IEC-60836. *Specifications for Unused Silicone Insulating Liquids for Electrotechnical Purposes*. International Electrotechnical Commission (IEC), Geneva, Switzerland, 2015.

[26] IEC-61099. *Insulating Liquids – Specifications for Unused Synthetic Organic Esters for Electrical Purposes*. International Electrotechnical Commission (IEC), Geneva, Switzerland, 2010.

[27] V. Segal, A. Hjortsberg, A. Rabinovich, D. Nattrass, and K. Raj. AC (60 Hz) and impulse breakdown strength of a colloidal fluid based on transformer oil and magnetite nanoparticles. In *Electrical Insulation, 1998. Conference*

Record of the 1998 IEEE International Symposium on, volume 2, pages 619–622, June 1998.

[28] F. M. O'Sullivan. *A model for the initiation and propagation of electrical streamers in transformer oil and transformer oil based nanofluids*. PhD thesis, Massachusetts Institute of Technology. Department of Electrical Engineering and Computer Science, 2007.

[29] J. G. Hwang, M. Zahn, F. M. O'Sullivan, L. A. A. Pettersson, O. Hjortstam, and R. Liu. Effects of nanoparticle charging on streamer development in transformer oil-based nanofluids. *Journal of Applied Physics*, 107(1):1–17, 2010.

[30] C. L. Wadhwa. *High Voltage Engineering*. New Age. New Age International, Delhi, 2007.

[31] V. Dahinden, K. Schultz, and A. Küchler. Function of solid insulation in transformers. *Transform 98*. München, 1998.

[32] H. P. Moser. *Transformerboard*. H. Weidmann AG, Rapperswil, Switzerland 1979.

[33] H. P. Moser. *Transformerboard II*. H. Weidmann AG, Rapperswil, Switzerland 1987.

[34] S. Tenbohlen, S. Coenen, M. Djamali, A. Müller, M. H. Samimi, and M. Siegel. Diagnostic measurements for power transformers. *Energies*, 9 (5):347, 2016.

[35] S. Tenbohlen, J. Jagers, G. Bastos, *et al*. Development and results of a worldwide transformer reliability survey. In *Proceedings of the CIGRE SC A2 Colloquium, Shanghai, China,* 20–25 September 2015.

[36] A. K. S. Jardine, D. Lin, and D. Banjevic. A review on machinery diagnostics and prognostics implementing condition-based maintenance. *Mechanical Systems and Signal Processing*, 20(7):1483–1510, 2006.

[37] T. V Oommen. Moisture equilibrium in paper-oil systems. In *Proceedings of the Electrical I Electronics Insulation Conference*, pages 162–166, 3–6 October 1983.

[38] Cigre. Ageing of cellulose in mineral-oil insulated transformers. In *CigrŽ Brochure 323*, page 46, 2007.

[39] F. Álvarez, F. Garnacho, J. Ortego, and M. Á. Sánchez-Urán. Application of HFCT and UHF sensors in on-line partial discharge measurements for insulation diagnosis of high voltage equipment. *Sensors*, 15(4):7360, 2015.

[40] I. Gottlieb. Practical Transformer Handbook: For Electronics, Radio and Communications Engineers. Butterworth-Heinemann, Woburn, MA, 1998.

[41] A. Cavallini, G. C. Montanari, A. Contin, and F. Pulletti. A new approach to the diagnosis of solid insulation systems based on PD signal inference. *IEEE Electrical Insulation Magazine*, 19(2):23–30, March 2003.

[42] R. W. Hurst and Canadian Electricity Forum Inc. *Electrical Transformer Testing Handbook*, Volume 5. The Electricity Forum, Ontario, 2008.

[43] D. A. Ward and J. L. T. Exon. Using Rogowski coils for transient current measurements. *Engineering Science and Education Journal*, 2(3):105–113, June 1993.

[44] M. Argüeso, G. Robles, and J. Sanz. Implementation of a Rogowski coil for the measurement of partial discharges. *Review of Scientific Instruments*, 76(6), 2005.

[45] V. Dubickas and H. Edin. High-frequency model of the Rogowski coil with a small number of turns. *IEEE Transactions on Instrumentation and Measurement*, 56(6):2284–2288, December 2007.

[46] G. Robles, J. M. Martínez, J. Sanz, B. Tellini, C. Zappacosta, and M. Rojas. Designing and tuning an air-cored current transformer for partial discharges pulses measurements. In *Instrumentation and Measurement Technology Conference Proceedings, 2008. IMTC 2008. IEEE*, pages 2021–2025, May 2008.

[47] G. Robles, J. M. Martínez-Tarifa, M. V. Rojas-Moreno, and J. Sanz-Feito. Inductive sensor for measuring high frequency partial discharges within electrical insulation. *IEEE Transactions on Instrumentation and Measurement*, 58(11):3907–3913, November 2009.

[48] M. V. Rojas-Moreno, G. Robles, B. Tellini, C. Zappacosta, J. M. Martínez-Tarifa, and J. Sanz-Feito. Study of an inductive sensor for measuring high frequency current pulses. *IEEE Transactions on Instrumentation and Measurement*, 60(5):1893–1900, May 2011.

[49] R. S. Simpson. *Lighting Control: Technology and Applications*. Focal Press, 2003.

[50] N. Ida. *Engineering Electromagnetics*. Springer International Publishing, Cham, 2015.

[51] M. V. Rojas-Moreno, G. Robles, R. Albarracín, J. Ardila-Rey, and J. M. Martínez-Tarifa. Study on the self-integration of a Rogowski coil used in the measurement of partial discharges pulses. *Electrical Engineering*, 99(3):817–826, 2017.

[52] L. E. Lundgaard. Partial discharge. XIV. Acoustic partial discharge detection-practical application. *IEEE Electrical Insulation Magazine*, 8(5):34–43, September 1992.

[53] H.-L. Liu. Acoustic partial discharge localization methodology in power transformers employing the quantum genetic algorithm. *Applied Acoustics*, 102:71–78, 2016.

[54] A. Hekmati. Proposed method of partial discharge allocation with acoustic emission sensors within power transformers. *Applied Acoustics*, 100:26–33, 2015.

[55] J. Ramírez-Niño and A. Pascacio. Acoustic measuring of partial discharge in power transformers. *Measurement Science and Technology*, 20(11):115108, 2009.

[56] X. Wang, B. Li, H. T. Roman, O. L. Russo, K. Chin, and K. R. Farmer. Acousto-optical PD detection for transformers. *IEEE Transactions on Power Delivery*, 21(3):1068–1073, 2006.

[57] J. Posada-Román, J. A. García-Souto, and J. Rubio-Serrano. Fiber optic sensor for acoustic detection of partial discharges in oil-paper insulated electrical systems. *Sensors*, 12(4):4793–4802, 2012.

[58] G. Robles, R. Albarracín, and J. L. Vázquez. Antennas in partial discharge sensing system. In Z. Chen, editor, *Handbook of Antenna Technologies*, pages 1–47. Springer, Singapore, 2015.

[59] M. D. Judd. Experience with UHF partial discharge detection and location in power transformers. In *Electrical Insulation Conference (EIC), 2011*, pages 201–205. IEEE, 2011.

[60] M. D. Judd, L. Yang, and I. B. B. Hunter. Partial discharge monitoring of power transformers using UHF sensors. Part I: Sensors and signal interpretation. *IEEE Electrical Insulation Magazine*, 21(2):5–14, 2005.

[61] M. D. Judd, L. Yang, and I. B. B. Hunter. Partial discharge monitoring for power transformer using UHF sensors. Part 2: Field experience. *IEEE Electrical Insulation Magazine*, 21(3):5–13, May 2005.

[62] A. Akbari, P. Werle, M. Akbari, and H. R. Mirzaei. Challenges in calibration of the measurement of partial discharges at ultrahigh frequencies in power transformers. *IEEE Electrical Insulation Magazine*, 32(2):27–34, March 2016.

[63] J. López-Roldán, T. Tang, and M. Gaskin. Optimisation of a sensor for onsite detection of partial discharges in power transformers by the UHF method. *IEEE Transactions on Dielectrics and Electrical Insulation*, 15(6): 1634–1639, December 2008.

[64] S. Tenbohlen, D. Denissov, S. M. Hoek, and S. M. Markalous. Partial discharge measurement in the ultra high frequency (UHF) range. *IEEE Transactions on Dielectrics and Electrical Insulation*, 15(6):1544–1552, 2008.

[65] H. Mirzaei, A. Akbari, E. Gockenbach, and K. Miralikhani. Advancing new techniques for UHF PD detection and localization in the power transformers in the factory tests. *IEEE Transactions on Dielectrics and Electrical Insulation*, 22(1):448–455, 2015.

[66] M.-K. Chen, J.-M. Chen, and C. –Y. Cheng. Partial discharge detection by RF coil in 161 kV power transformer. *IEEE Transactions on Dielectrics and Electrical Insulation*, 21(3):1405–1414, 2014.

[67] F. H. Kreuger. *Partial Discharge Detection in High-Voltage Equipment.* Butterworths, London, 1989.

[68] A. Cavallini, X. Chen, G. C. Montanari, and F. Ciani. Diagnosis of EHV and HV transformers through an innovative partial-discharge-based technique. *IEEE Transactions on Power Delivery*, 25(2):814–824, April 2010.

[69] A. Contin, A. Cavallini, G. C. Montanari, G. Pasini, and F. Puletti. Digital detection and fuzzy classification of partial discharge signals. *IEEE Transactions on Dielectrics and Electrical Insulation*, 9(3):335–348, 2002.

[70] A. Cavallini, M. Conti, A. Contin, and G. C. Montanari. Advanced PD inference in on-field measurements. II. Identification of defects in solid insulation systems. *IEEE Transactions on Dielectrics and Electrical Insulation*, 10(3):528–538, June 2003.

[71] J. M. Martínez-Tarifa, G. Robles, M. V. Rojas-Moreno, and J. Sanz-Feito. Partial discharge pulse shape recognition using an inductive loop sensor. *Measurement Science and Technology*, 21(10):105706, 2010.

[72] J. Ardila-Rey, J. Martínez-Tarifa, G. Robles, and M. Rojas-Moreno. Partial discharge and noise separation by means of spectral-power clustering techniques. *IEEE Transactions on Dielectrics and Electrical Insulation*, 20 (4):1436–1443, August 2013.

[73] R. Albarracín, G. Robles, J. M. Martínez-Tarifa, and J. Ardila-Rey. Separation of sources in radiofrequency measurements of partial discharges using time-power ratio maps. *ISA Transactions*, 58:389–397, 2015.

[74] R. J. Van Brunt. Stochastic properties of partial-discharge phenomena. *IEEE Transactions on Electrical Insulation*, 26(5):902–948, October 1991.

[75] G. C. Stone. The statistics of aging models and practical reality. *IEEE Transactions on Electrical Insulation*, 28(5):716–728, October 1993.

[76] A. Krivda. *Recognition of discharges: Discrimination and classification*. PhD thesis, Electrical Engineering, Mathematics and Computer Science, Delft University Press, 1995.

[77] A. Lapp and H.-G. Kranz. The use of the CIGRE data format for PD diagnosis applications. *IEEE Transactions on Dielectrics and Electrical Insulation*, 7(1):102–112, February 2000.

[78] A. Contin, G. C. Montanari, and C. Ferraro. PD source recognition by Weibull processing of pulse height distributions. *IEEE Transactions on Dielectrics and Electrical Insulation*, 7(1):48–58, February 2000.

[79] E. Gulski. Knowledge rules for partial discharge diagnosis in service. In *International Conference on Large High Voltage Electric Systems*, CIGRE Technical Brochure. CIGRE, 2003.

[80] E. Gulski and A. Krivda. Neural networks as a tool for recognition of partial discharges. *IEEE Transactions on Electrical Insulation*, 28(6):984–1001, December 1993.

[81] F. H. Kreuger, E. Gulski, and A. Krivda. Classification of partial discharges. *IEEE Transactions on Electrical Insulation*, 28(6):917–931, December 1993.

[82] R. E. James and B. T. Phung. Development of computer-based measurements and their application to PD pattern analysis. *IEEE Transactions on Dielectrics and Electrical Insulation*, 2(5):838–856, October 1995.

[83] W. J. K. Raymond, H. A. Illias, A. H. Abu Bakar, and H. Mokhlis. Partial discharge classifications: Review of recent progress. *Measurement*, 68: 164–181, 2015.

[84] M. Cacciari, A. Contin, and G. C. Montanari. Use of a mixed-Weibull distribution for the identification of PD phenomena [corrected version]. *IEEE Transactions on Dielectrics and Electrical Insulation*, 2(6):1166–1179, December 1995.

[85] A. Cavallini, M. Conti, G. C. Montanari, and A. Contin. Indexes for the recognition of insulation system defects derived from partial discharge measurements. In *Electrical Insulation, 2002. Conference Record of the 2002 IEEE International Symposium on*, pages 511–515, April 2002.

[86] G. Robles, E. Parrado-Hernández, J. Ardila-Rey, and J. M. Martínez-Tarifa. Multiple partial discharge source discrimination with multiclass support vector machines. *Expert Systems with Applications*, 55:417–428, 2016.

[87] K. Wang, J. Li, S. Zhang, *et al.* A hybrid algorithm based on s transform and affinity propagation clustering for separation of two simultaneously artificial partial discharge sources. *IEEE Transactions on Dielectrics and Electrical Insulation*, 22(2):1042–1060, April 2015.

[88] L. Hao and P. L. Lewin. Partial discharge source discrimination using a support vector machine. *IEEE Transactions on Dielectrics and Electrical Insulation*, 17(1):189–197, February 2010.

[89] J. Kennedy and R. Eberhart. Particle swarm optimization. In *Neural Networks, 1995. Proceedings., IEEE International Conference on*, volume 4, pages 1942–1948, November 1995.

[90] R. Poli, J. Kennedy, and T. Blackwell. Particle swarm optimization. *Swarm Intelligence*, 1(1):33–57, 2007.

[91] L. Hao, P. L. Lewin, and S. G. Swingler. Improving detection sensitivity for partial discharge monitoring of high voltage equipment. *Measurement Science and Technology*, 19(5):055707, 2008.

[92] H. de Oliveira Mota, L. C. D. da Rocha, T. C. de Moura Salles, and F. H. Vasconcelos. Partial discharge signal denoising with spatially adaptive wavelet thresholding and support vector machines. *Electric Power Systems Research*, 81(2):644–659, 2011.

[93] J. A. Hunter, P. L. Lewin, L. Hao, C. Walton, and M. Michel. Autonomous classification of PD sources within three-phase 11 kV PILC cables. *IEEE Transactions on Dielectrics and Electrical Insulation*, 20(6):2117–2124, December 2013.

[94] R. G. Stockwell, L. Mansinha, and R. P. Lowe. Localization of the complex spectrum: The S transform. *IEEE Transactions on Signal Processing*, 44(4):998–1001, April 1996.

[95] C.-C. Kuo. Artificial identification system for transformer insulation aging. *Expert Systems with Applications*, 37(6):4190–4197, 2010.

[96] M. Majidi and M. Oskuoee. Improving pattern recognition accuracy of partial discharges by new data preprocessing methods. *Electric Power Systems Research*, 119:100–110, 2015.

[97] S. Venkatesh and S. Gopal. Robust heteroscedastic probabilistic neural network for multiple source partial discharge pattern recognition – Significance of outliers on classification capability. *Expert Systems with Applications*, 38(9):11501–11514, 2011.

[98] H.-C. Chen, F.-C. Gu, and M.-H. Wang. A novel extension neural network based partial discharge pattern recognition method for high-voltage power apparatus. *Expert Systems with Applications*, 39(3):3423–3431, 2012.

[99] S. Venkatesh and S. Gopal. Orthogonal least square center selection technique. A robust scheme for multiple source – Partial discharge pattern recognition using radial basis probabilistic neural network. *Expert Systems with Applications*, 38(7):8978–8989, 2011.

[100] A. A. Mas'ud, R. Albarracín, J. A. Ardila-Rey, *et al.* Artificial neural network application for partial discharge recognition: Survey and future directions. *Energies*, 9(8), 2016.

Chapter 3

Moisture analysis for power transformers

Belén García[1], Alexander Céspedes[2] and Diego García[2]

3.1 Introduction

The life expectancy of power transformers is mainly dependent on the integrity of their cellulosic insulation. As cellulose ages, its tensile strength drops and therefore a transformer with a degraded solid insulation is less capable to withstand the dynamic efforts associated to certain operating conditions. The life of cellulosic insulation depends on the temperatures reached during transformer operation, and also on other factors as their moisture content, as water is a catalyst for the chemical reactions involved in cellulosic insulation ageing processes [1,2].

The presence of water inside a transformer may be the consequence of an external contamination or can be a by-product of the process of cellulose degradation itself. Because of the different hydrophobicity of oil and cellulose, most water remains adsorbed in solid insulation and just a small amount of it is dissolved in oil, although as will be explained in this chapter, the distribution between both media changes with temperature [3].

Being able to estimate the water content of transformer, solid insulation is highly desirable to take adequate decisions related to maintenance operations as to predict potential failure conditions in the transformer. Different methodologies have been proposed to determine the moisture content of transformer solid insulation. Some of them are based on the use of equilibrium charts [4,5], which allow the estimation of the water content of the solid insulation if the temperature and the water content of oil are known. Others are based on continuous monitoring of water in oil and dynamic modelling [6,7]. Alternatively, estimations based in the determination of the dielectric response of the transformer insulation have been pointed out by Cigré as the most suitable to assess the moisture content of paper and pressboard [8].

In this chapter, the problem of moisture in transformer insulation is tackled. The physical processes involved in moisture dynamics are first described, then the different methodologies that can be applied to estimate the value of this variable are

[1]Electrical Engineering Department, Universidad Carlos III de Madrid, Spain
[2]School of Electrical and Electronic Engineering, Universidad del Valle, Colombia

explained. Finally, the challenges that lie ahead regarding moisture monitoring are discussed.

3.2 Moisture in transformer insulation

3.2.1 Risks associated to the presence of high levels of moisture in transformers

Numerous studies have reported the negative impact of moisture in the life expectancy of transformer cellulosic insulation [1,9–11]. Experimental studies demonstrated [10] that for the same thermal stress, the life expectancy of a piece of insulation is halved if its moisture content is doubled.

Figure 3.1 shows the results of an ageing test in which Kraft paper was subjected to accelerated ageing in presence of mineral oil and copper at 130 °C. As can be seen, the degree of polymerisation, which is linked to the average length of the cellulose chains and thus to the remaining life of the cellulosic insulation, drops faster as the moisture content of the paper rises.

Cellulosic insulation in power transformers is mainly degraded by three chemical reactions: oxidation, hydrolysis and pyrolysis. Taking into account the typical activation energies of these reactions, the temperature profiles and the oxygen contents usually found in transformers, it seems clear that the hydrolysis reactions are the ones that require more attention.

Hydrolysis is a self-catalysed process, as the chain scission of cellulose molecules depends on the presence of carboxylic acids dissociated in water, as on water and carboxylic acids produced during the process [12].

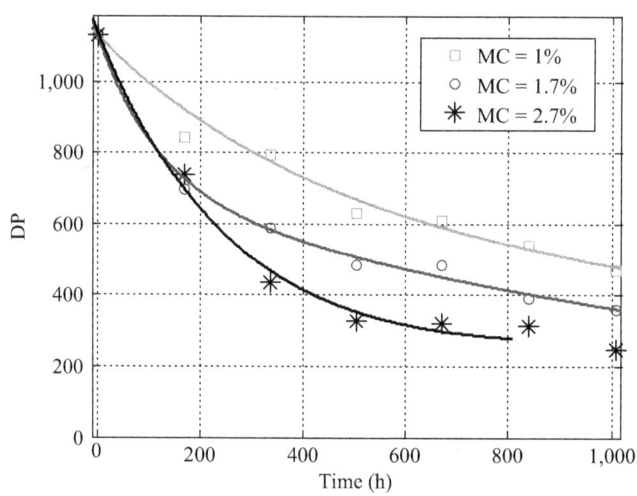

Figure 3.1 Evolution of the DP in samples with different initial MC in paper immersed in MO and aged at 130 °C for 6 weeks. © 2017 IEEE. Reprinted, with permission, from [12]

Table 3.1 Values of moisture contents of oil, expressed in relative saturation (RS) for different conditions of transformer, according to 62-1995 [1]

Moisture content of oil (RS)	Condition
0–5	Dry insulation
6–20	Moderate to wet insulation
21–30	Wet
>30	Extremely wet

Table 3.2 Values of moisture contents of solid insulation, expressed in per cent of dry weight, for different conditions, according to IEEE C57.91-1995 [1]

Moisture in paper (%)	Condition
0–2	Dry paper
2–4	Wet paper
>4.5	Excessively wet paper

Besides accelerating ageing reactions, water is harmful for transformers as it reduces the dielectric strength of the insulation and the partial discharge inception level. The presence of high levels of moisture also increases the risk of bubble formation at high operating temperatures [3].

International standard IEEE 62-1995 [1] classifies the moisture contents that can be present in transformers according to the condition of the equipment. These limits could be established in oil or in paper. In the case of moisture in oil, the reference values use to be expressed in relative saturation (RS) (Table 3.1) or parts per million of weight (ppm), while the advisable moisture content of solid insulation is generally expressed in per cent in weight (Table 3.2).

3.2.2 Sources of moisture contamination in transformers

There are three main sources which contribute to increasing the moisture content of transformer insulating systems: residual moisture, moisture ingress from the outside and moisture produced by the decomposition of the insulation [3]. These sources are described in the following subsections.

3.2.2.1 Residual moisture

Cellulose is a highly hydrophilic material and big amounts of water are adsorbed through the surface of the insulation when it is exposed to the atmosphere during transformer construction, installation or maintenance operations. After being manufactured, transformers are subjected to a drying process in factory to remove the water adsorbed during the construction stage. This process consists of the application of temperature and vacuum cycles to the active part of the transformer, and normally gets to reduce the moisture content of the insulation to levels below 0.5 per cent in weight.

Nevertheless, it must be considered that pieces with very different thicknesses and nature coexist in the transformer, which will require significantly different drying times to get the same drying level.

The thick pieces of wood or pressboard that are usually used in transformers for insulation and mechanical support, often retain non-negligible amounts of water even after the factory drying process. Part of this water may be released into the oil when the transformer starts its operation because of the moisture migration processes that will be described in a further section of this chapter.

3.2.2.2 Ingress of water from the atmosphere

There are different situations in which water ingress from the atmosphere could take place. The main ones are the following:

- Direct exposure of the insulation to the atmosphere during the process of installation or repair of the transformer.
- Moisture ingress through small pores on the tank in a phenomenon known as molecular or Knudsen that takes place when there is a difference on the vapour pressure of water between the inside and the outside of the transformer.
- Ingress of humid air in the tank because of a loss of sealing on the transformer.
- Adsorption of moisture through the transformer conservator. This contamination mechanism has a big impact on the moisture content of the transformer in 'free breathing' transformers.

3.2.2.3 Internal generation of water by the decomposition of cellulose and oil

Ageing of cellulose involves the cleavage of cellulose molecular chains (Figure 3.2) and a subsequent generation of by-products, such as water and furanic compounds. As can be seen in Figure 3.2, a water molecule is generated for each chain cleavage.

The rate of decomposition and generation of water is more intense near the so-called 'hot spots', which comprise about 5 per cent of the total solid insulation mass. Another mechanism in which water is generated is the process of oil oxidation [3].

3.3 Moisture dynamics in transformers

As was described before, an insulating fluid, usually mineral oil, and cellulosic insulation coexists inside transformer tank. Both materials have a very different behaviour with regards to water. While cellulosic materials are hydrophilic, oil is highly hydrophobic. In consequence, most of the water in a transformer remains stored in the solid insulation (about a 99 per cent of the water mass) and just a little

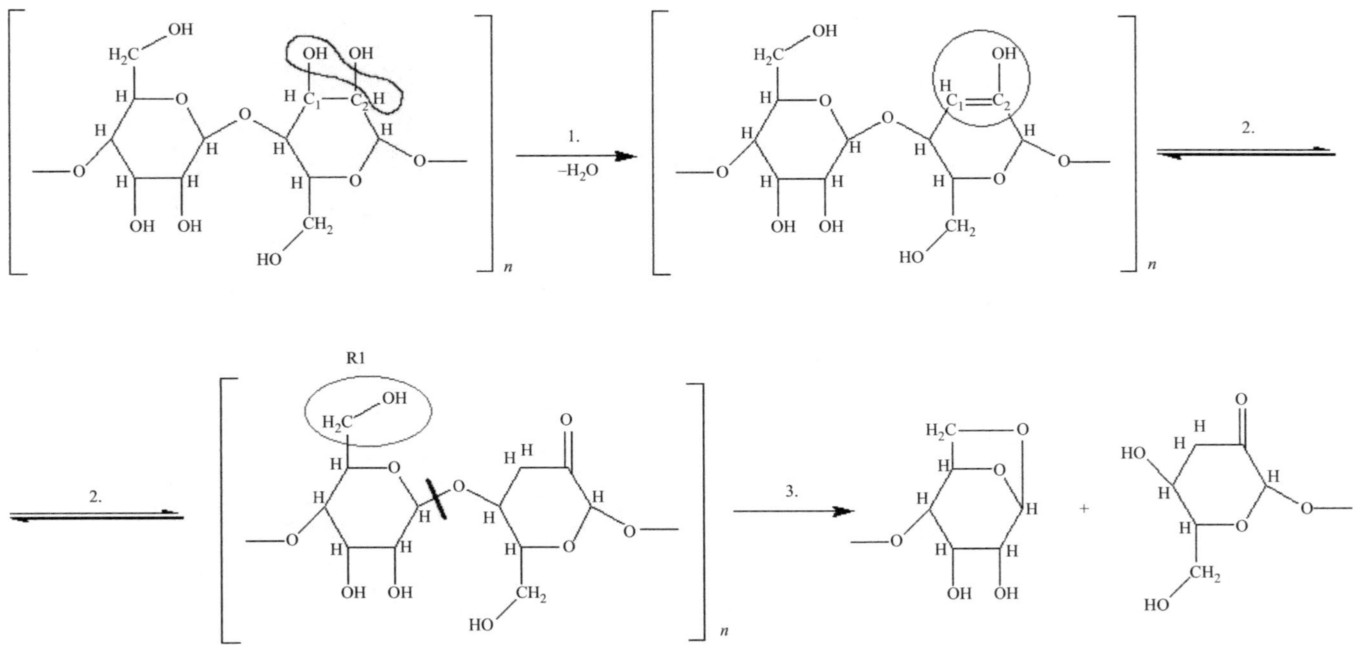

Figure 3.2 *Cleavage of cellulose chains.* *from [12]*

proportion of it is dissolved in oil (just 1 per cent of the mass). However, as will be described in this section, the distribution between oil and paper is not static but varies with the temperature cycles that take place during transformer operation.

3.3.1 Adsorption and desorption of moisture in cellulosic insulation

As Figure 3.3 illustrates, cellulosic insulation is composed of a maze of cellulose fibres and inter and intra-fibre spaces [13].

The adsorption of moisture in cellulosic insulation starts at the insulation surface and then water molecules penetrate into the cellulose moving through the so-called active sites on the micro-capillaries of the cellulose fibres. Active sites are polar structures, and so when water molecules in gaseous occupy one of these sites their movement is limited because of the electromagnetic attraction of the polar portions of cellulose. This phenomenon takes place until all the active sites in the cellulose structure are occupied by water molecules forming a layer of water molecules named as mono-layer.

If the concentration of water molecules in the surface of the insulation exceeds the number of active sites, the exceeding molecules will press the surfacial layer to the inner part of the insulation, forcing the internal cellulose molecules to absorb them and another layer of water molecules is formed in the surface of the insulation. This is the so-called multilayer stage of the process.

The process is then repeated in the same way until the equilibrium is reached. Figure 3.4 schematises the absorption of water molecules by the polar sites of the cellulose.

A similar process takes place when water is desorbed from the insulation to the environment, although unlike in the case of moisture adsorption, a certain activation energy is required to initiate the process, which is in general provided by

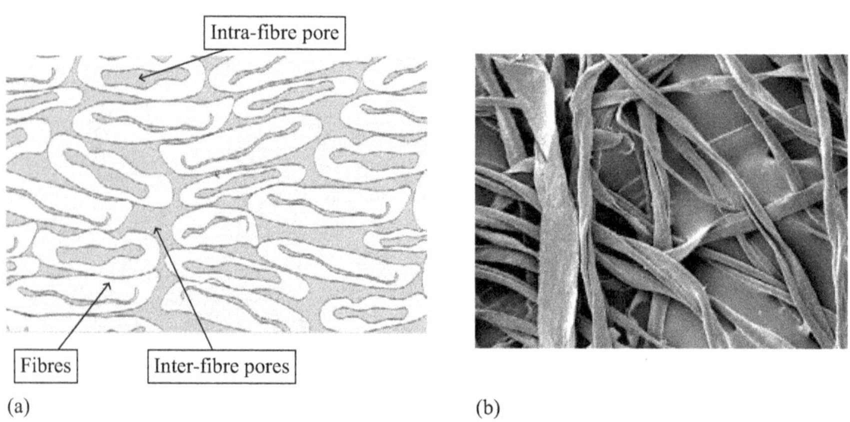

Figure 3.3 (a) Structure of cellulosic insulation and (b) cellulose nanofibre network (Taken from [14]. Work licenced under a Creative Commons Attribution)

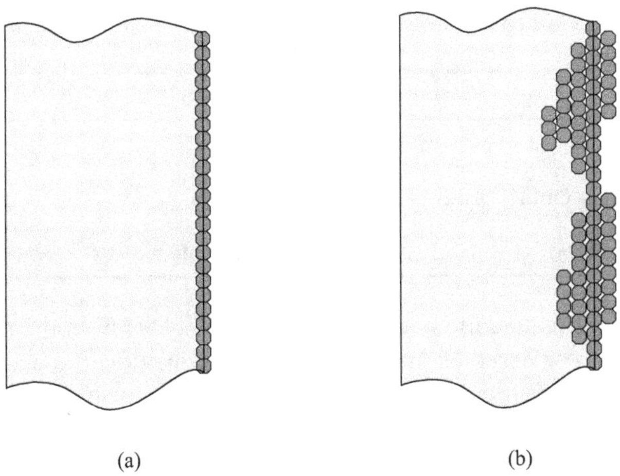

Figure 3.4 Adsorption of water molecules into a piece of cellulosic insulation: (a) mono-layer and (b) multilayer

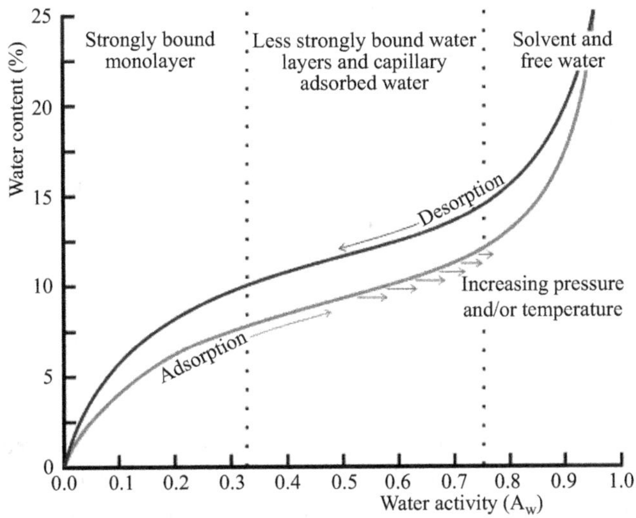

Figure 3.5 Isothermal sorption curve (Taken from [15]. Work licenced under a Creative Commons Attribution)

a temperature increase inside the material. For that reason, as temperature rises, the amount of water admitted by the material diminishes.

Both adsorption and desorption processes are well-defined by the so-called isothermal sorption curves, which in the case of cellulosic materials have an 'S' shape. As Figure 3.5 shows, there is a certain hysteresis in the adsorption and desorption processes. As explained before, the hysteresis has its origin in the

Table 3.3 Variation of the moisture-migration time constant with several factors

Factor	Variation	Time constant
Insulation thickness	Increases	Increases
Insulation density	Increases	Increases
Relative saturation on the surface	Increases	Diminishes
Temperature	Increases	Diminishes
Impregnation with oil	Impregnated	Increases

activation energy required to release the water molecules from the active sites of cellulose. As can be observed, the isothermal sorption curve presents three differentiated areas: The first and the second one correspond to single-layer and multi-layer processes, while the third correspond to high moisture level areas where water starts to appear in free form.

The time constant for the moisture migration process within cellulosic insulation depends on the properties of the material, but also on the conditions it is subjected to. Gasser [10] performed an experimental study to characterise how the moisture-migration time constant varies as a function of different factors. The results of the study are summarised in Table 3.3.

3.3.2 Moisture distribution within transformer solid insulation

Although the moisture distribution inside a transformer is not static, but subjected to continuous fluctuations, the hottest areas of cellulosic insulation typically have lower moisture contents while the cooler ones used to have higher concentrations. The different parts of transformer insulation are generally classified according to their thicknesses and operating temperature in three categories: thin hot insulation, thin cold insulation and thick cold insulation [3].

Typically, about a 60 per cent of the total mass of water in a transformer is stored in its thick insulation. These materials tend to have a low participation in moisture dynamics processes, as they used to be constituted by high density cellulosic materials, such as pressboard and wood, and present a little contact surface with oil in relation to their volume. Besides thick insulation is usually subjected to lower temperatures than other areas of the insulation and therefore, the time constant moisture diffusion of these elements tend to be extremely high, even in the order of years.

Hot thin insulation is typically subjected to higher temperatures and tend to have a lower moisture content. Density of these parts of the insulation used to be smaller and thus the mobility of water in them is high. Moreover, hot thin insulation is subjected to great temperature variations caused by transformer load changes and thus a constant moisture exchange with oil takes place in them.

The thin and cool insulation tends to have operating temperatures similar to that of oil and intermediate densities, so they use to be considered as the main store of water migrating water.

Figure 3.6 shows the typical masses of each type of insulation in a transformer and the typical water distribution among them.

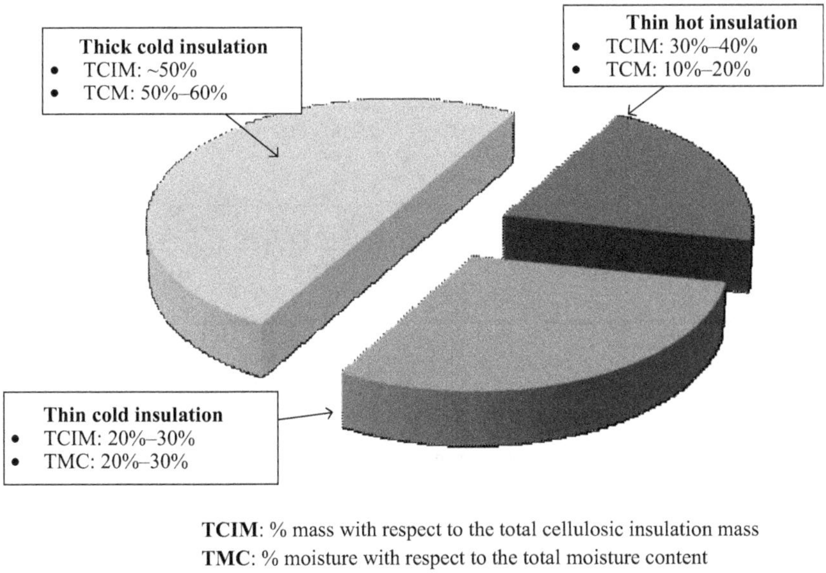

Figure 3.6 Typical masses and moisture contents of transformer thin and thick solid insulation

3.3.3 Solubility of water in oil

The most commonly used insulating fluid for power transformers is mineral oil. As it is well-known that liquid has a very low affinity for water and only admits a few parts per million of water in solution. If this amount of water, known as saturation limit, is exceeded then water in free state will appear in the transformer.

The saturation limit of oil changes with temperature and can be modelled according to Arrhenius law [eqs. (3.1) and (3.2)].

$$W_{s-oil} = 10^{(A-B/T_K)} \tag{3.1}$$

$$W_{s-oil} = e^{(A'-B'/T_K)} \tag{3.2}$$

where W_{s-oil} is the oil saturation limit (ppm), T_K is the oil temperature (K), A, B are the adjustment coefficients obtained experimentally and A', B' are the adjustment coefficients experimentally obtained.

Different values of the coefficients A, B, A' and B' have been proposed in the literature. Table 3.4 summarises some of these values.

Figure 3.7 represents the saturation limit of oil as a function of temperature, calculated with some of these parameters. As can be seen, oil saturation limit increases sharply with temperature.

As oil ages, polar by-products are generated and dissolved in it. The presence of such products will modify the water saturation limit of oil. To take into account the effect of oil ageing in saturation limits, Mladenov [16] proposed (3.3), which

Table 3.4 Coefficients for (3.1) and (3.2). Adapted from [4,6]

Author	A	B
Oommen	7.42	1,670
Griffin	7.09	1,567
Shell Company	7.3	1,630
Author	A'	B'
Davidov and Roizman	17.08	3,876
Fofana *et al.*	19.2	3,805

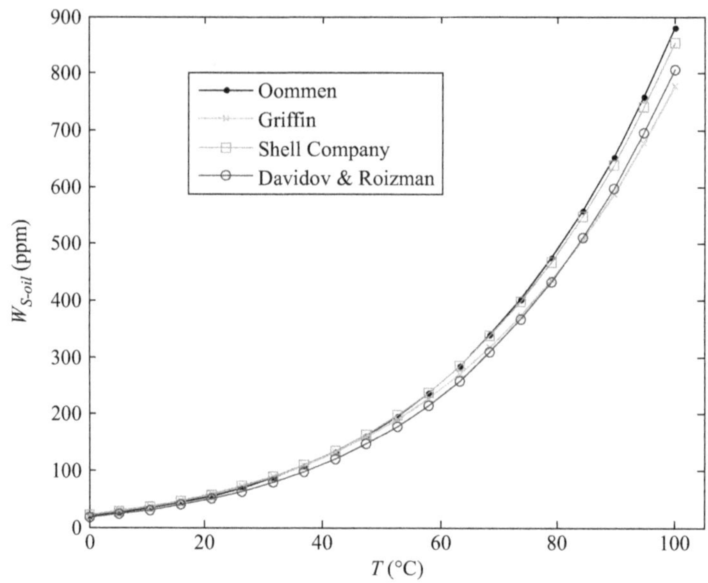

Figure 3.7 Saturation limit of mineral oil as a function of temperature

includes the dependency of the saturation limit of oil with the content of aromatic compounds and acidity.

$$W_{s-oil} = e^{(A'-B'/T_K)+C\cdot A_r+D\cdot A_n} \tag{3.3}$$

where A_r is the total aromatics content (%), A_n is the acidity (KOH/g) and C, D are the experimentally obtained constants.

3.3.4 Moisture equilibrium between paper and oil

As discussed in previous sections, the total mass of water inside a transformer is distributed between solid and liquid insulation being the amount of water admitted by each of the two media dependent on temperature.

To determine how moisture is distributed between paper and oil for certain conditions, it must be considered that if there is a thermodynamic equilibrium between a piece of cellulosic insulation and its surrounding media, an equilibrium between the partial pressures of water in both media and in between their relative humidity will be also established. In this way, the RS in the different materials must verify the relation of (3.4).

$$RS = \frac{p}{P_{sat}} = \frac{W_c}{W_{s-c}} = \frac{W_{oil}}{W_{s-oil}} \tag{3.4}$$

where RS is the relative saturation, P_{sat} is the saturation pressure of water vapour, p is the water partial pressure, W_c is the amount of water in the cellulosic insulation, W_{s-c} is the amount of water in saturation conditions in cellulosic insulation, W_{oil} is the water content in oil and W_{s-oil} is the amount of water in saturation conditions in oil.

Several authors have developed equilibrium charts, using the method proposed by Oommen in [17], which allow the calculation of the moisture content of cellulosic insulation and the oil for different temperatures under equilibrium conditions. As will be explained later, those charts could be a useful tool to estimate the moisture content of solid insulation, although their limitations should be understood to avoid misinterpretations.

A review of the different curves proposed in technical literature is presented in Du [4]. Figure 3.8 shows a reproduction of the curves proposed by Oommen, and Figure 3.9, another set of curves, calculated according to the methodology proposed by the author for higher moisture ranges.

Additionally, some authors have obtained equations to parameterise the equilibrium charts [18]. Fessler's eq. (3.5) expresses the water content in solid insulation as a function of the partial pressure of water vapour in the interphase

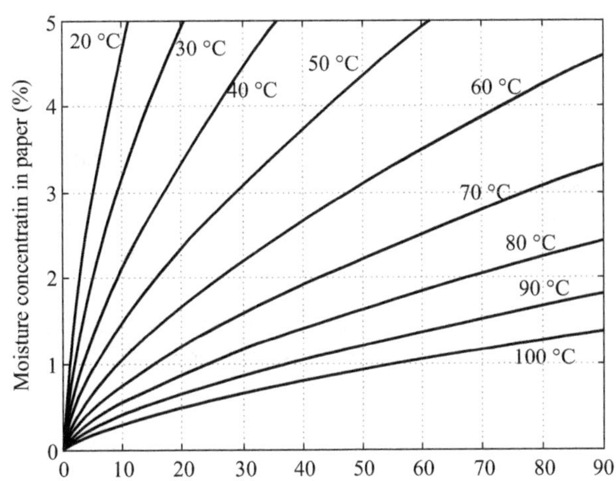

Figure 3.8 Reproduction of Oommen curves for low moisture region. Adapted from [17]

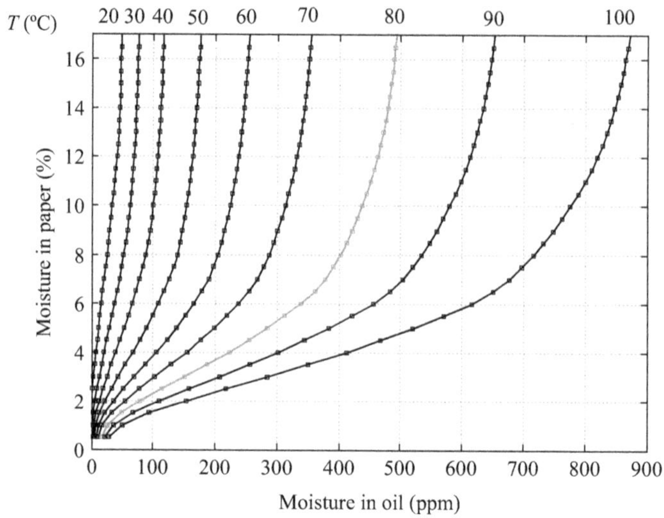

Figure 3.9 Recalculation of Oommen curves for high moisture levels. Adapted from [4]

between the solid insulation and the surrounding media under equilibrium conditions.

$$C_e = 2.173 \cdot 10^{-7} \cdot p^{0.6685} \cdot e^{4,725.6/T_K} \tag{3.5}$$

where C_e is the concentration of water in the insulation under equilibrium (g H_2O/g paper), p is the partial pressure of water (atm) and T_K is the absolute temperature (K).

3.3.5 Moisture equilibrium in alternative fluids

In the last years, ester-based fluids have been proposed as an alternative to mineral oil. Esters fluids are considerably more hydrophilic than mineral oils, and thus, different moisture equilibrium curves should be used to estimate the equilibrium in ester–cellulose systems.

Jovalekic in [19] obtained the water solubility and sorption curves of several liquid insulating fluids. The experiments were carried out on several samples of insulating liquids analysed at different temperatures. The relative humidity of oil was determined with a capacitive sensor until a constant value of the moisture was established. The parameters A and B of the saturation curve (3.6) were obtained for those liquids (Table 3.5).

$$W_s(T) = A \cdot e^{(-B/T)} \tag{3.6}$$

Other authors [20–24] have also obtained equilibrium curves for different types of insulating fluids and solid insulations. Figure 3.10 shows the curves

Table 3.5 Water saturation parameters for different insulating liquids. Adapted from [19]

Insulating fluid	*A*	*B*
FR3	$4.9{\cdot}10^5$	1,821
MIDEL 7131	$4.586{\cdot}10^5$	1,602
Lyra X	$6.013{\cdot}10^6$	3,396

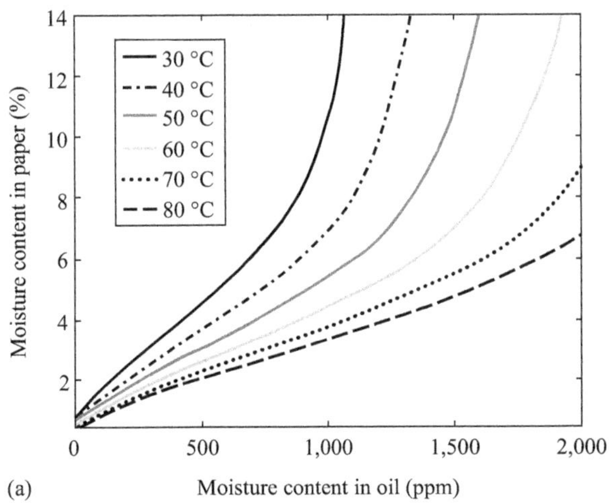

(a)

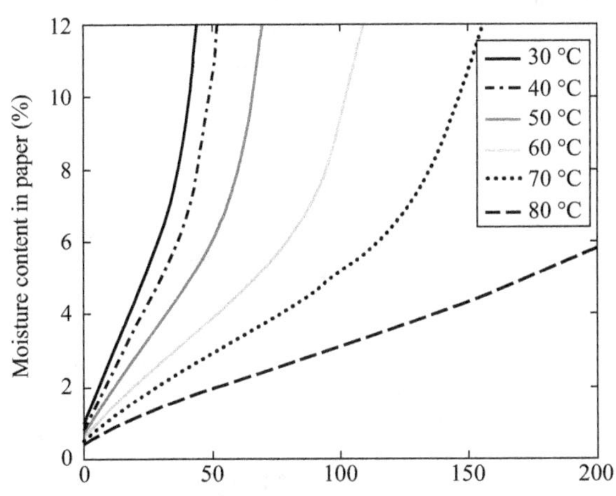

(b)

Figure 3.10 Equilibrium charts for a natural ester–Kraft paper system (a) and for a mineral-oil–Kraft paper system (b). Reproduced, with permission, from [20] determined by Villarroel [20] for Kraft paper and the natural ester Biotemp (Figure 3.10(a)) and Kraft paper and mineral oil (Figure 3.10(b)).

3.3.6 Moisture dynamics in a transformer under operation

The operation of a transformer involves a continuous thermal disequilibrium related with the temperature differences between solid and liquid insulation, and between the upper and lower parts of the transformer. Moreover, the temperature of the transformer is continuously changing with time because of the load and ambient temperature fluctuations it is typically subjected to. The dependency of transformer temperate with load has been widely studied and a good estimation of the temperature distribution can be obtained following the models proposed in international standards [25].

As was discussed before, the equilibrium of moisture between solid and liquid insulation depends on temperature of the insulation. In consequence, any change on the transformer temperature map is accompanied by a disequilibrium on the moisture distribution. When temperature rises, water saturation limit of oil increases, while water adsorption capacity of cellulose decreases. In this case, water will migrate from paper to oil. Conversely, if the temperature decreases, the migration of moisture will take place in the opposite direction. Figure 3.11 illustrates the phenomena.

Migration of water between paper and oil involves two stages: diffusion of moisture throughout the solid insulation and release or adsorption of water at paper–oil interface.

The typical time constants of these two processes are significantly different, and while the migration of moisture inside the solid is a very slow process, the exchange at the interface can be considered instantaneous.

The migration rate inside the transformer depends on the so-called Moisture diffusion coefficient [26,27], which depends on the insulation temperature, its

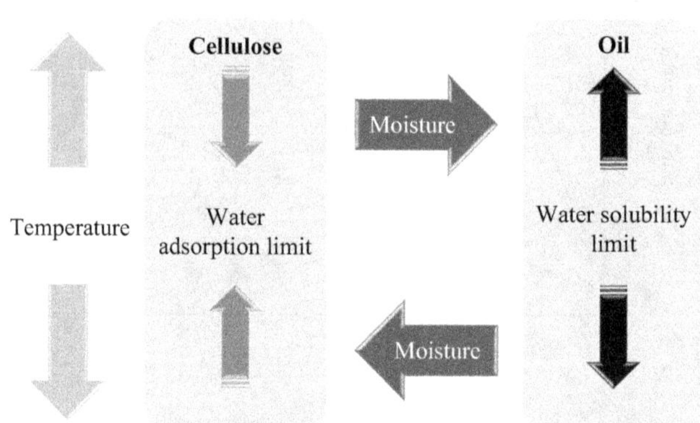

Figure 3.11 Moisture migration within oil–paper insulation

moisture concentration and the insulation properties, such as density, thickness and ageing condition and also on the type of oil that impregnates the solid insulation.

As a rule of thumb, it can be considered that, while in thin parts of the insulation operating at high temperatures the equilibrium of moisture could be reached within few hours, thick pressboard barriers or support pieces would need several months or even years to attain it [3]. Because of the different time constants of thermal and moisture migration phenomena in a transformer, the moisture distribution between liquid and solid insulation of a transformer is never in equilibrium when the transformer is in service.

3.4 Monitoring of moisture content in oil

Determination of the moisture content of oil used is to be included in routine physical–chemical testing of oil. Although this variable is subjected to continuous fluctuations caused by the transformer loading cycles and ambient temperature, and should not be interpreted as an isolated data, its determination is highly advisable to avoid potentially hazardous situations in the transformer and to help taking decisions related to transformer maintenance.

This section gives an overview of the different methods that can be applied to monitor the water content of transformer oil and also discusses how measurements might be interpreted.

3.4.1 Periodical sampling of oil

Samples of oil can be easily extracted from the transformer using the valves located at the bottom of the tank. Oil extracted from the tank is usually stored in dark crystal bottles to avoid potential degradation caused by light and taken to laboratory to be analysed. Three different types of analysis are typically carried out with these samples of oil: physicochemical analysis, focused in getting information on the condition of the oil, dissolved gas analysis, used to detect defects in the transformer, and furanic compounds and other markers focused to diagnose the condition of the solid insulation.

In order to determine the total mass of water dissolved in oil samples, measures using the Karl Fischer (KF) method can be carried out [28]. Figure 3.12 shows a picture of a KF coulometric titrator. As it can be seen, the KF measurement equipment consists of a main vessel, where the titration takes place, an oven, which is used to evaporate water adsorbed in solids or liquids, and a controller.

In general, a small sample of oil (1–2 g) is directly injected in the titration vessel to do the titration. As an alternative to direct injection, an oven can be used to heat the sample and evaporate the water contained in it. Evaporated water is then driven it to the titration vessel to be quantified.

KF method can also be used to determine the water content of samples of solid materials, as paper insulation.

The titration vessel is constituted as an electrolysis cell, with two compartments connected by a porous membrane. The vessel contains a solution formed by

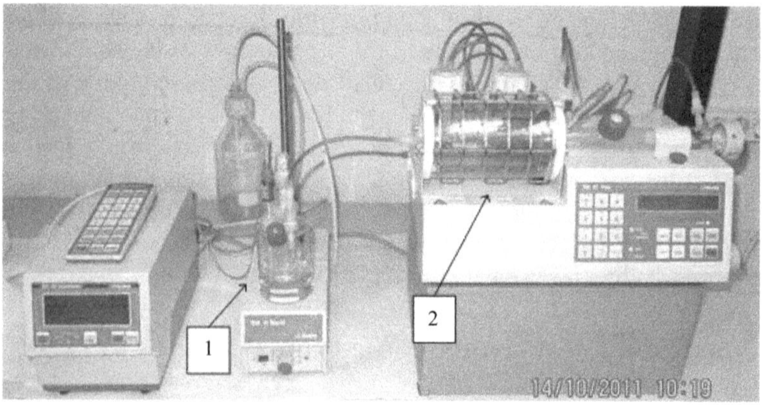

Figure 3.12 Karl Fischer coulometric titrator, 1 – titration vessel, 2 – oven

a base, an alcohol, iodine ions and SO_2. In KF coulometric method, the titration substance is iodine, which is generated by electrolysis from iodine ions, liberating electrons.

The following chemical reaction take place in the titration vessel in presence of water:

$$ROH + SO_2 + 3RN + I_2 + H_2O \rightarrow x(RNH) \cdot SO_4R + 2(RNH)I$$

The iodine that reacts with water is generated electrochemically by anodic oxidation in the coulometric cell according to the following half-reaction:

$$2I^- \rightarrow I_2 + 2e^-$$

The amount of water in the fluid that is being titrated is determined by measuring the electrical current resulting from the generation of the iodine. The analysis of the stoichiometry of the KF reaction leads to the result that the presence of 1 mg of water present in the reaction vessel corresponds to a consumption of electrical current of 10.712 C.

Although KF is considered a reliable measuring method, and although some factors should be taken into account, to avoid measuring errors, such as the contamination of the samples during their manipulation, moisture ingress in the titration vessel from the atmosphere, discrepancy between different extraction methods and influence of the heating temperature and times when the oven is used [3].

3.4.2 On-line measure of oil moisture with capacitive sensors

Capacitive sensors have been applied since the mid-90s to measure the moisture content of transformer oil in a continuous manner [3,29,30]. Continuous monitoring provides more information about the moisture content of the transformer insulation than periodical sampling, as it eases the interpretation of moisture fluctuations and their dependence with temperature [31].

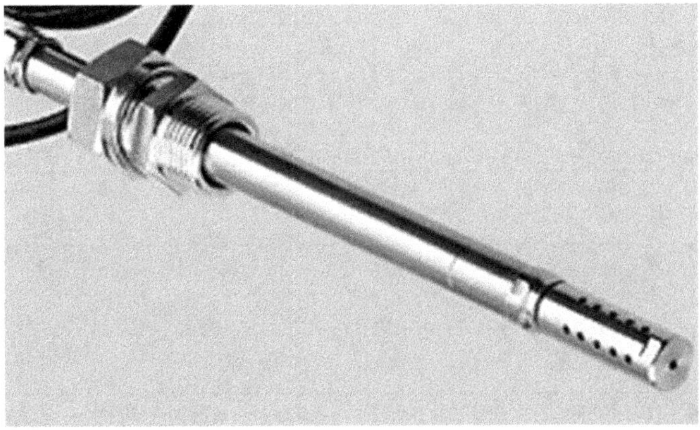

Figure 3.13 Measuring probe of a transformer moisture sensor

The measuring probe of capacitive sensors (Figure 3.13) consists of two electrodes with a hydroscopic dielectric polymer placed between them. Water molecules penetrate in the polymer and change the value of the measured capacitance in proportion to the RS of moisture of the fluid where the probe is immersed. Since capacitive probes measure RS, the type of oil and its condition do not have an influence in measures.

Most commercial moisture sensors provide a measure of the relative humidity of oil and incorporate a thermal sensor to measure oil temperature as well. Some sensors express the water content of oil in terms of the 'Water Activity' (Aw), which is defined as the relative humidity per unit [3] [eq. (3.7)].

$$\mathrm{Aw} = \frac{\mathrm{RH}}{100} \tag{3.7}$$

Most sensors provide a measure of the absolute concentration of moisture expressed in ppm of oil weight, which is calculated using the solubility equation of oil [eq. (3.8)].

$$W_{s-oil} = \exp\left(\frac{A - B}{T(K)}\right) \tag{3.8}$$

As was explained in Section 3.3.3, the parameters A and B are strongly dependent on the type of oil that is being analysed and on its ageing condition. In this way, a precise calculation of the ppm would require using the parameters of the fluid that is being measured.

Generally, moisture sensors are configured to calculate absolute concentrations of moisture (ppm) in new mineral oil. If the transformer that is being monitored is filled with an alternative fluid, or if oil inside the tank is degraded, different parameters A and B should be considered to get accurate ppm values. Most commercial sensors allow the user to configure the parameters used for the calculation.

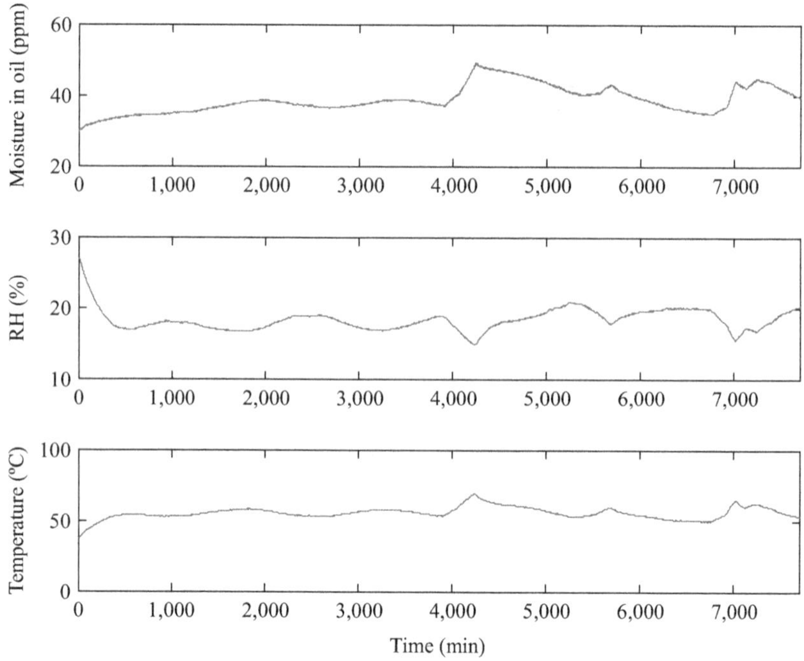

Figure 3.14 Evolution of temperature water in oil concentration and relative saturation in a transformer

Figure 3.14 plots show some measures of absolute and relative moisture in oil and temperature taken with a moisture capacitive sensor in an experimental transformer.

In relation to the installation of moisture sensors, manufacturers point out that their performance is better when the measuring probe is immersed into the transformer oil flow [32]. This will make possible to measure volumes of oil that are representative of the condition of the transformer and will lead to shorter response times. An ideal installation point would be the oil cooling circulation line; when this is not possible, the distance between the sensor and the oil flow should be minimised. Manufactures do not recommend to locate moisture sensors at the bottom of the transformer tank unless it is guaranteed that there is true oil exchange present at this point [32]. An image of an installed moisture sensors is shown in Figure 3.15.

3.4.3 *Interpretation of the moisture content of oil*

As was explained in Section 3.2, IEEE and IEC maintenance guides [1,33] provide values that allow to classify the dryness of a transformer according to the moisture contents of its oil and paper (Tables 3.1 and 3.2). IEEE Std C57.106-2006 [34] also

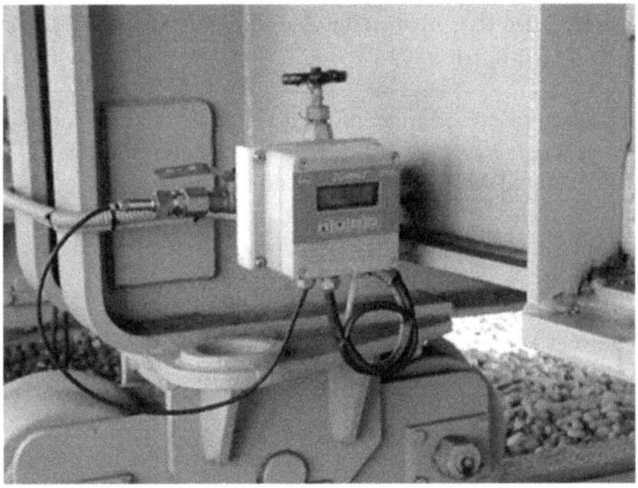

Figure 3.15 Installation of moisture capacitive sensors in transformers

Table 3.6 Moisture limits for continuous use in service-aged insulation. IEEE Std C57.106-2006 [34]

	Voltage class		
	Ur < 69 kV	**69 kV > Ur < 230 kV**	**Ur > 230 kV**
Moisture limit (ppm)	35	25	20

provides advisable limits for ppm in oil in service-aged insulation, according to the transformer voltage class (Table 3.6).

All those suggested values might help making decisions about the necessity of subjecting a transformer to a field drying process, or to programme different management actions. However, moisture values might not be considered in isolation, but should be analysed together with moisture historic recordings and with other key variables, and specially with temperature.

In [35], a study is presented to evaluate the importance of carrying out temperature recordings to ease the interpretation of the measures of moisture content of oil. The authors show the evolution of absolute moisture concentration (ppm) recorded on several transformers at different times of the year. The comparison of the measures performed on cold seasons showed clear discrepancies with those measured at warm seasons. Other sources also alert about that possible misinterpretation of the measured moisture content of oil in transformers [3,13,26,35–37].

3.5 Estimation of the moisture content of solid insulation from moisture in oil measures

As it is well-known, direct determination of the moisture content of paper is difficult, since extracting samples of solid insulation from a transformer is practically unfeasible, and so all the determinations should be done by indirect estimations. On the other hand, as was described before, determination of the moisture content of oil is quite simple, just requiring the extraction of an oil sample or the use of a moisture-in-oil on-line sensor.

Different approaches have been proposed to estimate the moisture content of solid using the measures of water in oil. Some of them are more complex than others, including the application of mathematical models to analyse the dynamic processes involved in the moisture migration within transformer insulation. An overview of these techniques is provided in this section.

3.5.1 Determination of moisture content of paper using the equilibrium charts

Equilibrium charts, such as those in Figures 3.8 and 3.9, have been widely used in order to estimate the moisture content of solid insulation on in-service transformers. These diagrams provide the amount of water that would be adsorbed in paper for a certain water content on oil and temperature under equilibrium conditions.

The approach used by several authors, and often applied by electrical companies, consists of measuring the water content of an oil sample and determining its temperature and then to introduce those data in the equilibrium charts to determine the water content of paper (Figure 3.16). Although a temperature determination is

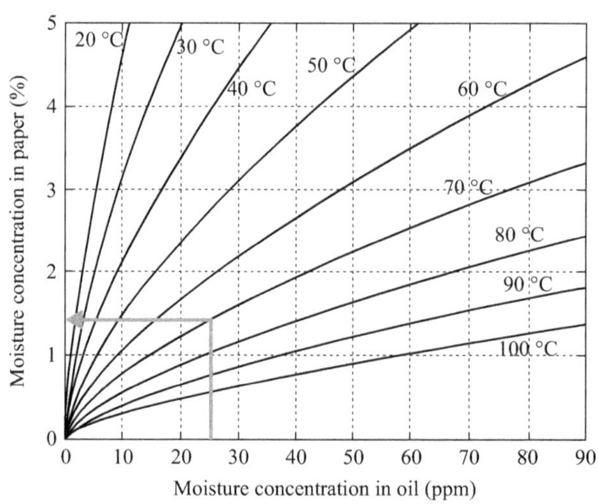

Figure 3.16 Determination of the moisture of solid insulation using equilibrium charts

sometimes carried out on the extracted sample by means of a thermal probe, the temperature recording is mostly based in the measure of the transformer thermometer.

This could apparently be an easy way to estimate the moisture content of solid insulation, although the procedure can lead to wrong conclusions, as the equilibrium charts are only valid under moisture equilibrium conditions. Taken into account the large time constants involved in moisture migration processes (Section 3.3.4), and the continuous temperature variation linked to the daily fluctuations of load and ambient temperature, it seems clear that moisture equilibrium between oil and paper equilibrium will hardly ever be attained on an in-service transformer.

Ward [38] identifies several sources of error that can affect the determination of the moisture content of paper using the equilibrium charts:

- The equilibrium charts are only valid under equilibrium conditions, and equilibrium of moisture is hardly ever attached in an on-service transformer.
- The charts are not accurate in certain regions, i.e. at low-temperature and low-moisture values.
- The estimated moisture contents differ if charts proposed by different authors are used.
- The distribution of moisture within transformer solid insulation is not homogeneous, and by using the equilibrium charts, a single value is estimated.
- The solubility of oil and the capacity of adsorption of water of cellulose vary as the transformer age, while most equilibrium charts were experimentally obtained on new materials.
- The equilibrium charts are only accurate for the material on which they were obtained.

The author estimates that errors of up to 200 per cent can affect the moisture values obtained by using moisture charts [34].

3.5.2 Improved methodologies to estimate the moisture content of paper from the measures of moisture content of oil

Different proposals have been done to get a more precise estimation of the moisture content of solid insulation when water content of oil is known.

IEE Std 62-1995 [2] proposes a methodology to estimate the moisture of paper in per cent of weight, when the moisture in oil is known. The standard provides a chart that allows the calculation of a so-called correlation multiplier, which relates the moisture content of oil and paper insulation. That coefficient is dependent on the temperature of the oil sample and can be obtained by entering in the provided chart with the temperature measured on a sample of oil plus 5 °C. The moisture content of solid insulation in per cent is then obtained as the product of that coefficient and the ppm measured in oil.

A similar method is proposed in IEC 60422:2013, in which a correction factor, f, is calculated using (3.9), where Ts is the temperature of the oil sample in °C.

The moisture content of paper is then obtained as the product of the correlation factor and the ppm of water in oil [eq. (3.10)].

$$f = 2.24 \cdot \exp(-0.04 \cdot Ts) \tag{3.9}$$

$$\% \text{ Water in paper} = f \cdot \text{ppm}_{\text{oil}} \tag{3.10}$$

Several published works implement more complex approaches to calculate the moisture content of solid insulation using the information provided by on-line moisture sensors to determine the moisture content of paper [39–46]. Some authors [39,40] propose the use of equilibrium charts in combination with those sensors to estimate the moisture content of solid insulation. Other researchers [20,41,42] have developed dynamic models that consider the physical phenomena involved in the moisture migration processes, combining the analysis of thermal and mass transfer phenomena.

3.6 Dielectric response methods for the estimation of moisture in solid insulation

Some of the most commonly accepted techniques to determine the moisture content of transformer solid insulation are those based in the measure and interpretation of the transformer dielectric response [47–50]. These methods characterise the response of the transformer insulation when it is subjected to an electric field. Some techniques analyse the response in the time domain, while others do it in the frequency domain. In this section, the main features of these techniques are reviewed.

3.6.1 Theoretical principles

Dielectric response measures are based in the fact that the behaviour of a dielectric material subjected to an electric field is strongly influenced by its condition, and particularly by its water content.

Unlike what happens in conductive materials, electric charges in dielectrics materials are trapped and cannot freely move. When an electric field is applied to a dielectric material, a polarisation process starts consisting in the aligning of positive and negative charges in the direction of the field.

Polarisation is characterised by a limited movement of the charges trapped in the material; in opposition to the conduction phenomena, where a flux of free loads arises [50]. From the mathematical point of view, the macroscopic polarisation can be defined as a vector proportional to the electric field:

$$P = \chi \varepsilon_0 E \tag{3.11}$$

where χ is the susceptibility of the material and ε_0 is the vacuum permittivity.

There are different mechanisms that contribute to the polarisation of the materials governed by different time constants, such as the atomic polarisation, ionic polarisation, orientational polarisation, interfacial polarisation and hopping; each of them is characterised by different time constants.

Dielectric response methods analyse the macroscopic polarisation of a material subjected to an electric field in the time or the frequency domain considering the materials to be isotropic, homogeneous and linear.

The dielectric response of the materials presents a time dependency different to that of the electric field that causes it. If a voltage $V(t)$ is applied to the plates of a capacitor separated by a dielectric material, the electric field will be proportional to the voltage, while the electric displacement $D(t)$, which represents the total induced load in the electrodes, will have an influence of the retarded response on the dielectric.

$$D(t) = \varepsilon_0 E(t) + P(t) \tag{3.12}$$

The dielectric response phenomena is characterised by the so-called dielectric response function, $f(t)$, which is a monotonously decreasing function that represent the response of a material to an impulse electrical field.

For an impulse of magnitude $E\Delta t$, the resulting polarisation would be

$$P(t) = \varepsilon_0 E\Delta t \cdot f(t) \tag{3.13}$$

For a field with a different time dependency, $E(t)$, the polarisation can be obtained as

$$P(t) = \varepsilon_0 \int_{-\infty}^{t} f(t-\tau)E(\tau)\partial\tau = \varepsilon_0 \int_{0}^{\infty} f(\tau)E(t-\tau)\partial\tau \tag{3.14}$$

The dielectric response function, $f(t)$, depends on certain properties of the material, such as its moisture content or its ageing condition, and thus, by measuring the response of the material to an electric field, information about those properties can be obtained [50].

Three methods based on the characterisation of the dielectric response of insulating materials are used to determine the moisture content of the transformer solid insulation: the measure of the polarisation and depolarisation currents (PDC), the recovery voltage method (RVM) and the frequency dielectric spectroscopy (FDS). PDC and RVM determine the response of the material in the time domain, while FDS does it in the frequency domain.

Pieces of equipment, specifically designed to monitor the condition of transformer insulation, are commercially available to perform RVM, PDC and FDS measures. These instruments are adequate to measure in field conditions, minimising the influence of electromagnetic disturbances or other influence factors. Most devices incorporate interpretation software which allows the user to estimate the moisture content of transformer insulation once its dielectric response is determined.

The following subsections provide a description of FDS, PDC and RVM techniques.

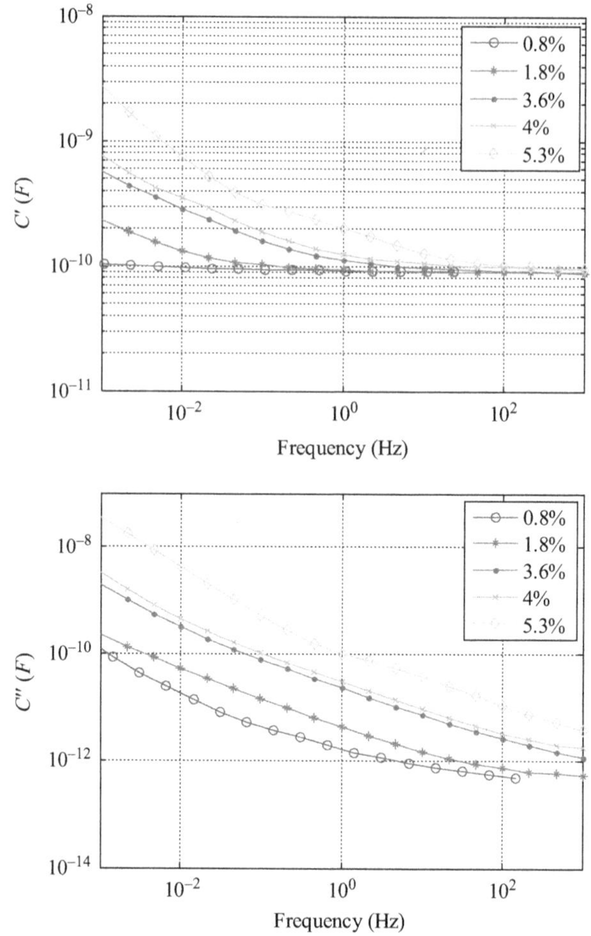

Figure 3.17 Variation of the real and imaginary part of the complex capacitance with the moisture content of the test object in a frequency range 1 MHz–1 kHz

3.6.2 Frequency dielectric spectroscopy

Frequency dielectric spectroscopy (FDS) is based in measuring the response of a material in the frequency domain. A sinusoidal voltage of variable frequency is applied to transformer insulation in a typical range of 1,000 Hz–0.1 MHz. The current flowing through the dielectric is measured, and the complex impedance of the object is determined for every characterised frequency. Relevant magnitudes, such as the tangent delta, complex permittivity or complex capacity, can be derived from these measures [49].

The moisture content of the insulation is a main influence factor on the evolution of these variables with frequency. Figure 3.17 shows the evolution of

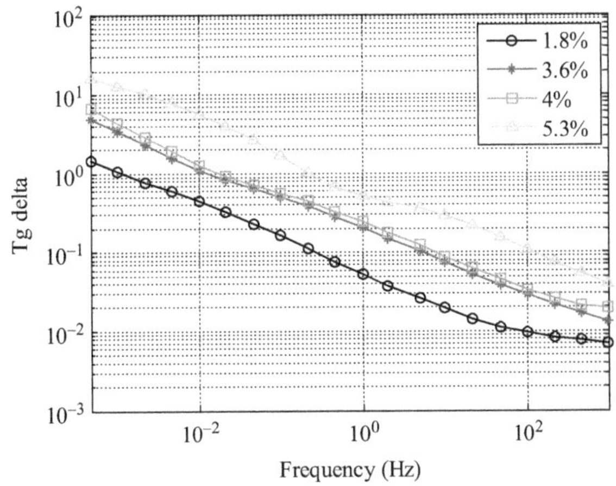

Figure 3.18 Variation of the tangent delta with the moisture content of the test object in a frequency range 1 mHz–1 kHz

Table 3.7 Different parts of transformer insulation that can be characterised by dielectric measures

Notation	**Insulation to be characterised**
CHL	Insulation between high-voltage winding and low-voltage winding
CH	Insulation between high-voltage winding and ground (tank)
CL	Insulation between low-voltage winding and ground (core)
CHT	Insulation between high-voltage winding and tertiary winding
CLT	Insulation between low-voltage winding and tertiary winding
CT	Insulation between tertiary windings and ground

complex capacity real and imaginary parts with the moisture content of a piece of solid insulation measured in laboratory conditions. As can be noted, the greater influence can be observed at low frequencies.

Figure 3.18 shows the dependence of tangent delta for a frequency range 1 mHz to 1 kHz at the same test specimen.

FDS measures can be used to characterise the moisture content of different parts of transformer insulation, generally named with the codes shown in Table 3.7. FDS measures do not require the application of high voltage levels. In fact, the voltage sources of most commercial provide voltages below 200 V.

Figure 3.19 shows the measuring scheme commonly used to carry out FDS measures on a transformer [50]. As can be observed, the transformer windings should remain short-circuited during the measures, and thus, the transformer under test should be out of service. Additionally, although studies have been done to analyse other measurement schemes [51], it is advisable that the transformer remains totally disconnected from the system during the tests. It should be

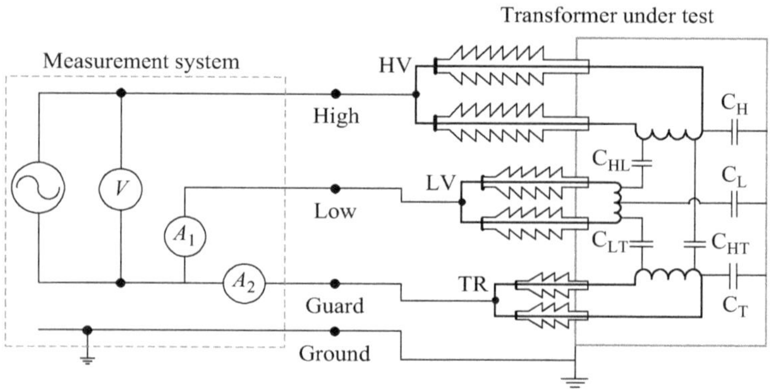

Figure 3.19 FDS measuring scheme in a field transformer

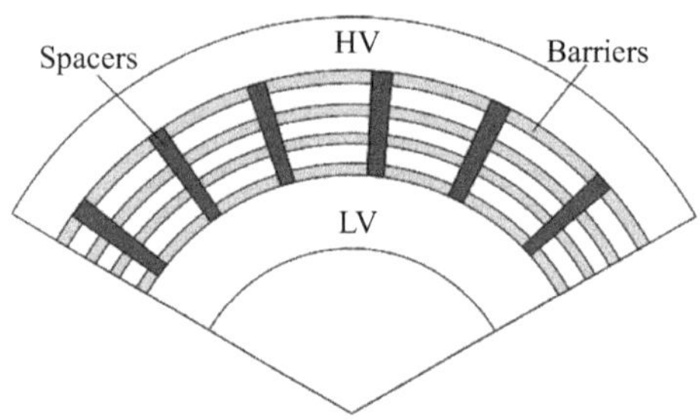

Figure 3.20 Section of main insulation in a core type transformer. © 2006 IEEE. Reprinted, with permission, from [51]

remarked that dielectric measures are generally time consuming, lasting from 1 to 3 h depending on the frequency ranges to be characterised.

By applying an adequate interpretation method, FDS measures can provide precise information related to the amount of water contained in the different parts of transformer solid insulation. A library containing data on dielectric properties of well-characterised solid insulation at different humidity content used to be used to this end. From this information, the dielectric response, i.e. its frequency dependent permittivity $\bar{\varepsilon}(\omega) = \varepsilon'(\omega) - j\varepsilon''(\omega)$, of the composite duct insulation is calculated and compared with results of the measurements.

There are several factors that influence the dielectric response of transformer insulation, e.g. the geometry of the duct, the dielectric response of oil (its conductivity and permittivity), the dielectric response of pressboard or paper and the temperature of the whole insulation system. To obtain the moisture content in the solid insulation, these variables should be known or estimated as parameters during the modelling. For modelling purposes, the transformer main insulation, consisting of cylindrical pressboard barriers in series with oil ducts and spacers (Figure 3.20), is represented by the so-called *X–Y* model, which is shown in Figure 3.21.

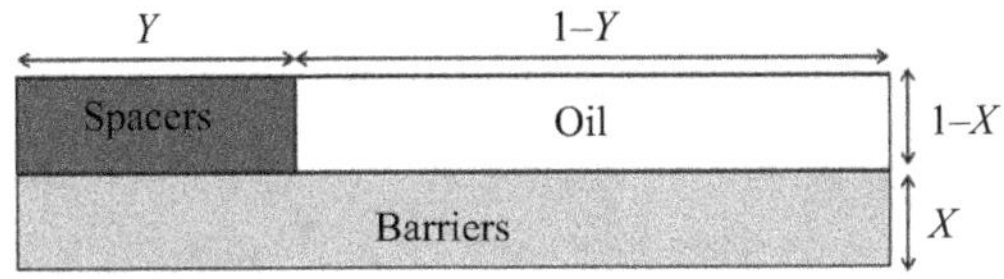

Figure 3.21 Representation of transformer main insulation by means of the X–Y model. © 2006 IEEE. Reprinted, with permission, from [51]

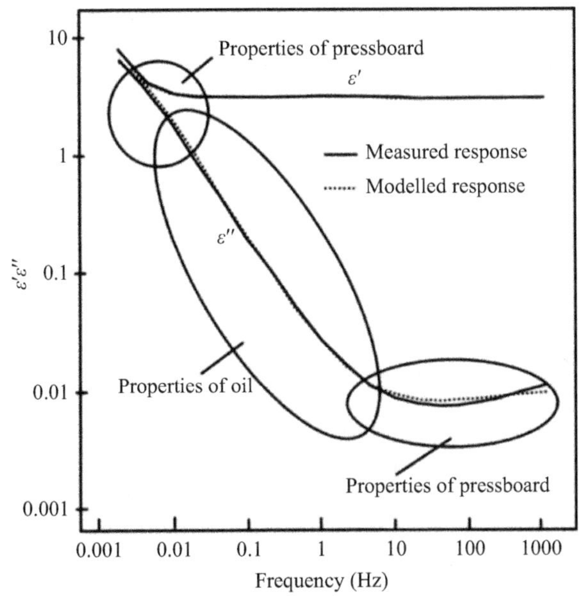

Figure 3.22 Measured vs. modelled dielectric response (ε' and ε'') in a real transformer. © 2006 IEEE. Reprinted, with permission, from [51]

Variables X and Y of the model are defined as

$$X = \frac{\text{radial thickness of total barriers}}{\text{radial width of the duct}} \tag{3.15}$$

$$Y = \frac{\text{total width of the spacers along periphery of the duct}}{\text{periphery of the duct}} \tag{3.16}$$

The interpretation of the results is based on a best-fit procedure, in which the measured and the modelled responses of oil–pressboard ducts are compared, as it is illustrated in Figure 3.22. In the figure, both responses fit together well. It is important to notice that different parts of the response, in frequency axis, are separately sensitive to properties of oil and solid parts of the insulation. At the very

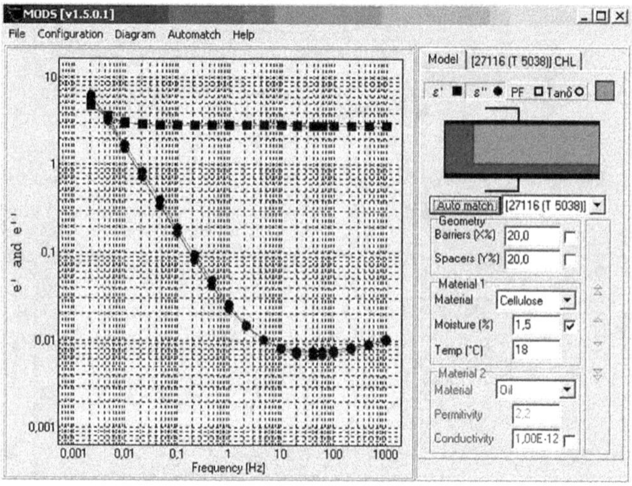

Figure 3.23 Fitting process to find the moisture content in a transformer with the software MODS included in the equipment IDA 200 [52]

low (<10^{-2} Hz) frequency range, the response is mainly influenced by the properties of the pressboard. The same is true for the higher frequency range (>10 Hz). The central part of the response is, on the other hand, influenced by the properties of the oil, mainly by its conductivity.

Different commercial FDS measuring devices are available nowadays, which generally incorporate their own interpretation software. As an example of those, Figure 3.23 shows how interpretation can be done using the software MODS, provided with the equipment IDA 200 [52]. In this case, the approximate moisture content of the insulation, between high and low-voltage windings, was estimated to be of 1.5 per cent. It can be noted that the software allows the user to introduce information about transformer geometry, conductivity of oil and temperature to be considered during the fitting process, what is highly advisable in order to get accurate estimations. Alternatively, these variables can also be estimated by the software if no information about them is available.

Finally, it must be remarked that dielectric measures are very sensitive to temperature variations. To get reliable estimations of the moisture content, it is basic to determine the temperature of the transformer during the measurements and to apply adequate temperature corrections to refer FDS results to a reference temperature.

The temperature dependence of dielectric measurements has been widely studied [3,48], observing that changes of temperature cause a shifting of the dielectric response frequency spectra, although in most dielectric materials, such as oil–paper insulation, temperature does not affect significantly the shape of the dielectric response. This allows to normalise the data measured at different temperatures into one single curve called master curve.

The shifting of the dielectric response curves with temperature can be expressed with an Arrhenius factor and may be calculated using the equation proposed by Ekanayake in [49]:

$$\text{shift} = \log(\omega_1) - \log(\omega_2) = -\frac{E_a}{k \cdot 2.3258}\left(\frac{1}{T2} - \frac{1}{T1}\right) \tag{3.17}$$

where k is the Boltzmann's constant and E_a is the activation energy of the material (typically 1 eV in oil–paper insulation).

3.6.3 Recovery voltage method

As described before, FDS method, is based in measuring the response of transformer insulation when an excitation of variable frequency is applied, i.e. it analyses the dielectric response in the frequency domain. On the other hand, methods are also available which use a DC excitation to determine the response of the transformer solid insulation. These techniques are called Time domain methods, and within them, two measuring schemes are currently used: the RVM and the measure of PDC. In both cases, the insulation is subjected to a DC voltage, and the response of the insulation is characterised by using different approaches.

RVM is based in the measuring scheme shown in Figure 3.24, which allows the characterisation of the so-called recovery voltage (RV) of an insulation. The method consists of four steps [53]:

1. The insulation is charged by subjecting it to a DC voltage during a well-established charging time (tc).
2. Then the test object is short-circuited for a certain discharging time (td), being the relation between charging and discharging time: tc = 2td.
3. Once the discharging time has elapsed, the short-circuit is released, and the voltage between the terminals of the object is measured. The measured voltage is the so-called RV, which is originated by Dielectric Relaxation processes appearing in a polarised material. The maximum RV (Urmax) and the initial slope of the RV (dUr/dt) are key parameters to interpret measures, as will be later explained.
4. Finally, the object is short-circuited again to remove all remaining polarisation in the material before starting a new measuring cycle.

When RVM is applied to diagnose a transformer, several measuring cycles are needed using different charging times in a typical range 0.1–2,000 s. These measures characterise the different polarisation phenomena of the material which are governed by different time constants.

To estimate the moisture content of transformer insulation, RVM uses the so-called polarisation spectra of the insulation, which is a representation of the maximums of the RV vs. charging times measured in the transformer. Figure 3.25 shows the polarisation spectra of several transformers with different moisture contents. As can be observed, an increase of moisture content implies a shifting of the curve to the left.

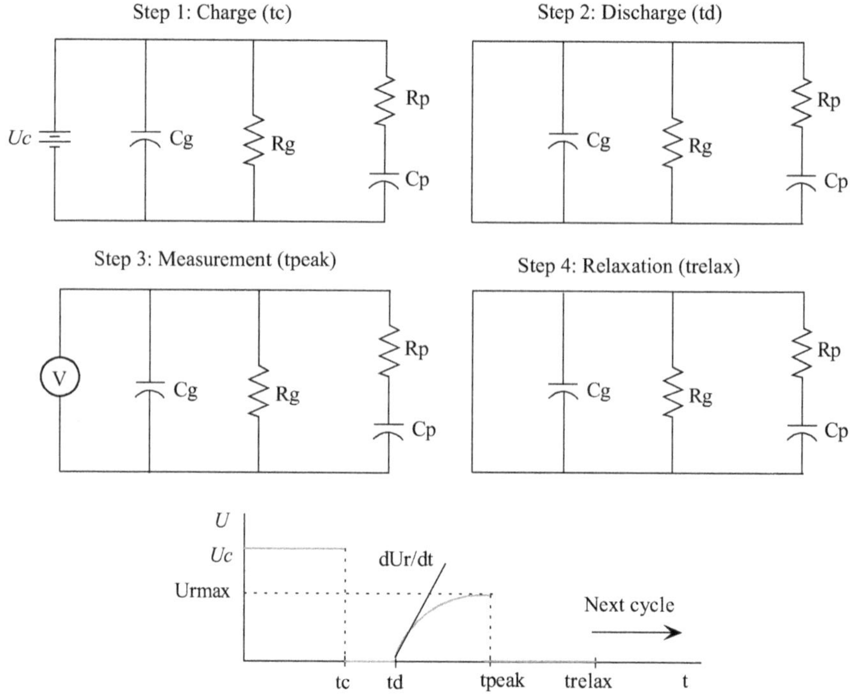

Figure 3.24 Steps for RVM measures. Cg, geometrical capacitance; Rg, insulation resistance; Rp, polarisation resistance and Cp, polarisation capacitance

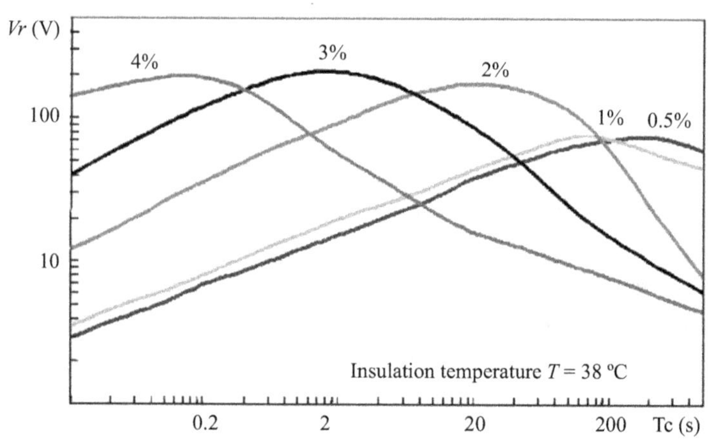

Figure 3.25 RVM measurements on five transformers with various moisture contents in solid insulation. © 2008 CIGRE. Reprinted, with permission, from [54]

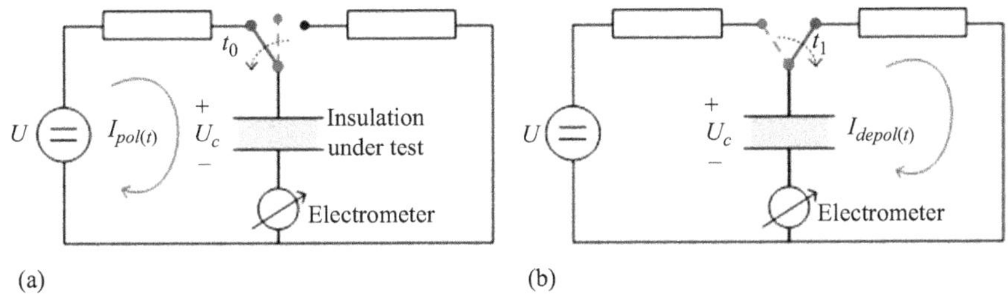

Figure 3.26 PDC measuring scheme: (a) measure of the polarisation current and (b) measure of the depolarisation current. Adapted from [55]

The moisture content of the specimen under study can be estimated by comparing the determined polarisation spectra with those on a library obtained in the lab for samples with controlled moisture contents.

Some authors [48] have criticised this method for not taking into account the geometry of the transformer and the condition of the oil in the interpretation of the measures. Skipping these factors could lead to overestimations of the MC, especially in new units.

Cigré proposed an alternative interpretation method for RVM measures, based in the representation of the maximum value of the RV vs. its initial slope. This representation, known as Guinic representation, seems to gather up in a clearer way the influence on the transformer geometry [48] and is commonly included in the interpretation software of commercial equipment.

3.6.4 Polarisation and depolarisation currents

PDC method is based on characterising the currents associated to the polarisation and depolarisation processes in dielectric materials [48]. These currents are influenced by the water content of the dielectric under test, and then an estimation of the moisture content of a piece of insulation can be obtained from the measure of their values.

In order to apply PDC method, two steps need to be followed:

1. A slope voltage is applied to the object under test and kept for a time long enough to guarantee the completion of the polarisation processes. During this time, the polarisation current flowing through the dielectric is measured. Polarisation current will decrease with time, as the different polarisation processes taking place in the material are completed [Figure 3.26(a)].

   The voltage level typically applied to measure PDC in transformers is around 1,000 V. That voltage should not be too high to avoid non-linear effects, but neither too low, to allow the measure of the currents, which are in the range of pA.
2. The object under test is then short-circuited, and the depolarisation current is measured [Figure 3.26(b)]. This current is caused by relaxation of polarisation processes at the material.

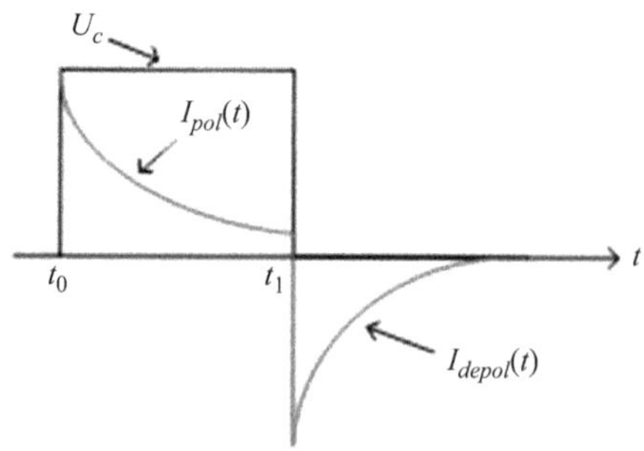

Figure 3.27 Typical shape of polarisation and depolarisation currents

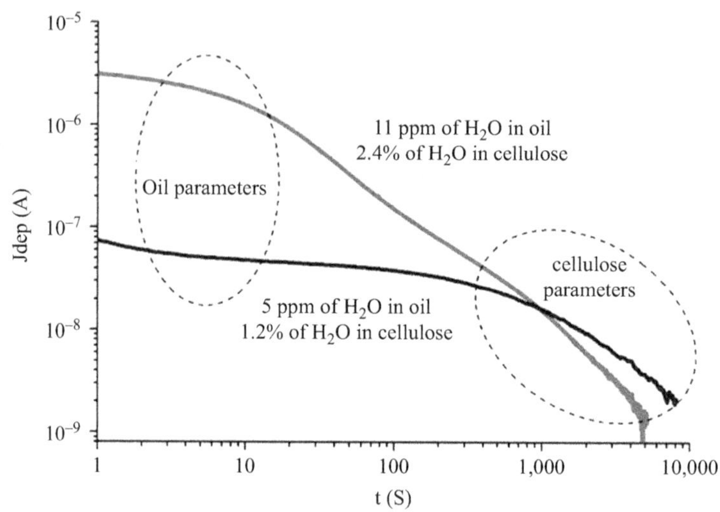

Figure 3.28 Depolarisation current measured at transformers with different moisture contents. © 2008 CIGRE. Reprinted, with permission, from [54]

PDC depend on the applied charging times, and thus, to avoid misinterpretation of the results, applied charging times should be 5–10 times higher than discharging times.

Figure 3.27 shows the typical evolution of PDC. One of the advantages of PDC method is that, as conduction processes are involved in the polarisation current but not in the depolarisation one, it allows a direct determination of the DC conductivity of the insulation.

Figure 3.28 shows the dependency of depolarisation currents with the moisture content of the solid and liquid insulation.

The interpretation of PDC measures is based upon their comparison with the so-called Master Curves, which are obtained by applying PDC method on laboratory models of oil–paper insulation with well-characterised moisture contents [48]. Commercial pieces of equipment use to provide interpretation software including libraries of curves, which allows the user the interpretation of field measures.

Although PDC is agreed to be a reliable method which can provide fair estimations of the moisture content of transformer solid insulation, it must be noted that moisture is not the only factor that influences the measures. Other variables, such as the conductivity of oil or the ageing condition of solid insulation must be included in the interpretation process of the measures in order to avoid wrong diagnosis.

As in the case of FDS and RVM, PDC measures are highly influenced by temperature, and the registers obtained in field transformers must be referred to a reference temperature before interpreting them.

3.7 Conclusions, future trends and challenges

This chapter reviews the main issues related with the presence of moisture in transformers. The chapter comprises two main blocks. The first one includes basic aspects, such as the explanation of the origin of water in transformers, its admissible levels, the risks associated to its presence and distribution of moisture between liquid and solid insulation in an in-service transformer. The second one reviews the main methods that can be used to monitor the water content of transformers.

As was discussed through the chapter, the main challenge for moisture content monitoring in transformers is to get insight into the moisture content of solid insulation. The measure of the moisture content of oil is relatively straightforward and can be even done in a continuous manner using capacitive sensors. However, the mass of water contained in oil supposes only about a 1 per cent of the total mass of water in the transformer, and it is greatly influenced by the loading profile of the transformer and other factors, such as the ambient temperature, and thus, the information that can be extracted from its measure is limited. As an example, in a transformer with mass of oil 45 T and mass of solid insulation 4.5 T, and with an average moisture content in solid insulation of 2.2 per cent, the measured moisture content of oil will change from 1.3 to 26 ppm when it is off service and when it works at full load.

Different methods are currently used to estimate the moisture content of solid insulation. Some of them are based in doing chemical determinations while others are based in modelling or in electric measures. Each method has its own advantages and weaknesses, but all of them can be useful to get insight on the real condition of the transformer. The future trend in moisture monitoring should go towards the combination of different on-line and off-line techniques and interpretation methods to obtain a reliable diagnosis of this variable.

As has been described in this chapter, knowing the moisture content of a transformer is fundamental to determine its maximum admissible load and to programme maintenance operations in the transformer. It seems clear that moisture monitoring must be included in modern management systems to increase the transformer reliability and availability and to optimise the investment in transformer maintenance.

References

[1] Cigré T.D1.01.10, "Brochure 323 – Ageing of cellulose in mineral-oil insulated transformers," 2007.

[2] "IEEE Guide for Diagnostic Field Testing of Electric Power Apparatus – Part 1: Oil Filled Power Transformers, Regulators, and Reactors," in *IEEE Std 62-1995*, pp. 1–68, Dec 1995. doi: 10.1109/IEEESTD.1995.122640

[3] CIGRÉ, WG A2.30, "Brochure 349 – Moisture equilibrium and moisture migration within transformer insulation systems," 2008. Available: http://c1.cigre.org/content/download/15206/539983/version/1/file/A2.30_Moisture_TutorialID55VER20.pdf.

[4] Y. Du, M. Zahn, B. C. Lesieutre, A. V. Mamishev and S. R. Lindgren. 1999, Moisture equilibrium in transformer paper–oil systems. *IEEE Electrical Insulation Magazine* 15(1), pp. 11–20.

[5] F. M. Clark. 1942, Factors affecting the mechanical deterioration of cellulose insulation. *Transactions of the American Institute of Electrical Engineers* 61(10), pp. 742–749.

[6] L. Zhou, G. Wu and J. Liu. 2008, Modeling of transient moisture equilibrium in oil–paper insulation. *IEEE Transactions on Dielectrics and Electrical Insulation* 15(3), pp. 872–878.

[7] M. Scala, "Moisture model for transformer online monitoring," in International Conference on Power Transformers, Transformer 01, Bydgoszcz (Poland), 2001.

[8] S. Gubanski, P. Boss, G. Csepes, *et al.* 2003, Dielectric response methods for diagnostics of power transformers. *IEEE Electrical Insulation Magazine* 19(3), pp. 12–18.

[9] V. Sokolov, P. Griffin, B. Vanin, Moisture Equilibrium and Moisture Migration within Transformer Insulation Systems. Report of CIGRE WG12.18 Life Management of Transformers, 2001.

[10] H. Gasser, C. Krause and T. Prevost, "Water absorption of cellulosic insulating materials used in power transformers," In *Solid Dielectrics, 2007. ICSD'07. IEEE International Conference on*, 2007, pp. 289–293.

[11] B. V. Vanin, "Oil-impregnated cellulosic insulation: Moisture diffusion and equilibrium in view of interfacial adsorption water vapor by cellulose in insulation microcapillaries," in *CIGRÉ Session*, Paris, August 2000.

[12] B. García, T. García, V. Primo, J. C. Burgos and D. Urquiza. 2017, Studying the loss of life of natural-ester-filled transformer insulation: impact of moisture on the aging rate of paper. *IEEE Electrical Insulation Magazine* 33(1), pp. 15–23.

[13] B. V. Ramarao, A. Massoquete, S. Lavrykov and S. Ramaswamy. 2003, Moisture diffusion inside paper materials in the hygroscopic range and characteristics of diffusivity parameters. *Drying Technology* 21(10), pp. 2007–2056.

[14] 'Cellulose nanofiber network.jpg' by Qspheroid4 is licensed under CC BY-SA 4.0. Accessible online at: https://commons.wikimedia.org/w/index.php?title=File:Cellulose_nanofiber_network.jpg&oldid=224873435.

[15] 'Water Structure and Science' by Martin Chaplin is licensed under CC BY-NC-ND 2.0. Accessible online at: http://www1.lsbu.ac.uk/water/water_structure_science.html.

[16] E. Mladenov, S. Staykov and G. Cholakov. 2009, Water saturation limit of transformer oils – [Feature article]. *IEEE Electrical Insulation Magazine* 25(1), pp. 23–30.

[17] T. V. Oommen, "Moisture equilibrium in paper–oil insulation systems," in *1983 EIC 6th Electrical/Electronical Insulation Conference*, 1983, pp. 162–166.

[18] W. A. Fessler, T. O. Rouse, W. J. McNutt and O. R. Compton. 1989, A refined mathematical model for prediction of bubble evolution in transformers. *IEEE Transactions on Power Delivery* 4(1), pp. 391–404.

[19] M. Jovalekic, D. Kolb, S. Tenbohlen, L. Bates and R. Szewczyk, "A methodology for determining water saturation limits and moisture equilibrium diagrams of alternative insulation systems," in *2011 IEEE International Conference on Dielectric Liquids*, Trondheim, 2011, pp. 1–5.

[20] R. D. Villarroel Rodríguez, "Moisture dynamics in transformers insulated with natural esters," PhD Thesis, Departamento de Ingeniería Eléctrica, Universidad Carlos III de Madrid. Accessible online at: https://orff.uc3m.es/handle/10016/21660. 2015.

[21] Z. Zhang, J. Li, S. Grzybowski and P. Zou, "Moisture equilibrium in vegetable oil and paper insulation system," in *2011 Annual Report Conference on Electrical Insulation and Dielectric Phenomena*, 2011, pp. 428–431.

[22] P. Przybylek. Jun. 2016, Water saturation limit of insulating liquids and hygroscopicity of cellulose in aspect of moisture determination in oil–paper insulation. *IEEE Transactions on Dielectrics and Electrical Insulation* 23(3), pp. 1886–1893.

[23] R. Liao, Y. Lin, P. Guo, H. Liu and H. Xia. Apr. 2015, Thermal aging effects on the moisture equilibrium curves of mineral and mixed oil–paper insulation systems. *IEEE Transactions on Dielectrics and Electrical Insulation* 22(2), pp. 842–850.

[24] H. Gasser, C. Krause and T. Prevost, "Water absorption of cellulosic insulating materials used in power transformers," in *2007 IEEE International Conference on Solid Dielectrics*, 2007, pp. 289–293.

[25] IEC 60076-7:2005, "Loading guide for oil-immersed power transformers", 2005.

[26] D. F. Garcia, B. Garcia and J. C. Burgos. 2013, A review of moisture diffusion coefficients in transformer solid insulation – Part 1: Coefficients for paper and pressboard. *IEEE Electrical Insulation Magazine* 29(1), pp. 46–54.

[27] D. F. Garcia, R. Villarroel, B. Garcia and J. C. Burgos. 2013, A review of moisture diffusion coefficients in transformer solid insulation – Part 2: Experimental validation of the coefficients. *IEEE Electrical Insulation Magazine* 29(2), pp. 40–49.

[28] IEC 60814-1997, "Insulating liquids – Oil-impregnated paper and pressboard – Determination of water by automatic coulometric Karl Fischer titration", 1997.
[29] S. Leivo and J. Leppänen, "Transformer's moisture assessment with online monitoring," in *23rd International Conference on Electricity Distribution*, Lyon, 2015, pp. 1–5.
[30] T. Islam, Md F. A. Khan and S. A. Khan, "Moisture measurement of transformer oil using thin film capacitive sensor," in *Power India International Conference (PIICON), 2014 6th IEEE*, Delhi, 2014, pp. 1–5.
[31] W. Li and C. Gao, "Mechanisms and testing of moisture content measurement of transformer pressboard," in *Electronic Measurement & Instruments, 2009. ICEMI'09. 9th International Conference on*, Beijing, 2009, pp. 2-352–2-356.
[32] Vaisala, white paper, "The Effect of Moisture Sensor Location on Reliable Transformer Oil Monitoring," August 2017, accessible online at: http://www.vaisala.com/Vaisala%20Documents/White%20Papers/CEN-TIA-G-whitepaper-transformer_moisture-sensor-B211462EN-A-LOW.pdf.
[33] IEC Std 60422:2013, "Mineral insulating oils in electrical equipment – Supervision and maintenance guidance," 2013.
[34] "IEEE Guide for Acceptance and Maintenance of Insulating Oil in Equipment," in IEEE Std C57.106-2006 (Revision of IEEE Std C57.106-2002). pp. 1–36, Dec. 2006. doi: 10.1109/IEEESTD.2006.371041
[35] S. J. Tee, Q. Liu, Z. D. Wang, *et al.* 2016, Seasonal influence on moisture interpretation for transformer aging assessment. *IEEE Electrical Insulation Magazine* 32(3), pp. 29–37.
[36] P. J. Griffin, "Water in transformers – So what!" in *National Grid Condition Monitoring Conference*, May 1996.
[37] L. Lewand. 2002, Understanding water in transformer systems. *NETA World*, Spring 2002, pp. 1–4.
[38] B. W. Ward, T. V. Oommen and J. A. Thompson, "Moisture estimation in transformer insulation," in *Technical Presentation for the IEEE/PES Transformers Committee, Spring 2004 Meeting*, 9 Mar. 2004.
[39] D. Martin, C. Perkasa and N. Lelekakis, Jul. 2013, Measuring paper water content of transformers: A new approach using cellulose isotherms in nonequilibrium conditions. *IEEE Transactions on Power Delivery* 28(3), pp. 1433–1439.
[40] M. Koch, S. Tenbohlen and T. Stirl. Oct. 2010, Diagnostic application of moisture equilibrium for power transformers. *IEEE Transactions on Power Delivery* 25(4), pp. 2574–2581.
[41] B. Garcia, J. C. Burgos, A. M. Alonso and J. Sanz.Apr. 2005, A moisture-in-oil model for power transformer monitoring – Part I: Theoretical foundation. *IEEE Transactions on Power Delivery* 20(2), pp. 1417–1422.
[42] J. Li, Z. Zhang, S. Grzybowski and M. Zahn. Oct. 2012, A new mathematical model of moisture equilibrium in mineral and vegetable oil–paper insulation.

IEEE Transactions on Dielectrics and Electrical Insulation 19(5), pp. 1615–1622.

[43] L. J. Zhou, G. N. Wu, Y. F. Wang and H. Tang, "Calculating method of moisture in oil–paper insulation at arbitrary temperature," in *2007 Annual Report Conference on Electrical Insulation and Dielectric Phenomena*, Vancouver, 2007, pp. 723–726.

[44] C. Sumereder, M. Muhr and F. Musai, "Moisture determination in solid transformer insulation on the basis of capacitive oil sensors," in *Performance of Conventional and New Materials for High Voltage Apparatus – CIGRE SC D1 – Colloquium in Hungary Budapest 2009*, Budapest, 2009, pp. 1–4.

[45] S. Leivo and J. Leppänen, "Transformer's moisture assessment with online monitoring," in *23rd International Conference on Electricity Distribution*, Lyon, 2015, pp. 1–5.

[46] T. Islam, Md F. A. Khan and S. A. Khan, "Moisture measurement of transformer oil using thin film capacitive sensor," in *Power India International Conference (PIICON), 2014 6th IEEE*, Delhi, 2014, pp. 1–5.

[47] CIGRé Task Force, "Dielectric response methods for diagnostics of power transformers," *IEEE Electrical Insulation Magazine*, 19(3), pp. 12–18, 2003.

[48] Cigré TF D1.01.09, "Brochure 254 – Dielectric response methods for diagnostics of power transformers," 2004.

[49] C. Ekanayake, "Diagnosis of moisture in transformer insulation – Application of frequency domain spectroscopy," PhD. Thesis. Chalmers University of Technology, 2006.

[50] Cigré WG D1.01 (TF 14), "Brochure 414 – Dielectric response diagnoses for transformer windings," 2010.

[51] J. Blennow, C. Ekanayake, K. Walczak, B. Garcia and S. M. Gubanski. Apr. 2006, Field experiences with measurements of dielectric response in frequency domain for power transformer diagnostics. *IEEE Transactions on Power Delivery* 21(2), pp. 681–688.

[52] General Electric, "IDA 200: Insulation Diagnostics System." [Online]. Available: http://mldt.pl/files/ida200.PDF. [Accessed: 28 Apr. 2017].

[53] H. Hipotronics, "Advanced Automatic Recovery Voltage Meter for Diagnosis of Oil Paper Insulation." [Online]. Available: http://www.haefely-hipotronics.com/_em-asset/5462_DS.pdf. [Accessed: 28 Apr. 2017].

[54] R. Malewski, J. Subocz, M. Szrot, J. Płowucha and R. Zaleski, "Condition assessment of medium power transformers using diagnostic methods: PDC, FDS, FRA to support decision to modernize or replace service aged units," in *Cigré Conference*, 2008.

[55] W. Zaengl. Sep./Oct. 2003, Dielectric spectroscopy in time and frequency domain for HV power equipment, Part I: Theoretical considerations. *IEEE Electrical Insulation Magazine*, 19(5), pp. 5–19.

Chapter 4

Assessing DP value of a power transformer considering thermal ageing and paper moisture

Ricardo David Medina Velecela[1], Andres Arturo Romero Quete[2], Enrique Esteban Mombello[2], Giuseppe Rattá[2] and Diego Xavier Morales Jadán[1]

Abstract

Current power transformer (PT) ageing models are approximations to reality based on experimental and theoretical evidence; these models are simplifications of complex interactions inside units, because of that loss-of-life results could be presented in a wide range. In PT decision-making context—investment, replacement or maintenance, this amplitude could lead inadequate actions and, in extreme cases, might compromise technical operation and profitability.

Solid insulation failure is the leading cause of end-of-life of PTs, insulation paper is composed by long fibers of cellulose; the average length of these fibers is called degree of polymerization (DP), it is widely accepted that DP value is a good index of PT loss-of-life.

Considering the above, this chapter presents a holistic methodology for solid insulation ageing assessing based on all thermal degradation process (oxidation, hydrolysis and pyrolysis) and the influence of paper moisture dynamics.

Paper moisture is estimated using as input external variables such as hot-spot temperature, transformer technical data and measurements regarding oil moisture, in order to consider uncertainty in oil moisture growing, arithmetic-Brownian-motion (ABM) algorithms are presented.

For illustrating the application of this method, DP value of two units is assessed; available information for this example are hourly profiles of load and ambient temperature and yearly moisture in oil samples for a period over 15 years.

[1]Smart Grid Research Group and Industrial Informatics Research Group, Universidad Católica de Cuenca, Ecuador

[2]Instituto de Energía Eléctrica, Universidad Nacional de San Juan, Argentina

4.1 Introduction and preliminary issues

Power transformers (PTs) are efficient, reliable and very important assets in the power system, with a low failure rate and a long useful life—around 40 years. Moreover, a well-maintained transformer can operate normally some years more. There is a large portion of installed units near the end of its operative life.

Solid insulation failure is the leading cause of the end-of-life of PTs [1], therefore, a measure of the state of winding solid insulation, i.e., paper, is accepted as an indicator of the condition of the unit [2]. Paper is composed by long chains of glucose rings that build the cellulose polymer molecule; the average length of these chains is termed degree of polymerization (DP). During the PT life cycle, DP decreases due to ageing processes [3], and consequently insulating paper loses its dielectric and mechanical properties. Main ageing processes are oxidation, hydrolysis and pyrolysis that are activated by agents as heat, oxygen, acids and water.

DP can be measured directly from a sample of paper, this action could be addressed during an overhauling process; but for operative units, it is unfeasible as this operation requires a disconnection and manipulation of paper. In addition, as a thumb-rule says: the worst paper sample is the hardest to take.

4.2 State of the art

PT solid insulation has been studied extensively. In the following section, a brief description of the state of the art is presented.

In IEEE C57.91 [4], a noninvasive method to determinate PT loss-of-life was proposed. This method assesses the aging of paper caused by temperature in the hot spot (θ_{HS}) during a time period, and assumes that humidity and acidity in the oil–paper insulating system keeps constant. In such a method, and others similar, the well-known Arrhenius relation, presented in [5], is employed to model the aging of paper, but it is important to remark that the author states the noninclusion of water and oxygen makes current methods "not satisfactory" and, it is necessary to know the complete PT operation history to make realistic life consumption estimations.

In [6], an updated method to calculate remaining life of PT solid insulation was presented. The research was conducted for 3 years over three single-phase 4 kVA – 2 kV/230 V transformers that ran at high load and reached the end of insulation life. In order to compare estimated and real degradation, samples of oil and paper were taken along the experiment; DP value, water content and gas content of the oil were measured. Findings can be summarized as follows: (1) degradation constants proposed in [7] are probably right; (2) proposed equations relating oil oxygen content, paper moisture with paper ageing are in a good agreement with the experimental evidence and (3) it is improper to use a numerical average of temperature profiles because the relationship between the temperature and the reaction rate is not linear—as was probed on [8]. It must be noted that paper moisture influence is emphasized, but no practical evaluation method was proposed.

In [9], a ratio of the mass of adsorbed water to the mass of dry cellulose, C, was proposed and experimentally proved. DP was calculated using C and θ_{HS}. Advantages of this proposal are: direct relation between DP and C, the value of C is stable over time and not affected by hourly/daily quick oscillations. Some restrictions are: this method requires water-activity probes installed in the units, isotherm curves are not accurate above 2–3 percent of water content in paper and the coefficient used in the formulation changes when oil and paper ages.

In [10], a postmortem evaluation of a PT was performed in order to validate the Arrhenius models. Main conclusion of this work is that the available expression is not sufficiently accurate to estimate DP, because they do not include the combined effects of moisture, acidity, and oxygen.

From the above review, it is concluded that an advanced model for estimation of the insulation paper degradation of a PT in operation must suitably consider dynamics in both the paper temperature, θ_{HS}, and the chemical environment inside the tank.

4.3 Theoretical framework

4.3.1 Paper as power transformer solid insulation system

Oil–paper insulation systems are widely used in electrical machinery. In oil-cooled PTs, paper is used as insulating material due to its outstanding characteristics, e.g., excellent dielectric and mechanical properties. Oil degradation can be managed, there are many oil-treatment methods: (1) dehumidification, (2) purification, (3) filtration and (4) in extreme cases oil replacement. However, there are no paper recovery methods; it is no possible to bring maintenance to it.

Paper comprises lignin, hemicellulose and cellulose. Long chains of glucose rings build the cellulose molecules. As was mentioned in Section 4.1, DP is the average length of cellulose chains and has direct relation with the mechanical strength of the paper. Consequences of paper degradation are (1) reduction of mechanical strength, in case of a solicitation, the paper may rupture exposing windings conductors and losing its dielectric function, (2) sludge formation hinders cooling capacity of the unit and (3) formation of water and polar particles.

When paper is exposed to water, oxygen, heat and acids, cellulose molecular chains are cleaved by chemical reactions breaking hydrogen bridges, and then, reducing paper mechanical strength, it is accepted that DP value at the beginning of the paper life is around 1,200–1,000 units, after the dying process DP value falls around 950–900. In the range between 1,000 and 500, paper mechanical strength keeps constant, from 500 to 200, mechanical strength reduces proportional to DP reduction, when DP is lower than 200–150 paper is not able to support a mechanical stress, and the end of life criterion has been reached [11].

4.3.2 Paper degradation process

Paper aging processes act simultaneously and synergistically; consequently, the mathematical model of aging is a very complex task. However, for practical purposes, most researchers and organizations assume three independent degradation

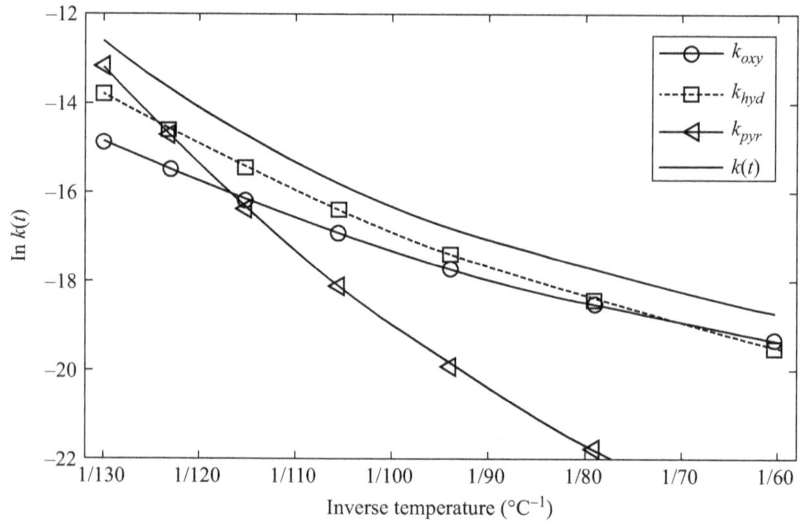

Figure 4.1 Arrhenius plot of degradation rates of oxidation (k_{oxy}), hydrolysis (k_{hyd}) and pyrolysis (k_{pyr}) and total degradation rate k(t)

processes: oxidation, hydrolysis and pyrolysis, each one acts in specific temperature range of prevalence, the total degradation becomes the sum of all process [12]. Figure 4.1 presents an Arrhenius plot of all after-mentioned processes and the total degradation rate $k(t)$.

4.3.2.1 Oxidation

Oxidation is the process that combines paper and oil with oxygen, this reaction increases the acidity of the medium, it is predominant at temperatures below 60–75 °C, this means a unit working undercharged was mainly degraded by acidification. Lundgaard *et al.* proposed in [13] and verifies experimentally in [14] the activation energy of oxidation is lower than hydrolysis. Oxidation subproducts are CO, CO_2 and H_2O. According to [15] for oxidation point of view, thermal upgraded papers (TUP) has no big advantages than non-TUPs (No-TUP).

4.3.2.2 Hydrolysis

Hydrophilic process requires heat, water and an acid environment to trigger paper degradation. Inside PT the main acid source is oil oxidation. Hydrolysis is the main process between 70 and 120 °C. It produces H^+ ions, CO, CO_2 and furans in no thermally upgraded papers. Thermally upgraded paper ages 1.5–3 times slower than no upgraded ones [15].

4.3.2.3 Pyrolysis

Pyrolytic degradation is produced exclusively by heat; its activation energy is 1.4–2 times greater than hydrolysis. Beyond 130 °C is the main degradation process, at 140 °C starts an self-accelerated reaction due H_2O and O_2 generation [16].

4.3.3 *Degradation accelerators*

4.3.3.1 Heat

Heat inside the transformer is not uniformly distributed; the hot-spot temperature θ_{HS} is placed commonly at the top of the transformer over the windings. θ_{HS} is a function of load, ambient temperature and constructive features of the unit. Consequently, the functional age of the transformer due to thermal aging can be roughly assessed through the θ_{HS} estimation. In an operating transformer, internal temperatures increase as load and ambient temperature rise, and vice versa.

θ_{HS} in a PT can be measured by using fiber-optical thermocouples installed at the windings or estimated using, e.g., a set of differential equations. Susa in [17] proposed a dynamic θ_{HS} estimation method based in oil temperature, and the solution of two differential equations: first one (4.1) is associated with top oil temperature θ_{TO}. Second one (4.2) computes hot-spot temperature, θ_{HS}.

$$\left(\frac{(1+R\cdot K^2)}{(1+R)}\cdot \mu_{pu}^{n}\cdot \Delta\theta_{TO,R}\right)=\left(\mu_{pu}^{n}\cdot \tau_{TO,R}\cdot \frac{d\,\theta_{oil}}{d\,t}\right)+\frac{(\theta_{TO}-\theta_{amb})^{n+1}}{\Delta\theta_{TO,R}{}^{n}} \tag{4.1}$$

$$\left(\left\{K^2\cdot P_{W,pu}(\theta_{HS})\right\}\cdot \mu_{pu}^{n}\cdot \Delta\theta_{HS,R}\right)=\left(\mu_{pu}^{n}\cdot \tau_{W,R}\cdot \frac{d\,\theta_{HS}}{d\,t}\right)+\frac{(\theta_{HS}-\theta_{TO})^{m+1}}{\Delta\theta_{HS,R}{}^{m}} \tag{4.2}$$

where R is the relation between load and no-load losses, K is the load factor, μ_{pu} is oil viscosity per unit, n is a constant that depends on the oil circulation, m is an empirical constant that models nonlinear behavior of coils [4], $\Delta\theta_{TO,R}$ is the rated top-oil θ_{TO} rise over ambient, $\Delta\theta_{HS,R}$ is the rated hot-spot temperature rise over top-oil, $\tau_{TO,R}$ rated thermal time constant, $\tau_{W,R}$ is the rated winding time constant and θ_{amb} is the ambient temperature.

Solving simultaneously (4.1) and (4.2) for each i time, $\theta_{HS,i}$ is obtained and will be used to evaluate the depolymerization process.

4.3.4 *Paper humidity*

Paper has high hydrophilicity and accumulates water during its lifespan. Oil has low water affinity and acts as water transport. It is correct to assume the moisture flows from oil to paper, there are many oil-drying methods, but there are no efficient paper-drying methods.

Water accelerates paper degradation, in fact, mechanical strength reduces half when paper moisture duplicates [18], another risk of high moisture is bubble formation on windings surface at high temperatures.

Table 4.1 presents a paper categorization according to moisture content. It is remarkable that from humilities beyond 4–5 percent, degradation rate is very high. In [19], authors consider that if the paper moisture content keeps lower than 2 percent, the degradation rate is almost constant; beyond this limit, degradation increases exponentially.

Table 4.1 Paper moisture ranges

Moisture (%)	Status
0.5	New oil, start of life
1–2	Acceptable, normal operation
2–5	Degraded, accelerated ageing
>5	End of life

Main humidity sources are: ingress from atmosphere, oil–paper decomposition and residual humidity not removed during factory drying procedure. Following a bibliographical review, there are many paper moisture estimation/measurement procedures, e.g., direct measurement taking-off a paper sample, frequency response analysis as presented in [20] and indirect estimation using oil humidity measures and θ_{HS}. The rest of this section presents a brief resume of indirect estimation of moisture in paper.

Paper water sorption is a dynamic process that depends on temperature and oil humidity, on the other hand, desorption is a slow process as it requires some activation energy. There are two sorption ways in a porous material as paper: absorption and adsorption; in the first one, water moves freely and becomes part of the substance, second one is just a superficial addition of water. Adsorption is practically instantaneous, antagonist process of desorption. For deeper treatment of cellulose sorption and desorption please refer to [21].

Paper moisture is proportional to oil humidity and inverse to temperature [22], many methods have been proposed to correlate these variables [23]. Most of them fit well when oil–paper are in hydric equilibrium [24], which means the humidity migration between paper and oil is null in short time periods (hours); in order to fulfill hydric equilibrium, it is indispensable to reach thermal equilibrium.

Thermal equilibrium is not reachable in real world; PTs operates under variable load and ambient temperature which varies along the day. According to reference [25], time to reach equilibrium is around many hours to few days. It is evident that those methods are unsuitable for an operating unit.

In [26,27], "dynamic equilibrium" has been proposed. PT works under daily and weekly cyclic load profile, this situation produces cyclic θ_{HS} profiles, supposing that, moisture adsorption/desorption oscillates around an equilibrium point or region as oil humidity varies slowly. This idea was probed in [28] for the range of operative temperatures of the unit. Koch in [29] establish a method to create vapor-saturation-based "equilibrium curves," that assumes: cellulose sorption is fast and desorption takes long time and requires high temperatures, i.e., water flows from oil to paper, paper moisture variations are slow, no hourly variations exists.

In [30], the company ABB presents a paper humidity formula, shown in (4.3) and plotted in Figure 4.2, works reasonably well keeping in mind "dynamic equilibrium"; hence, this equation will be used in this work.

$$H_{paper} = 2.06915 \cdot e^{(-0.0297 \cdot \theta_{HS})} \cdot (H_{oil})^{0.4089 \cdot \theta_{HS}^{\,0.09733}} \tag{4.3}$$

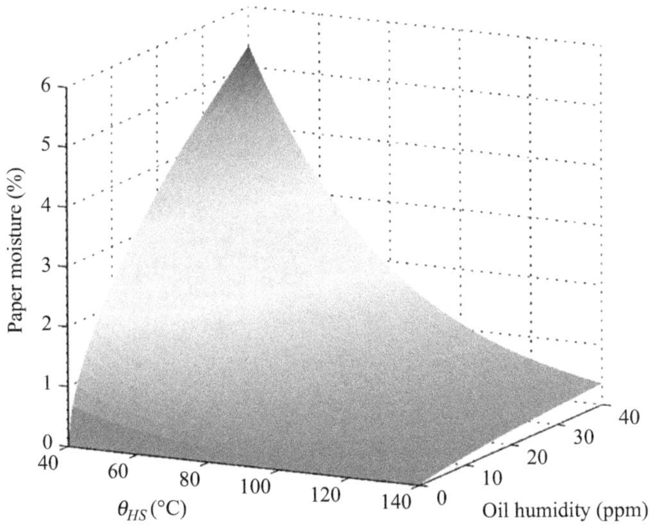

Figure 4.2 Paper moisture using ABB formula

where H_{paper} is paper moisture expressed in percent, H_{oil} is oil humidity expressed in parts per million and θ_{HS} is hot-spot temperature in centigrade.

4.3.5 Assessing of depolymerization process

Depolymerization process was widely analyzed by Dakin [31] and other researchers. The estimation of the DP value is based on Dakin–Arrhenius relationship shown in (4.4). In [12], degradation processes were disaggregated into hydrolysis, oxidation and pyrolysis according to (4.5).

$$\frac{1}{DP(t)} - \frac{1}{DP(t_0)} = \sum_{t_0}^{t} k(t) \cdot \Delta t \tag{4.4}$$

with:

$$\begin{aligned} k(t) = \big(&A_{oxy}(t) \cdot e^{-E_{a,oxy}/R\cdot(273+\theta_{HS})} + A_{hyd}(t) \cdot e^{-E_{a,hyd}/R\cdot(273+\theta_{HS})} \\ &+A_{pyr}(t) \cdot e^{-E_{a,pyr}/R\cdot(273+\theta_{HS})}\big) \end{aligned} \tag{4.5}$$

where $DP(t_0)$ and $DP(t)$ are DP values at the start (t_0) and at the end (t) of the time interval Δt; $k(t)$ is the degradation rate; E_a is the activation energy of the reactions, expressed in (J/mol); A is the preexponential factor and represents the environmental influence, it is expressed in (h^{-1}); R is the gas constant, equal to 8.314 (J/mol/K); θ_{HS} is the hot-spot temperature. *Oxy, Hyd* and *Pyr* subindices correspond to oxidation, hydrolysis and pyrolysis, respectively. *Ea* and *A* are closely related [32].

In (4.4) and (4.5), the preexponential factor, A, represents the environmental influence on the degradation rate $k(t)$. Many accelerated ageing experiments have been performed in order to quantify the relationship between humidity, heat and paper condition. For instance, in [33,34], the following quadratic equations that relates A_{hyd} value, paper moisture and oxygen were presented: for no TUP (4.6), and (4.7) for thermally upgraded paper. Oxygen ranges are: low content (<7,000 ppm), medium (7,000–14,000 ppm) and high content (>16,500 ppm).

$$A_{hyd} = \begin{cases} 1.78 \cdot 10^8 \cdot H_p^2 + 1.1 \cdot 10^8 \cdot H_p + 5.28 \cdot 10^7 \rightarrow \text{Low } O_2 \\ 2.07 \cdot 10^8 \cdot H_p^2 + 5.61 \cdot 10^8 \cdot H_p + 2.31 \cdot 10^8 \rightarrow \text{Med } O_2 \\ 2.29 \cdot 10^8 \cdot H_p^2 + 9.78 \cdot 10^8 \cdot H_p + 3.86 \cdot 10^8 \rightarrow \text{High } O_2 \end{cases} \tag{4.6}$$

$$A_{hyd} = \begin{cases} 6.92 \cdot 10^7 \cdot H_p^2 + 2.61 \cdot 10^8 \cdot H_p + 1.03 \cdot 10^7 \rightarrow \text{Low } O_2 \\ 2.64 \cdot 10^8 \cdot H_p^2 + 7.32 \cdot 10^8 \cdot H_p + 2.37 \cdot 10^9 \rightarrow \text{Med } O_2 \\ 4.29 \cdot 10^8 \cdot H_p^2 + 2.03 \cdot 10^9 \cdot H_p + 4.27 \cdot 10^9 \rightarrow \text{High } O_2 \end{cases} \tag{4.7}$$

where H_p is the paper moisture, expressed in percent.

It is desirable to maintain low oxygen levels for units in service [12]. If oxygen content is high, there are some possible actions as: add oxygen inhibitors into oil or to conduct a degasification process. For most common service conditions, it is recommendable to use low oxygen equation.

Values for $E_{a,ox}$ and A_{ox} were obtained by Feng [35] and Lelekakis [33]. These values are reported in Table 4.2. For pyrolysis, $E_{a,pyr}$ and A_{pyr} values can be found in the works of Conesa [36], Capart [37] and Kashiwagi [38] and, these are presented in Table 4.3.

Table 4.2 $E_{a,oxi}$ and A_{oxi} values

Author	**No-TUP Feng and Lelekakis**	**TUP Lelekakis**
$E_{a,oxi}$ (kJ/mol)	89	82
A_{oxi}	4.6×10^5	3.2×10^4

Table 4.3 $E_{a,pyr}$ and A_{pyr} values

Author	**Conesa**	**Capart**	**Kashiwagi**
$E_{a,pyr}$ (kJ/mol)	215.7	255	220
A_{pyr}	4×10^{17}	5.8×10^{20}	7.2×10^{20}

4.4 Proposed method

4.4.1 Problem description

Direct measurement of θ_{HS} and paper moisture are not available in old units and, considering that these two variables are needed to estimate solid insulation state of degradation by means of Dakin–Arrhenius relation. In the following section, a robust methodology for indirect DP value estimation is presented.

4.4.2 Oil moisture estimation

Oil humidity (H_O) increases continuously with time, it depends on the transformer sealing quality and environment conditions. If the unit owner have historic data of water content in the oil, it is possible to analyze tendencies and forecast humidity increase. If such information is not available, and assuming that the unit has a good sealing, it is possible to estimate the amount of water contamination inside the tank using empirical values to create probable profiles.

Oil humidity variations inside PTs satisfy the following conditions: (1) stochasticity, (2) continuity, (3) temporal independence, (4) self-similarity and (5) it is a memory-less process. According to this, oil humidity can be modeled using the generalized Wiener Process, so called ABM [17]. The corresponding mathematical formulation is presented in (4.8), where $Z_t^{(m)}$ is the t^t term of the m path, μ is the expected value, σ is the standard deviation, Δt is time $[0,T]$ divided in N intervals and X_n is Gaussian white noise $N(0,1)$. For all paths, there are only one origin value, that is, there exists exactly one Z_0 for all m paths. This is a diffusive process for which uncertainty grows over time.

$$Z_t^{(m)} = Z_{t-1}^{(m)} + \mu \cdot \Delta t + \sigma \cdot \sqrt{\Delta t} \cdot X_n \tag{4.8}$$

If there are two or more known points, it is possible to create a Brownian Bridge (BB), assuming that the uncertainty is zero in its nodes and maximum in the middle of two known points. Increments in BB are not independent. The corresponding equation is (4.9), where $ZB_t^{(m)}$ is the tth term of the m path between the two known points W_n and W_{n+1} with ($t_n<t<t_{n+1}$), σ is the standard deviation and X_n is the Gaussian white noise $N(0,1)$. There are common points for all paths, that is, there exists exactly the same W_n for all paths.

$$ZB_t^{(m)} = \left[W_n + \frac{t - t_n}{t_{n+1} - t_n}(W_{n+1} - W_n)\right] + \sigma \cdot \sqrt{\frac{(t - t_n) \cdot (t_{n+1} - t)}{t_{n+1} - t_n}} \cdot X_n \tag{4.9}$$

ABM will be used when there are little or no water in oil content information; on the other hand, BB considers uncertainties when some data exist.

4.4.3 New approach for degree or polymerization assessing

The proposed methodology for indirect estimation of DP value of an operative PT is presented in Figure 4.3, the five steps are detailed as follows:

1. **Step 1. Generation of random oil humidity profiles for the analysis period**
   This information is needed to model moisture dynamics and water migration from oil to paper. If there is not an oil moisture monitoring system, there are two forecast alternatives:
   (a) If there are some water in oil, historical records use (4.9); otherwise,
   (b) use (4.8) with statistical information.

2. **Step 2. Calculation of hot-spot temperature**
   If some θ_{HS} measuring system exists, use those profiles, otherwise solve simultaneously (4.1) and (4.2); required inputs are: transformer characteristics, hourly load and ambient temperature profiles.
3. **Step 3. Estimation of hot-spot paper moisture**
   To estimate the paper moisture use (4.3), using as inputs: θ_{HS} profile and oil humidity profiles. It is important to consider that estimated paper moisture corresponds to the most degraded point of the windings, where the temperature is the highest. Once the paper moisture profile is estimated, using a smoothing method to accomplish the pseudo-equilibrium conditions is proposed. In [27], a long-term average is proposed for reaching a mathematical equilibrium that balances the daily load cycles. In [29], a daily value is calculated by averaging the maximum and minimum temperature value in order to correctly model the

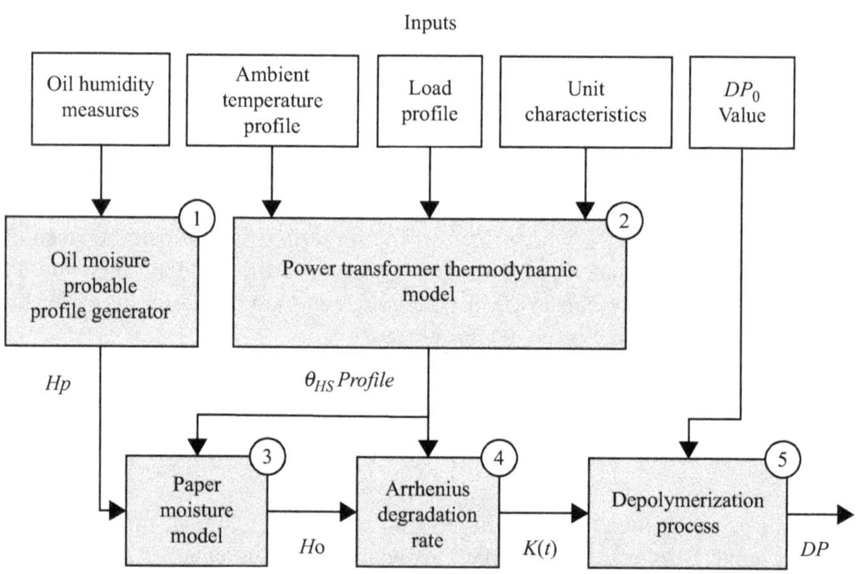

Figure 4.3 Proposed loss-of-life methodology

paper moisture dynamics. In [39], a moving average method is used with a data window of one day.

4. **Step 4. Compute degradation rate $k(t)$**
   Degradation rate is computed using (4.5), and A_{oxi} and $E_{a,oxi}$ values are taken from Table 4.2. Considering that A_{pyr} and $E_{a,pyr}$ values were experimentally obtained, the average value describes well pyrolysis degradation, mean values are taken from Table 4.3, A_{hyd} value is estimated with (4.6) for No-TUP and (4.7) for TUP. $E_{a,hyd}$ is 111 (kJ/mol).
5. **Step 5. Assessment of depolymerization**
   Final DP value is computed using (4.4) where $DP_0 = 1{,}000$ corresponds to a new unit; for old units it is possible to assume a DP_0 value derived from other sources, e.g., furan analysis or direct sample measure.

   Consider that in oil moisture forecast, some "possible paths" were proposed, consequently same number of DP profiles must be computed; final DP value is obtained from an average of the final DP value of all profiles.

4.5 Casestudy

Final DP value of two PTs, termed T1 and T2, is assessed. Technical characteristics are presented in Table 4.4, the operator began to store load and ambient temperature data on Jan. 20, 2001 (installation date of both units); this study comprises the period from that date until Sep. 01, 2016. Therefore, 24 values of both load and ambient temperature for each of the 5,825 days listed in the database are available. Units T1 and T2 were installed in the same substation; the ambient temperature profile is presented in Figure 4.4; load profiles and the computed hot-spot temperature are shown in Figures 4.5 and 4.6(a) for T1 and T2, respectively.

Oil humidity measurements were performed yearly since 2008. The results are presented on Table 4.5. As previous history is not available; initial oil moisture value was assumed as 0.5 ppm on the installation date. Figures 4.7 and 4.8 show 1,000 oil and paper profiles from Jan. 2001 to Sep. 2016 generated with AMB and BB tool for unit T1 and T2, respectively; as mentioned above, the constants $\sigma = 0.00709$ and $\mu = 0.0068$ were computed from the average previous growing rates, these values are consistent with [21]. Note all paths in oil moisture passes

Table 4.4 Characteristics of studied units

	T1	T2
Rated power (MVA)	15	30/30/20
Voltage (kV)	132/34.5	132/34.5/13.8
Connections	YNy0	Yy0/Yd11/Yd11
Cooling class ONAN/ONAF (%)	0–70/70–100	0–70/70–100
Paper class	No-TUP	TUP
Year of manufacture	2000	2000

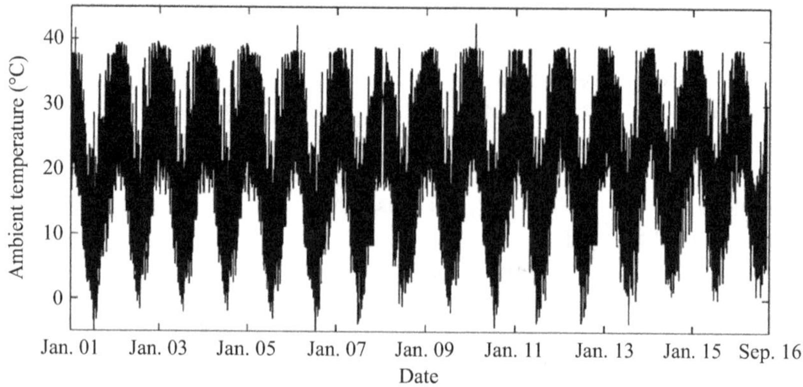

Figure 4.4 Ambient temperature of the substation for the period Jan. 2001–Sep. 2016

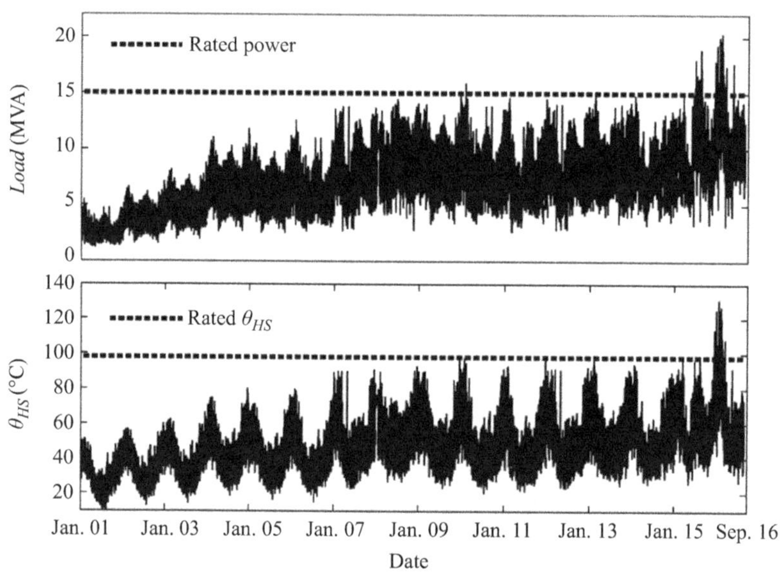

Figure 4.5 Load and θ_{HS} of Unit T1 for the period Jan. 2001–Sep. 2016

through the known points—identified with circles. Uncertainty in paper moisture is proportional to uncertainty in oil moisture, which is evident in the period Jan. 2001–Jan. 2008.

4.5.1 Results

DP values were computed according the previously outlined methodology. To this effect, the same hot-spot temperature profile was used, and in order to consider oil moisture uncertainty, 1,000 oil humidity profiles were proposed for each unit.

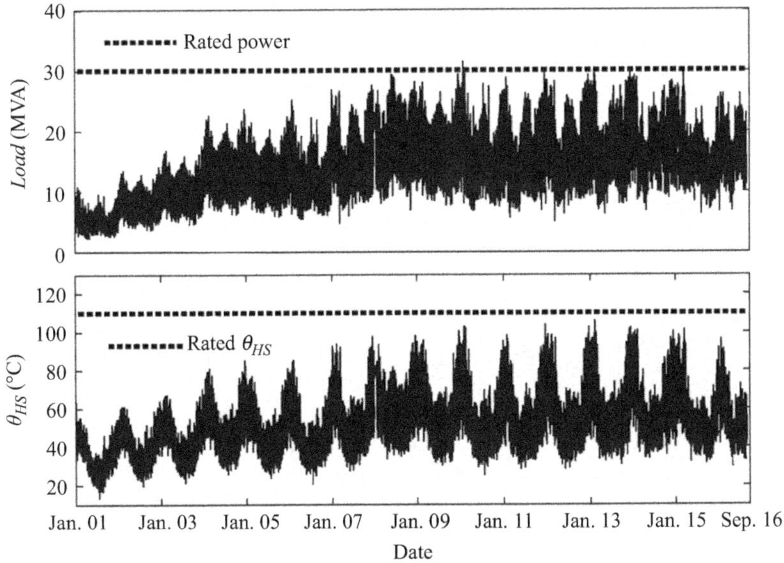

Figure 4.6 Load and θ_{HS} of Unit T2 for the period Jan. 2001–Sep. 2016

Table 4.5 Reported oil moisture content, in (ppm)

Date (H_O)	T1	T2
Jan. 2001	0.5*	0.5*
Jan. 2008	5	4.8
Jul. 2008	5.4	5
Jul. 2009	6.1	5.2
May 2011	7.6	5.7
Jan. 2013	7.1	8
Feb. 2014	8.2	7.6
Jan. 2015	8.8	7.4
May 2016	9	7

* Assumed value.

Figure 4.9 presents assessed DP profiles for the unit. There are small dispersions in final DP values this is because the unique uncertain source was oil moisture.

Hot-spot temperature profile follows an annual cycle, summer (from December to March) presents peaks that could reach recommended operative temperature, and accelerates the depolymerization process.

The average final DP value was 498 and 568 for T1 and T2, respectively. The analysis was performed over 5,825 days and it represents 139,800 h (around 90 percent of normal life—150,000 h), considering that units operate for a

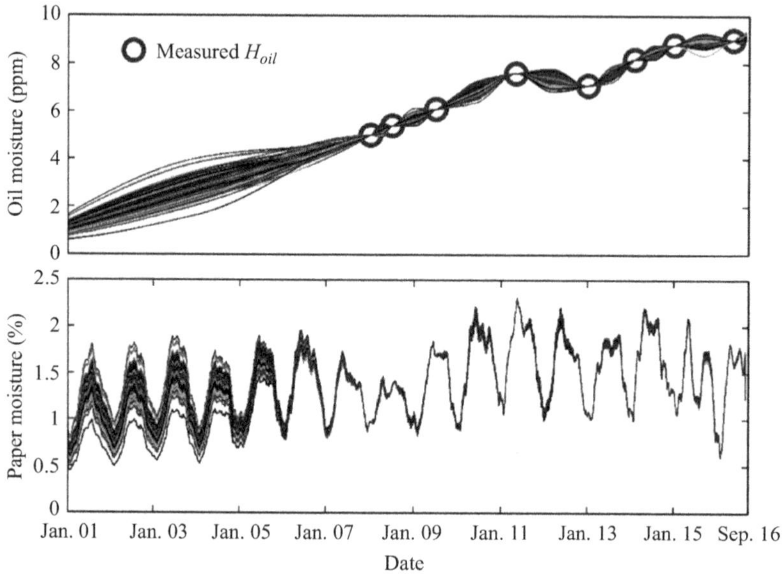

Figure 4.7 Thousand random oil and paper moisture profiles for T1

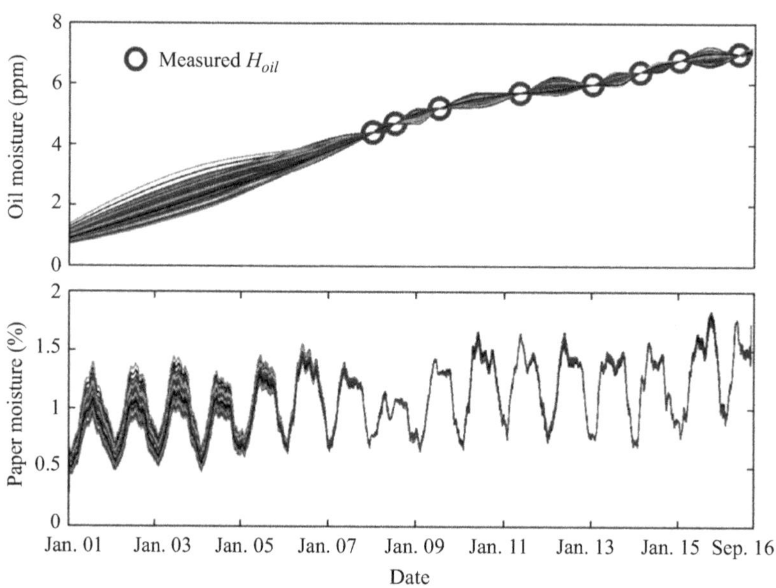

Figure 4.8 Thousand random oil and paper moisture profiles for T2

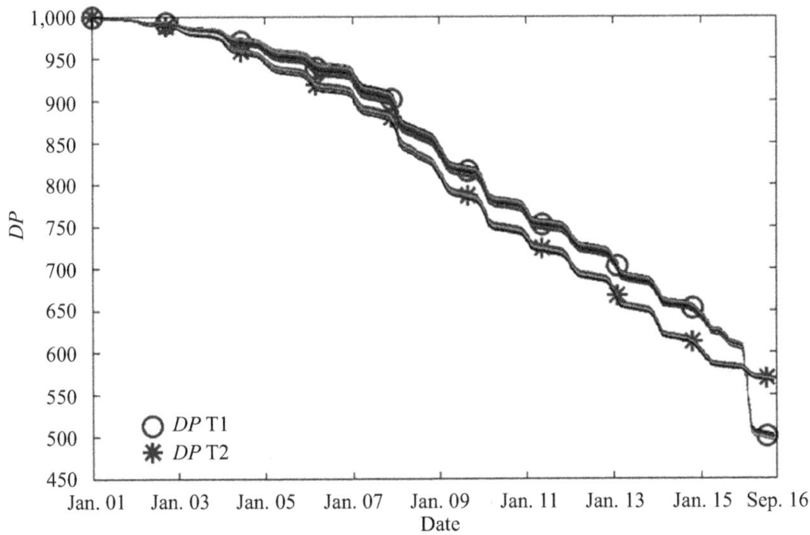

Figure 4.9 Computed thousand DP profiles for units T1 and T2

long period. Under its nominal load, the reduction in DP value is not significant except for the period of summer 2016, where unit T1 works overcharged, this condition increases heat over windings and consequently accelerate depolymerization process. On the other hand, unit T2 works under its nominal load and hot-spot temperature, this situation causes the progressive and constant degradation of solid insulation.

Respect to oil moisture, it is important to remark that the substation is located in a desert region and water ingress inside units is lower, the difference in water-in-oil content between units can be attributed to cellulose degradation in unit T1, oxidative and pyrolitic degradation of cellulose produces free water and H^+ ions.

4.6 Conclusions

This chapter presents a holistic methodology to assess degradation of PT solid insulation considering thermal aging and dynamics in oil moisture. Advantages of this method can be summarized as follows:

- Proposed method is supported by state-of-the-art methods,
- works for thermally upgraded paper papers as well for no thermally upgraded units,
- moisture in paper is estimated using external information as oil moisture, load and ambient temperature and,
- uncertainty in moisture is addressed with mathematical tools.

References

[1] W. Bartley. (2004). *An Analysis of International Transformer Failures, Part 1.* Available: http://www.hsb.com/TheLocomotive/Article.aspx?id=157&terms=transformer.

[2] W. Flores, E. Mombello, G. Ratta, and J. A. Jardini, "Life of power transformers immersed in oil.—State-of-the-art—Part I. Correlation between life and temperature," *IEEE Latin Am Trans*, vol. 5, pp. 50–54, Mar. 2007.

[3] A. Abu-Siada and S. Islam, "A new approach to identify power transformer criticality and asset management decision based on dissolved gas-in-oil analysis," *IEEE Trans Dielectr Electr Insul*, vol. 19, pp. 1007–1012, Jun. 2012.

[4] IEEE, "IEEE Guide for Loading Mineral-Oil-Immersed Transformers and Step-Voltage Regulators," *IEEE Std C57.91-2011 (Revision of IEEE Std C57.91-1995)*, pp. 1–123, 2012.

[5] A. M. Emsley and G. C. Stevens, "Review of chemical indicators of degradation of cellulosic electrical paper insulation in oil-filled transformers," *IEE Proc Sci Meas Technol*, vol. 141, pp. 324–334, Sep. 1994.

[6] D. Martin, Y. Cui, C. Ekanayake, H. Ma, and T. Saha, "An updated model to determine the life remaining of transformer insulation," *IEEE Trans Power Delivery*, vol. 30, pp. 395–402, Jan. 2015.

[7] L. E. Lundgaard, W. Hansen, D. Linhjell, and T. J. Painter, "Aging of oil-impregnated paper in power transformers," *IEEE Trans Power Delivery*, vol. 19, pp. 230–239, Jan. 2004.

[8] R. D. Medina, A. A. Romero, E. E. Mombello, and G. Ratta, "Comparative study of two thermal aging estimating methods for power transformers," *IEEE Latin Am Trans*, vol. 13, pp. 3287–3293, 2015.

[9] D. Martin, T. Saha, T. Gray, and K. Wyper, "Determining water in transformer paper insulation: effect of measuring oil water activity at two different locations," *IEEE Electr Insul Mag*, vol. 31, pp. 18–25, May 2015.

[10] N. Lelekakis, G. Wenyu, D. Martin, J. Wijaya, and D. Susa, "A field study of aging in paper-oil insulation systems," *IEEE Electr Insul Mag*, vol. 28, pp. 12–19, Jan. 2012.

[11] T. K. Saha, "Review of modern diagnostic techniques for assessing insulation condition in aged transformers," *IEEE Trans Dielectr Electr Insul*, vol. 10, pp. 903–917, Oct. 2003.

[12] CIGRE, "323 Ageing of cellulose in mineral-oil insulated transformers," vol. 323, ed: CIGRE, 2007.

[13] L. E. Lundgaard, W. Hansen, and S. Ingebrigtsen, "Ageing of mineral oil impregnated cellulose by acid catalysis," *IEEE Trans Dielectr Electr Insul*, vol. 15, pp. 540–546, 2008.

[14] K. B. Liland, M. Kes, M. H. G. Ese, L. E. Lundgaard, and B. E. Christensen, "Study of oxidation and hydrolysis of oil impregnated paper insulation for

transformers using a microcalorimeter," *IEEE Trans Dielectr Electr Insul* vol. 18, pp. 2059–2068, 2011.

[15] CIGRE, "A393 Thermal performance of transformers," ed: CIGRE Group A2.24, 2009.

[16] CIGRE, "A227 Life Management Techniques for power transformers" vol. A227, ed: CIGRE Work Group A2.18, 2003.

[17] D. Susa, "Dynamic thermal modelling of power transformers," Doctoral Dissertation, Department of Electrical and Communications Engineering, Helsinki University of Technology, 2005.

[18] L. J. Zhou, G. N. Wu, Y. F. Wang, and H. Tang, "Calculating method of moisture in oil-paper insulation at arbitrary temperature," in *CEIDP*, 2007, pp. 723–726.

[19] A. Krontiris, "Fuzzy systems for condition assessment of equipment in electric power systems," Shaker, 2012.

[20] T. K. Saha and P. Purkait, "Understanding the impacts of moisture and thermal ageing on transformer's insulation by dielectric response and molecular weight measurements," *IEEE Trans Dielectr Electr Insul*, vol. 15, pp. 568–582, Apr. 2008.

[21] CIGRE, "Moisture equilibrium and moisture migration within transformers insulation systems," vol. A2.30, ed: CIGRE, 2008.

[22] T. Gradnik, B. Cucek, and M. Koncan-Gradnik, "Temperature and chemical impact on determination of water content in dielectric liquids by capacitive moisture sensors," in Dielectric Liquids (ICDL), *2014 IEEE 18th International Conference on*, 2014, pp. 1–5.

[23] J. A. Thompson, "A moisture diffusion model for transformer oil and paper," in *Power and Energy Society General Meeting, 2011 IEEE*, 2011, pp. 1–3.

[24] S. K. Ojha, P. Purkait, and S. Chakravorti, "Understanding the effects of moisture equilibrium process on dielectric response measurements for transformer oil-paper insulation systems," in *Power and Energy in NERIST (ICPEN), 2012 1st International Conference on*, 2012, pp. 1–6.

[25] Y. Du, M. Zahn, B. C. Lesieutre, A. V. Mamishev, and S. R. Lindgren, "Moisture equilibrium in transformer paper-oil systems," *IEEE Elect Insul Mag*, vol. 15, pp. 11–20, Jan./Feb. 1999.

[26] D. Martin, C. Perkasa, and N. Lelekakis, "Measuring paper water content of transformers: a new approach using cellulose isotherms in nonequilibrium conditions," *IEEE Trans. Power Delivery*, vol. 28, pp. 1433–1439, 2013.

[27] J. Abdallah, "Investigating the transformer operating conditions for evaluating the dielectric response," in *Presented at The World Academy of Science, Engineering And Technology*, Issue 0051 March 2011, 2011.

[28] W. Wei, W. Xin, M. Zhiqing, *et al.*, "Disequilibrium of moisture in oil-pressboard insulation under transient temperature condition," in *Condition Monitoring and Diagnosis (CMD), 2012 International Conference on*, 2012, pp. 894–896.

[29] M. Koch, S. Tenbohlen, and T. Stirl, "Diagnostic application of moisture equilibrium for power transformers," *IEEE Trans Power Delivery*, vol. 25, pp. 2574–2581, Oct. 2010.

[30] Pahlavanpour, M. Martins, and Eklund, "Study of moisture equilibrium in oil-paper system with temperature variation," in *Properties and Applications of Dielectric Materials, 2003. Proceedings of the 7th International Conference on*, 2003, vol. 3, pp. 1124–1129.

[31] T. W. Dakin, "Electrical insulation deterioration treated as a chemical rate phenomenon," *Trans Am Inst Electr Eng*, vol. 67, pp. 113–122, Jan. 1948.

[32] A. M. Emsley, "The kinetics and mechanisms of degradation of cellulosic insulation in power transformers," *Polym Degrad Stab*, vol. 44, pp. 343–349, Jan. 1994.

[33] N. Lelekakis, D. Martin, and J. Wijaya, "Ageing rate of paper insulation used in power transformers Part 1: Oil/paper system with low oxygen concentration," *IEEE Trans Dielectr Electr Insul*, vol. 19, pp. 1999–2008, Dec. 2012.

[34] N. Lelekakis, D. Martin, and J. Wijaya, "Ageing rate of paper insulation used in power transformers Part 2: Oil/paper system with medium and high oxygen concentration," *IEEE Trans Dielectr Electr Insul*, vol. 19, pp. 2009–2018, Dec. 2012.

[35] D. Y. Feng, Z. D. Wang, and P. Jarman, "Modeling thermal life expectancy of the UK transmission power transformers," in *ICHVE*, 2012, pp. 540–543.

[36] J. A. Conesa, J. Caballero, A. Marcilla, and R. Font, "Analysis of different kinetic models in the dynamic pyrolysis of cellulose," *Thermochim Acta*, vol. 254, pp. 175–192, Apr. 1995.

[37] R. Capart, L. Khezami, and A. K. Burnham, "Assessment of various kinetic models for the pyrolysis of a microgranular cellulose," *Thermochim Acta*, vol. 417, pp. 79–89, Jul. 2004.

[38] T. Kashiwagi and H. Nambu, "Global kinetic constants for thermal oxidative degradation of a cellulosic paper," *Combust Flame*, vol. 88, pp. 345–368, Mar. 1992.

[39] R. D. Medina, A. Romero, E. Mombello, and G. Rattá, "Assessing degradation of power transformer solid insulation considering thermal stress and moisture variation," *Electr Power Syst Res*, vol. 151, pp. 1–11, 2017.

Chapter 5
Frequency response analysis

Mehdi Bagheri[1] *and Toan Phung*[2]

Abstract

Transformers can be treated as a time-invariant system which can be characterized by its response to the Dirac's delta function. This impulse response is completely unique and should remain so over the system life. Therefore, it can be used as a kind of indicator to check if the internal compartments, elements or parameters have physically changed during service life. The system response is the convolution of the system impulse response with the excitation signal. This time-domain response has its equivalent response in the frequency domain. Thus, the frequency response of a time-invariant system should also remain unchanged if system parameters are unchanged. This concept can be utilized to evaluate the mechanical structure integrity as well as diagnosis of transformers and rotating machines. It is called frequency response analysis. This chapter is specifically focused on frequency response measurement and analysis of transformers.

5.1 Introduction

Power transformers are commonly considered as the heart of the transmission and distribution power system networks. They are in service under different environmental, electrical and mechanical conditions and may be subject to enormous hazards during the course of operation. Monitoring their condition and diagnosing faults are important parts of the maintenance function. Utility engineers strive to keep power transformers in service, and to prevent even short-term outages. Failure of a transformer can cause extensive damage to equipment owned by consumers and/or the utility.

For many years, short-circuit impedance (SCI) measurement has been used as a simple technique to detect transformer winding deformation and core displacement. It is still implemented in many countries for transformer diagnosis. As a more sensitive method, frequency response analysis (FRA) for transformer was

[1]Department of Electrical and Computer Engineering, Nazarbayev University, Kazakhstan
[2]School of Electrical Engineering and Telecommunications, University of New South Wales, Australia

introduced by Dick and Erven in 1978 [1]. Nowadays, it has increasingly become a popular technique for assessment of the mechanical integrity of transformers. FRA is considered a highly accurate, fast, economical and nondestructive method of detecting winding defects and damage in the transformer core. Transformer winding deformation, turn-to-turn short-circuit, faulty grounding core or screens, core deformation and movement, conductor rupture, radial or axial winding displacements, partial winding failure or collapse, internal winding connection problems and even tap-changer contact disconnection or bad contacts can be recognized through FRA. FRA is a comparison-based method, i.e., the newly acquired measurement data are compared with the historical data recorded in the past. Any discrepancy between new and old measured data would indicate some physical change in the transformer. This change could be due to mechanical damage of transformer active parts or insulation deterioration. Both will be discussed in this chapter.

FRA measurement setup, results interpretation, factors affecting FRA, practical considerations, and online FRA (OFRA) implementation will be discussed in this chapter. To facilitate better understanding, different industrial examples on different transformers are provided, and transformer frequency responses are interpreted. This chapter also attempts to explain reasons causing transformer active part damage and discusses diagnosis solutions.

5.2 Transformer winding deformation

5.2.1 Deformation types and short-circuit current

Different types of transformer windings are designed and manufactured based on voltage level and electromagnetic relations. Spacers, barriers, dense woods and other materials are used to provide mechanical support for the winding. Spacers are employed to separate one disk from another disk in order to provide easy heat dissipation as well as mechanical support. They can also help as a part of the cooling system by providing appropriate distance between two adjacent disks for oil flow. Larger space between the disks may result in oil flow speed reduction and inefficient cooling, whereas small space can impede oil flow and make cooling less effective. Hence, there is an optimum disk space based on transformer winding design and its physical configuration and also oil viscosity.

On the one hand, the spacers can be placed in layers to provide suitable disk-to-disk distance; on the other, they can be arranged horizontally in a disk to provide mechanical support. In this regard, the number of spacers per disk (horizontally) is given by

$$N_{spacer} = \frac{\pi D_{ave}}{\theta_s + W_{spacer}} \tag{5.1}$$

where D_{ave} is the average diameter of winding, W_{spacer} is the spacer width and θ_s represents the desirable angle between spacers based on mechanical support concepts.

Mechanical defects in transformers can occur due to various disturbances such as short-circuit currents, severe explosion of combustible gas in transformer oil, earthquake or even improper transportation. Winding deformations are reportedly happening between spacers and barriers. Statistical evidence has shown that well-placed spacers can prevent winding deformation during short-circuit, especially for the inner winding subjected to radial inward forces. In fact, winding sections between spacers are considered the weak points of a winding during short-circuit events.

Of all the possible causes of transformer failure, mechanical deformation of windings as a result of large short-circuit currents is probably the most common. Such currents may generate radial, axial or combined forces acting on a transformer winding. The result could be radial, axial or angular deformation of the windings, or conductor rupture. Transformer winding deformation may be categorized as follows [2]:

- Radial forces
  - Forced buckling
  - Free buckling (hoop buckling)
  - Hoop tension (stretching)
  - Relaxation buckling
- Axial forces
  - Tilting (cable-wise tilting, strand-wise tilting)
  - Conductor bending between radial spacers
- Combined forces
  - Spiraling
  - Telescoping
  - Twisting

In all cases, the force F acting on the transformer winding is given by

$$F = \int_{\lambda} I_{sh} \cdot d\lambda \times B \tag{5.2}$$

where I_{sh} denotes the short-circuit current, B is the vector of magnetic induction and λ is the winding length.

5.2.1.1 Radial forces

The radial forces produced by the axial leakage field act outward on the outer winding, stretching the winding conductors and causing hoop stress. On the other hand, they cause the inner winding to experience radial compressive stress [3]. The radial force due to the axial leakage flux in the gap between the two windings is calculated as follows [4]:

$$F_{radial} = \pi D_{ave}\left(\frac{\sqrt{2}\mu_0 NI}{2H_w}\right)\left(\sqrt{2}NI\right), \tag{5.3}$$

where H_w is the winding height, D_{ave} is the average winding diameter, NI denotes RMS winding's ampere-turns value and μ_0 represents vacuum permeability.

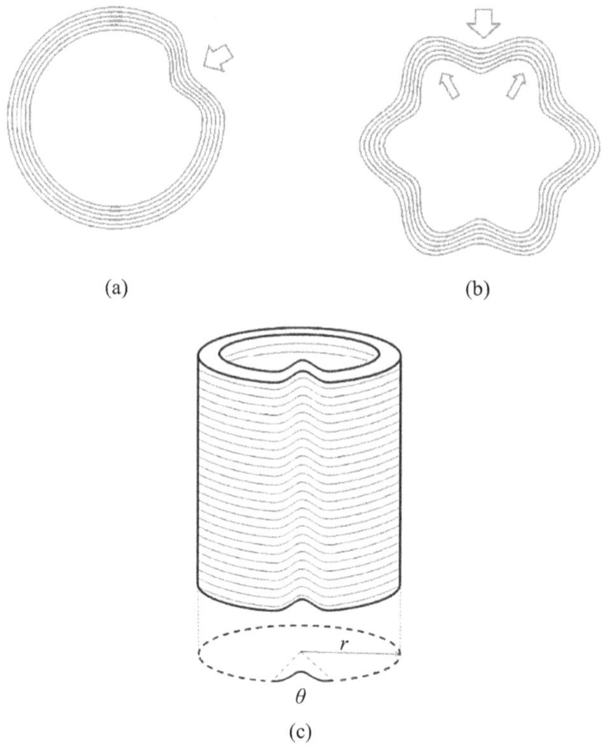

Figure 5.1 Winding deformation (buckling): (a) free buckling (top view), (b) forced buckling (top view), (c) free buckling (side view)

Various types of deformation due to the radial forces were mentioned in the previous sub-section. Amongst all of them, the buckling type of deformation is the most often reported in transformer windings. Radial forces in concentric windings lead to free buckling and forced buckling as illustrated in Figure 5.1.

5.2.1.2 Axial forces

Axial forces due to short-circuit current are produced by the radial magnetic flux. Axial forces lead to tilting or bending of conductors and will be more dangerous when the windings are not placed symmetrically. Any ampere-turn mismatch between LV and HV windings will strengthen axial forces. Titling and bending of conductors between spacers due to axial forces are shown in Figure 5.2. Any small displacement during transformer transportation or due to earthquake would result in intensified axial forces when short-circuit occurs.

5.2.2 Transformer transportation causing active part displacement

Transformer transportation is one of the major causes in transformer active part displacement and winding deformation. Transformer must be carefully transported,

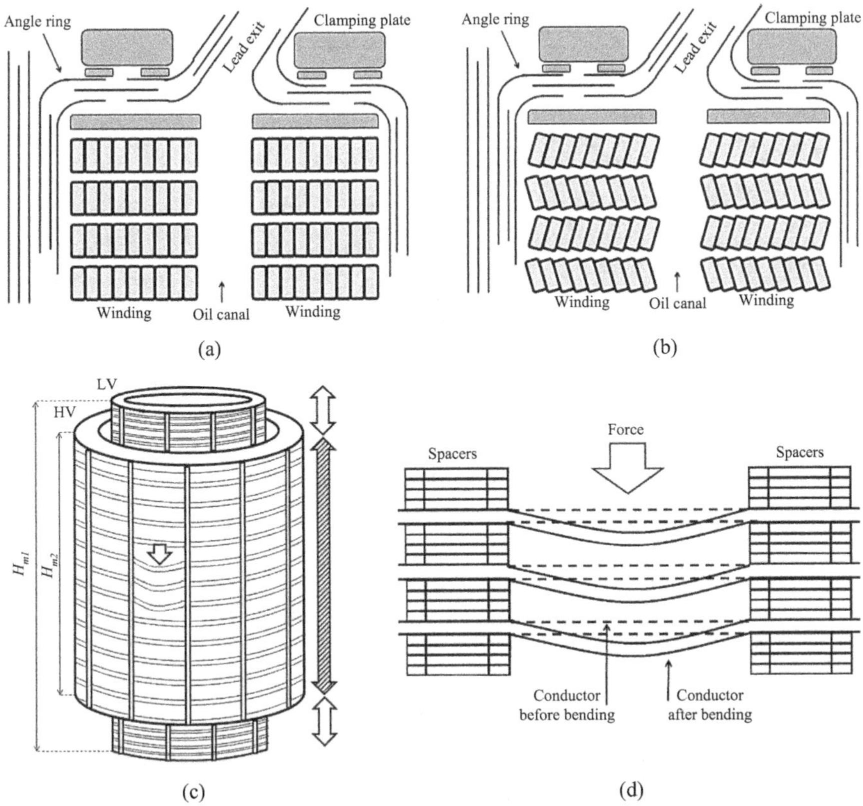

Figure 5.2 Winding deformation: (a) before tilting, (b) after tilting, (c) bending (side view), (d) bending (close view)

especially in the case of large power transformers. In this regard, there are different means to transport a transformer: truck, railroad, sea carrier, air carrier.

From the safety point of view, the approach chosen should produce the lowest vibration and high reliability even if it is slow and time consuming. While a transformer is transported, the main tank is filled with dry air, nitrogen or any other gas which does not cause chemical reactions on transformer major and minor insulation systems and at the same time prevents the core and windings from absorbing moisture until final on-site oil filling. Therefore, the pressure of the gas inside the transformer tank must be higher than the ambient. Also, the bushings are disassembled for easier transport. A typical arrangement for transformer transportation is shown in Figure 5.3.

Mechanical forces which can potentially have serious effect on the transformer tank during transportation need to be considered when it is settled on the portable support surface. Hence, all of ramps, winding routes, wind force, stop shock as well as bumps should to be taken into account in calculating the mechanical forces.

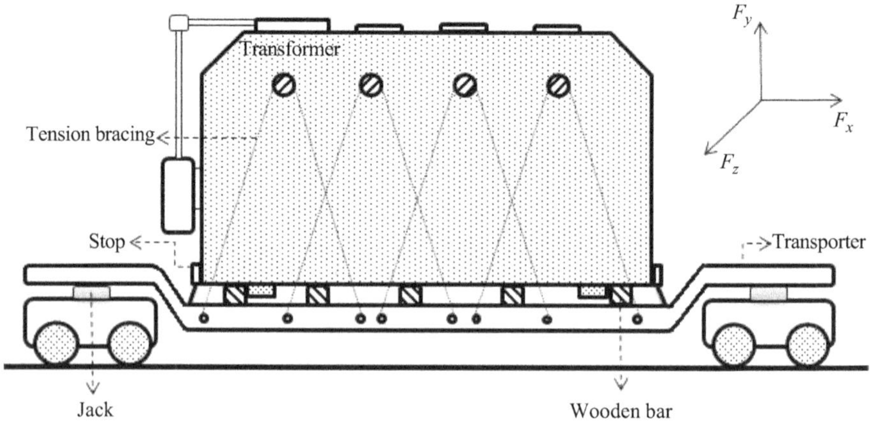

Figure 5.3 Transformer transportation

All effective acting forces must be less than the static friction force between the transformer tank and portable surface plus stop forces. In this regard, the overall acting force can be calculated as

$$\sum F_{radial} + \sum F_{wind} + \sum F_{predicted} < \mu_s mg + \sum F_{stop} \tag{5.4}$$

where F_{radial} is the forces imposed during transportation, F_{wind} is the wind force, $F_{predicted}$ denotes other predictable forces, F_{stop} is the security bracing or stop forces, m is the total transformer mass and μ_s represents the static friction constant between the transformer tank and portable surfaces and wooden bars if used.

5.3 Methods to recognize winding deformation

5.3.1 Short-circuit impedance

5.3.1.1 Short-circuit impedance concept

The SCI may be considered as a parameter which highlights imperfect magnetic coupling between primary and secondary windings. It contains resistive and inductive terms, the latter being more important than the former. The SCI (or leakage inductance) can be represented as an additional inductance in series with the transformer primary inductance, as shown in Figure 5.4. A high SCI value leads to a high voltage-drop across the transformer terminals and thus affects network voltage regulation, while a low value influences the network short-circuit fault current.

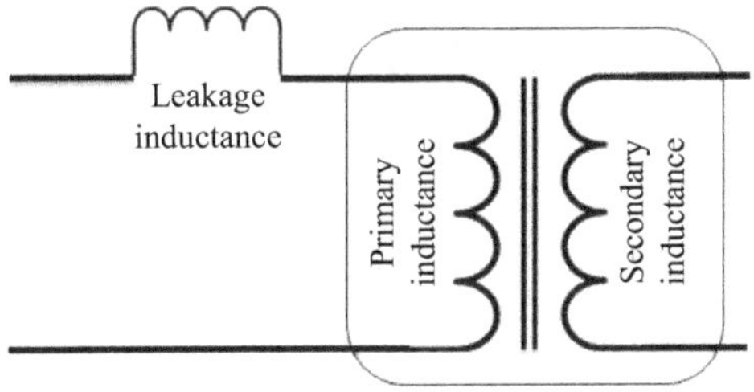

Figure 5.4 Schematic model of primary, secondary and leakage inductances of a transformer

The distances between the HV winding, the LV winding and the core of the transformer have considerable influence on the SCI value, see Figure 5.5. The SCI is given by

$$U_x = 0.248 \frac{SC_xD_x}{2(V/N_t)^2 H_m N_B} K_R \left(\frac{f}{50}\right) \tag{5.5}$$

where U_x is the short-circuit impedance, N_B is the number of transformer core limbs surrounded by HV and LV windings, K_R is the Rogowsky coefficient taken as 1 for most HV and LV winding arrangements [5] but can be calculated if necessary [6], f is the operating frequency, $H_m = (H_{m1} + H_{m2})/2$, where H_{m1} is the height of the LV winding and H_{m2} is the height of the HV winding, V is the nominal voltage of the winding, N_t is the number of winding turns and S is the apparent power quoted on the transformer nameplate. In addition, $D_x = D_{C2} + (B_{W2} - B_{W1})/3$ and $C_x = C_2 + (B_{W1} + B_{W2})/3$, where the distances C_2, D_{C2}, B_{W1} and B_{W2} are shown in Figure 5.5(b). Clearly, winding deformation (changes in the geometrical factors) will result in a change of U_x.

The measured SCI for a transformer should be compared to the value printed on the nameplate or quoted in factory test results. Winding displacement that may have occurred since the factory tests were performed may then be detected. According to [7], changes of more than $\pm 3\%$ should be considered as indicating winding deformation or core displacement. However, changes should not exceed $\pm 1\%$ for transformers with capacities above 100 MVA [8].

5.3.1.2 Short-circuit impedance measurement setup

SCI measurements can be performed on single or three-phase transformers, usually on the HV winding, with the LV winding short-circuited. The cross-sectional area of the cable used to short-circuit the LV winding must be at least 30% greater than that of the winding conductor, and it must be as short as possible. The resistance of the connection between the LV terminals and the shorting cable must also be as small as possible. The SCI measurement test setup for single and three-phase transformers is shown in Figure 5.6(a) and (b), respectively.

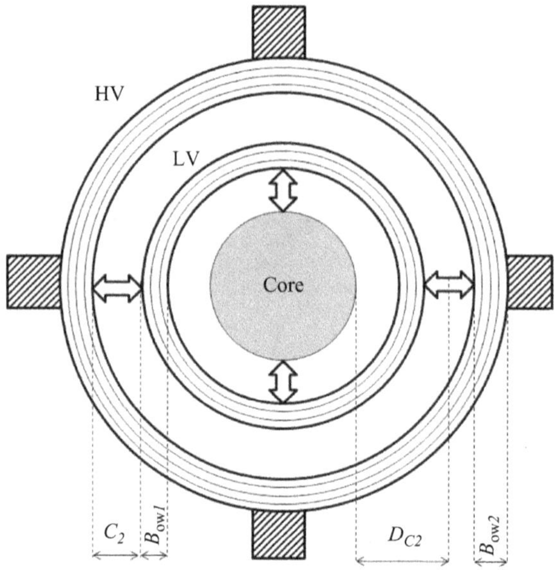

(a)

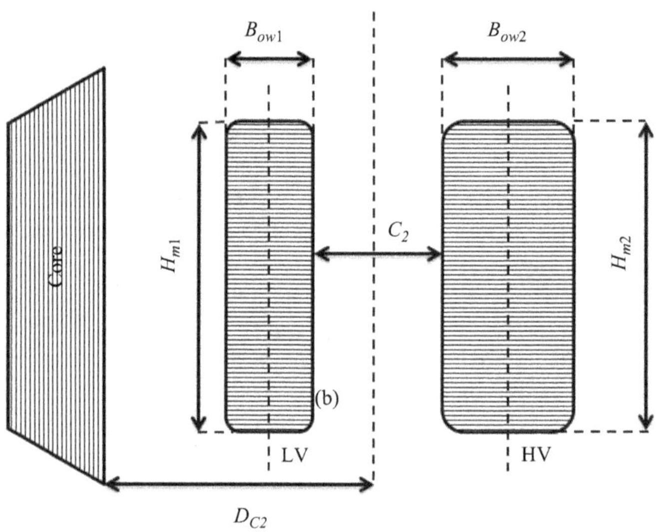

(b)

Figure 5.5 Schematic of transformer core and windings: (a) top view and (b) side cut view

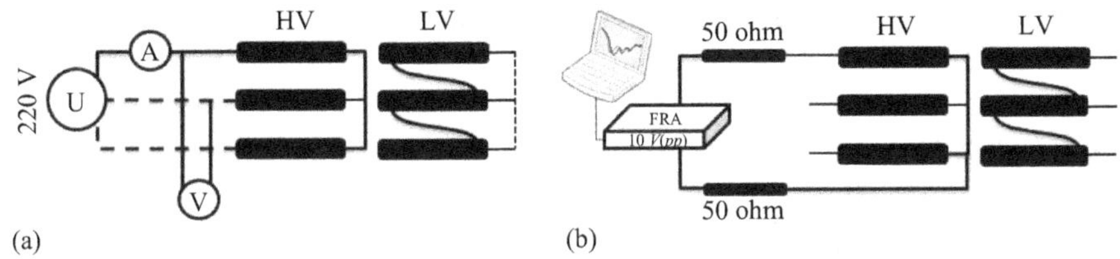

Figure 5.6 Short-circuit impedance measurement setup: (a) single-phase transformer and (b) three-phase transformer. The test is made by short-circuiting the line-leads of the low-voltage windings and applying a single-phase voltage at rated frequency to two terminals of the other winding. Three successive readings are taken on the three pairs of leads [7]. If the neutral terminal is available, the measurement can be conducted through the line-lead and the neutral-lead

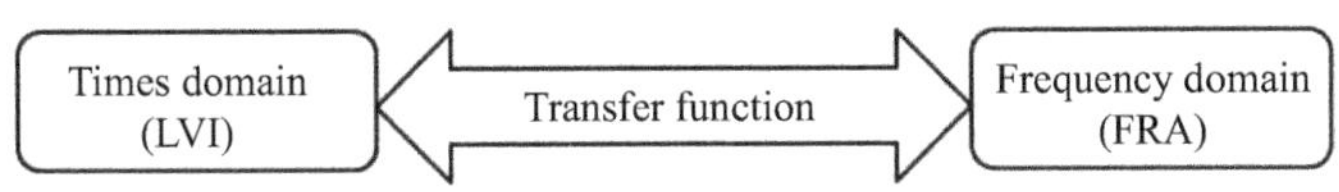

Figure 5.7 Transfer function measurement techniques

5.3.2 Transfer function

Transfer function is basically a way of describing the input–output behavior of a system. It is an established convention for quantifying the system response as well as other means such as differential equations. Transfer functions encapsulate information about frequency response and time response given certain input signals, and system characteristics can be picked out through the transfer function as well. There are two popular methods that can be used for transfer function measurement (see Figure 5.7). The low-voltage impulse (LVI) is a time-domain approach, whereas the FRA is a frequency domain equivalent.

In the time-domain method, the impulse waveform is injected into the test object input and the test object output signal is recorded. Once the time-domain measurement data is obtained, the transfer function in frequency domain can be determined by using fast Fourier transform (FFT) technique. With the frequency-domain method, the system response is determined by excitation at one frequency at a time and then continued in the same fashion to build up the response spectrum. Technically speaking, the transfer function determined by using the frequency domain method is not identical to that calculated by employing time-domain measurement and then utilizing FFT technique. For consistency, it is recommended to choose one method and use it in all the measurements to enable comparisons. LVI is generally faster than FRA, while FRA is more accurate. Based on industry practice, the general consensus favors the latter.

5.4 Sweep frequency response analysis

Frequency domain measurement is performed by injecting a swept sinusoidal waveform within a predetermined frequency band:

$$U(t) = A\sin(\omega t) = A\sin(2\pi f_{sweep}t), \quad f_{\min} \le f_{sweep} \le f_{\max} \tag{5.6}$$

where A is the sinusoidal signal amplitude and f_{sweep} denotes the variable frequency. It is worth noting that for commercially available FRA equipment, A varies from 1 to 25 V_{pp} for different manufacturers.

FRA data are commonly presented as magnitude Bode diagrams, with the x-axis for frequency and the y-axis for the response magnitude. In some cases, more information is also provided in the form of the FRA phase diagram, although this does not seem to yield much information on the mechanical integrity of transformer. Thus, most of literatures are just concentrated on the magnitude. The phase diagram is useful for checking how exactly the FRA spectrum (transformer winding behavior) is changed from inductive to capacitive, and vice versa. Occasionally, analyzing the zero crossing in the FRA phase diagram is also useful for transformer diagnosis.

FRA is considered as a kind of offline test technique because the transformer should be out of service and all bushing terminals disconnected from the overhead lines or feeders during FRA measurements.

In general, FRA measurement is performed in the frequency band 20 Hz–1 MHz for transformers with highest voltage of >72.5 kV, and in the range of 20 Hz–2 MHz for transformers with highest voltage of ≤ 72.5 kV [9]. To be on the safe side that all transformer resonant frequencies are observed in FRA spectrum, FRA measurement can be performed over the range 20 Hz–2 MHz for all transformers irrespective of their voltage rating. However, in the case of special transformers or reactors, the upper limit may be shifted to even higher frequencies. For instance with air-core reactors, this limit could be increased up to 20 MHz. It should be noted that on-site FRA measurement beyond 2 MHz is likely to experience undesirable oscillations or additional fluctuations.

In transformer winding studies, the FRA could be measured for HV windings or LV windings. In three-phase transformers, the windings are interconnected (e.g., wye, delta or zigzag); the FRA data can be extracted collectively or independently for the windings. In fact, the FRA can be measured for any proposed circuit when the transformer is out of service. Note that the same connection configuration must be used to enable comparison of the test results obtained at different dates.

FRA has been widely used as a comparative diagnosis method for some years now. The initial FRA measurements during factory testing serve as the winding fingerprint (reference, baseline or original trace). Changes in winding configuration would almost certainly cause some changes in the frequency response trace. It is also useful that the initial measurements of the distributed resistance, capacitance and inductance of a winding are compared with the same measurements following a transformer maintenance, repair or transport. Three types of comparisons can be considered: (1) comparison with earlier baseline FRA spectrum, (2) comparison with a sister (twin) transformer and (3) comparison between individual phases (exploiting winding symmetry).

5.5 Standard connection methods

Independent measurement of the frequency response for each individual winding is recommended as it is more convenient for future comparison. So for Y_n winding connection, the phase and neutral bushing leads would be the input and output, respectively. When the transformer includes isolated wye or delta connection, the two phase leads would be used instead. Based on this, different test setups for FRA measurement are given as follows [9,10].

5.5.1 End-to-end measurement

In this setup, FRA measurements can be made on a winding by injecting a preset signal V_{in} at the line-lead, and detecting the response V_{out} at the neutral-lead of the transformer, as shown in Figure 5.8(a). The frequency response magnitude K_{mag} (the voltage attenuation in dB) is given by

$$K_{mag} = 20\log_{10}\left(\frac{V_{out}}{V_{in}}\right) = 20\log_{10}\left(\frac{Z_{out}}{Z_{out} + Z_w}\right) \tag{5.7}$$

Z_{in} and Z_{out} are illustrated in Figure 5.8.

5.5.2 Inductive interwinding measurements

In this setup, FRA measurements can be made on a winding by injecting a preset signal V_{in} at the line-lead (for instance HV winding, phase *U*), and detecting the response V_{out} at the line-lead of the corresponding concentric winding (for instance LV winding, phase *u*), where the neutral-lead of both windings are grounded as shown in Figure 5.8(b).

5.5.3 Capacitive interwinding measurements

In this case, FRA measurements can be made on a winding by injecting a preset signal V_{in} at the line-lead (for instance HV winding, phase *U*), and detecting the response V_{out} at the line-lead of the corresponding concentric winding (for instance LV winding, phase *u*), where the neutral-lead of both windings are left open circuit as shown in Figure 5.8(c).

5.5.4 End-to-end short-circuit measurements

In this case, FRA measurements can be made on a winding by injecting a preset signal V_{in} at the line-lead, and detecting the response V_{out} at the neutral-lead of the transformer, where all terminals of the other side are short-circuited as shown in Figure 5.8(d). Table 5.1 and Figure 5.9 also provide a guide for terminal labels and connections, as well as detailed FRA setups for both star and delta connections (star and delta connections do not influence illustrated terminals in Table 5.1).

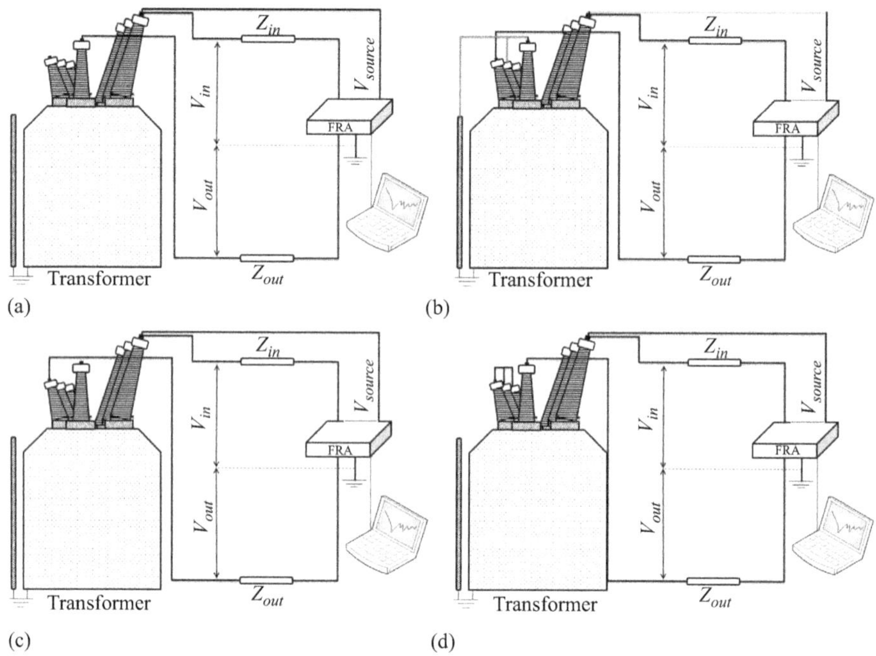

Figure 5.8 FRA standard test setups: (a) end-to-end measurement, (b) inductive interwinding measurement, (c) capacitive interwinding measurement, (d) end-to-end short-circuit measurement

Table 5.1 FRA measurement connections (phases A, B and C). See Figure 5.9 to find connection labels

Setup	Source (V_{in})	Response (V_{out})	Terminals earthed	Terminals connected together
End-to-end	A_1	A_2	None	None
Inductive interwinding	A_1	a_1	A_2 and a_2	None
Capacitive interwinding	A_1	a_1	None	None
End-to-end short-circuit	A_1	A_2	None	a_1–a_2–b_1–b_2–c_1–c_2
End-to-end	B_1	B_2	None	None
Inductive interwinding	B_1	b_1	B_2 and b_2	None
Capacitive interwinding	B_1	b_1	None	None
End-to-end short-circuit	B_1	B_2	None	a_1–a_2–b_1–b_2–c_1–c_2
End-to-end	C_1	C_2	None	None
Inductive interwinding	C_1	c_1	C_2 and c_2	None
Capacitive interwinding	C_1	c_1	None	None
End-to-end short-circuit	C_1	C_2	None	a_1–a_2–b_1–b_2–c_1–c_2

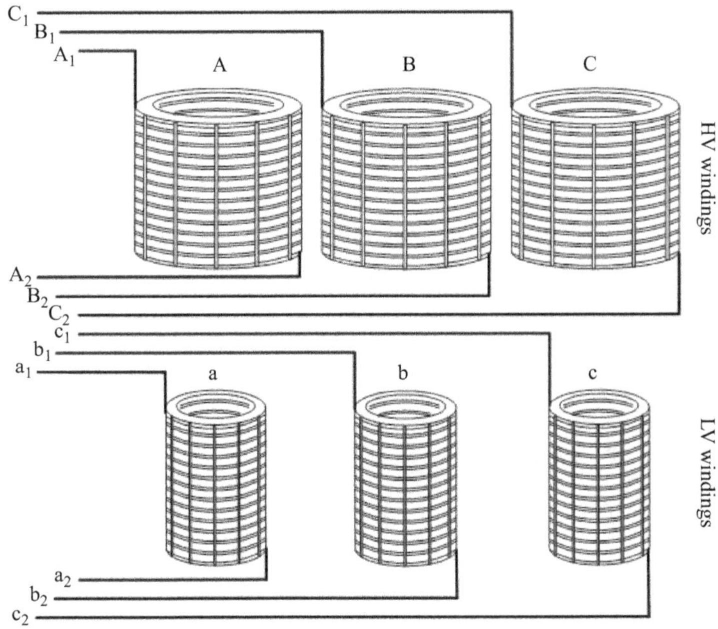

Figure 5.9 FRA test setups (detailed connections)

5.6 FRA signature assessment

Precise FRA spectrum interpretation requires that detail modeling of transformer active part is carried out and the response can then be determined. This requires detail information of transformer internal dimensions, insulation thickness, winding size and oil insulation parameters. Usually, such information is not available. Hence, FRA evaluation through "visual FRA signature assessment" and "statistical indicators" are introduced and recommended in practice.

5.6.1 Visual assessment of FRA signature

It is common practice to make visual assessment of the FRA spectrum based on physical concepts and expertise knowledge. To perform FRA signature interpretation and assess transformer condition, the frequency response spectrum of transformer should be classified. To facilitate classification, the frequency response data may be divided into three frequency bands: namely low-, medium-, and high-frequency bands, see Figure 5.10(a). The data are dominated by the transformer core at low frequencies (LFs), by the winding structure at medium frequencies (MFs), and by the connection leads at high frequencies (HFs). However, the boundaries between the bands are not universally agreed.

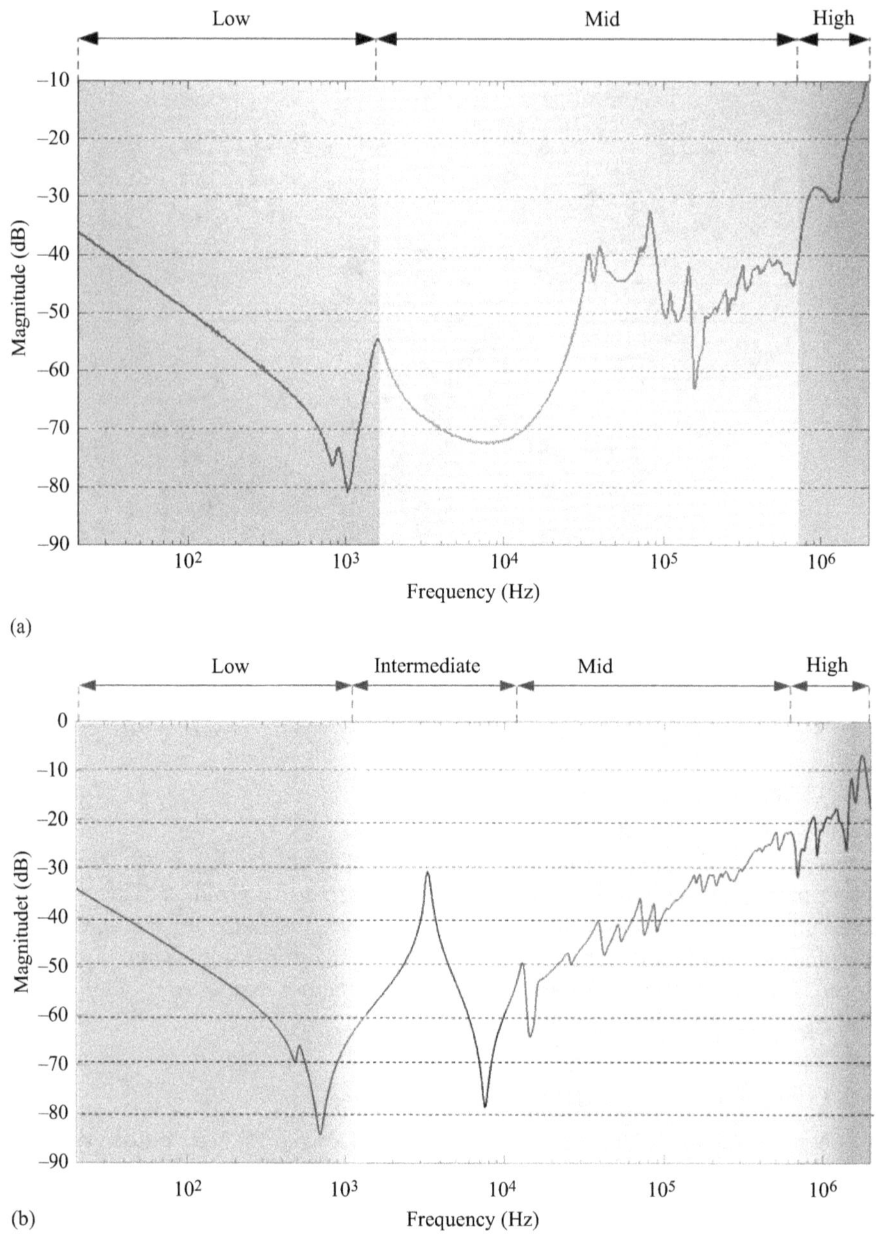

Figure 5.10 Typical FRA frequency bands: (a) three frequency bands, measured on 400-MVA transformer, (b) four frequency bands, measured on 73-MVA transformer

The transformer characteristics would be different in different frequency bands and so the FRA spectrum will vary in each frequency region. Thus, interpretation depends on which frequency band the winding is going to be studied.

From physical point of view, winding inductances as well as capacitances play a major role in the formation of the frequency response trace. In LFs, transformer winding inductances play a dominant role in the low-frequency range. Hence, in the Bode diagram, the frequency response magnitude follows a falling trend as frequency increases, see (5.7). Figure 5.10(a) demonstrates the descending trend of the frequency response trace in LFs, up to $\sim$1 kHz for the transformer under investigation.

For inductive reactance, the transformer winding self-inductance is greater than the mutual inductance. Therefore, the LF behavior is affected mainly by the self-inductance (L_m) which can be obtained as

$$L_m = \frac{N^2}{\Re} \tag{5.8}$$

where N is the number of coil turns and $\Re$ is the core reluctance.

With increasing frequency toward the mid-frequency range, the magnitude of the self-inductance will become large as compared to the capacitive reactance (in parallel) and can be neglected. Thus, the capacitive reactance gradually becomes a major player in the formation of frequency response trace. On the other hand, the mutual inductances become significant. Mid-frequency fluctuations of the frequency response are controlled by winding capacitances including series and shunt (ground) capacitances as well as mutual inductances. Series capacitances consist of turn-to-turn and disk-to-disk capacitances, whereas shunt capacitance is the capacitance of winding with respect to the core and to the electrostatic screens/tank. The mid-frequency band is generally up to some hundreds of kHz and strongly depends on the transformer size and active part configuration.

At HFs, the inductive reactance associated with the mutual inductances becomes large enough to be neglected. Therefore, the series and shunt capacitances will become dominant at HFs and hence the trace will follow an ascending trend, see Figure 5.10(a).

In summary, it can be concluded that variations in the LFs are associated with the transformer core. The response in the MFs is strongly influenced by the winding capacitances and mutual inductances. In the HF band, the behavior is dominated by connection leads as well as ground and leakage capacitance effects.

Furthermore, some literatures suggest that FRA spectrum can be divided into four frequency bands, see Figure 5.10(b), in which the new intermediate frequency band is influenced by transformer winding coupling and interaction between primary and secondary windings. Also, the winding connections as well as arrangement (delta connection, single or three-phase or auto-transformer) can affect this frequency band [9]. Deviation of the FRA spectrum in any frequency band can be certainly related to one of the mentioned parameters. Table 5.2 provides some guidance about which frequency band will change due to transformer defect. This table is quite useful for FRA interpretation. However, it does not cover entire transformer winding or core deformation defects.

Table 5.2 Possible changes in different frequency bands of FRA spectrum due to various transformer defects

	Low-frequency band	**Intermediate frequency band**	**Mid-frequency band**	**High-frequency band**
Core deformation	✓			
Winding rupture	✓	✓	✓	✓
Winding deformation		✓	✓	
Winding axial displacement			✓	
Winding radial displacement		✓	✓	
Tap-changer connection problems				✓
Turn-ratio problems	✓			
Core grounding problem	✓			
Turn-to-turn short-circuit	✓	✓		
Partial winding failure			✓	
Winding internal connection problems				✓
Tap-changer bad contact				✓
Winding resistance problems				✓
Winding insulation problems		✓	✓	✓
Winding huge water content		✓	✓	✓
Winding clamping pressure loss		✓	✓	
Turn-to-turn insulation deterioration	✓	✓		

Example 5.1 A brand-new 73-MVA transformer was installed in a substation 2 years ago; however, it was never energized. Transformer specifications are provided in Table 5.3. Recently, it was decided to energize this transformer; therefore, all commissioning tests were performed on it. All the test results show proper values except for the oil–water content (>4%) and the no-load current in one of the transformer phases (imbalanced and high). Hence, it was decided to drain the transformer oil, dry-out transformer active part and readjust clamping forces on the windings. Following the repair, commissioning tests were performed

Table 5.3 Specification of transformer in Example 5.1

Power rate	55/73 MVA	Limb no.	3
Voltage	230/33 kV	Winding no.	3
Freq.	50 Hz	Phase no.	3
VG	$Y_N y_n 0$(d1)	Max. Insult.	1,050 kV
Cooling	ONAN/ONAF	Max. Amb. Temp.	60 °C

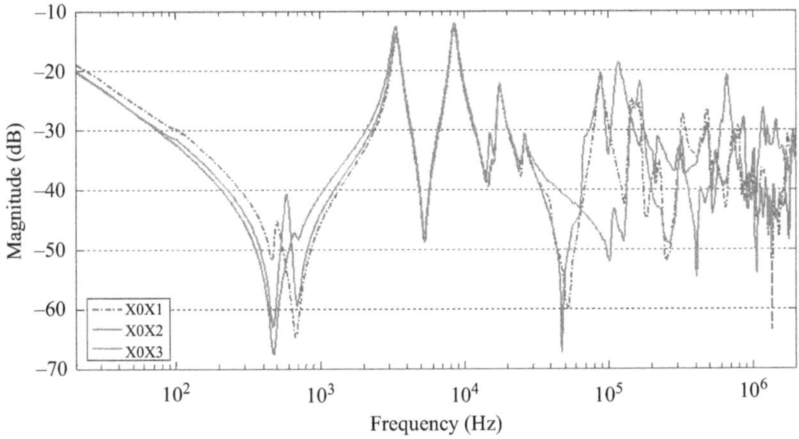

Figure 5.11 Frequency response spectra of LV side

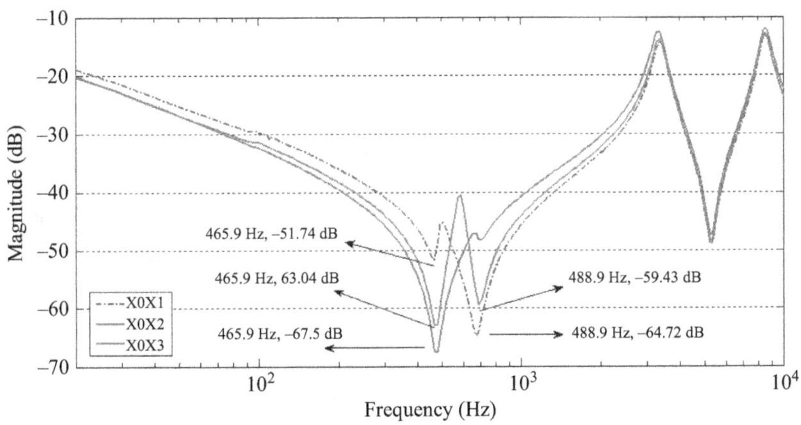

Figure 5.12 Frequency response spectra of LV side (20 Hz–10 kHz)

to double-check all the results once more. It was decided to perform FRA tests on transformer HV and LV phases to check mechanical integrity.

LV side terminal names: u = X1, v = X2, w = X3, neutral = X0
HV side terminal names: U = H1, V = H2, W = H3, neutral = H0

- ***Assessment***

The manufacturer did not provide the original FRA information, so this was the first time that the FRA spectra were recorded. Based on standard connection setups, end-to-end FRA measurement was conducted at 40 °C for each LV and HV winding, individually, while other windings were left open circuit. Figure 5.11 shows the LV winding traces for phases a, b and c (denoted as X1, X2 and X3). X0 represents the neutral terminal in LV side. Figures 5.12 and 5.13 provide the

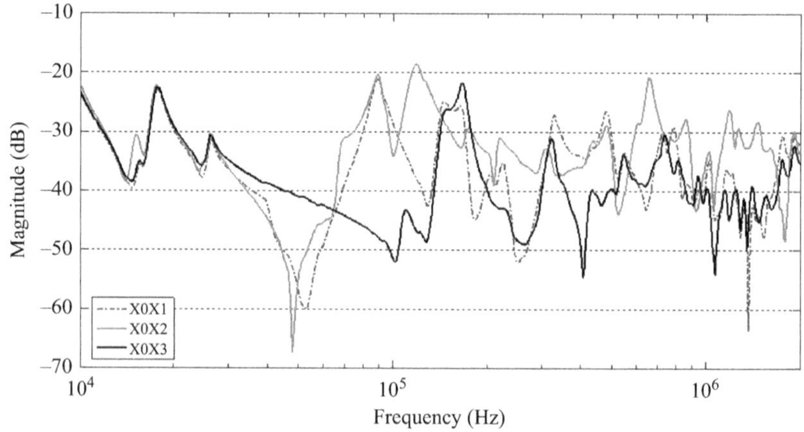

Figure 5.13 Frequency response spectra of LV side (10 kHz–2 MHz)

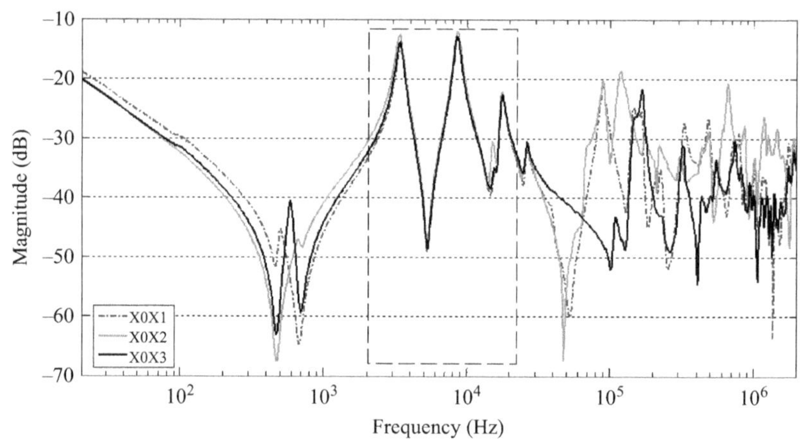

Figure 5.14 Frequency response spectra of LV side (the well-matched frequency band)

close-up view for the frequency bands of 20 Hz–10 kHz and 10 kHz–2 MHz, respectively. In Figure 5.14, the frequency band in which all the traces align is highlighted by dash-line. Note that only the magnitude response is presented here; the phase response does not convey useful information in this particular case.

In the low-frequency range, it can be seen that the trace for the middle winding (phase b) in LV side experiences a single antiresonance as compared to double antiresonance for the other phases. The lateral windings (phases a and c) show similar characteristic in terms of magnitude, trend (decreasing/increasing) as well as number of antiresonance. However, in the results (specifically in Figure 5.12), it is quite clear that the frequency response spectra for phases a and c show different magnitudes even though their falling and rising trends are similar. Phase a (X0X1) demonstrates less inductance than other phases, and phase c shows very close inductance to the middle winding. It is obvious that the transformer limb reluctance of phase a magnetic core is larger than other phases. All spectra are well-matched in the frequency band 2–20 kHz, and it means that the shunt capacitances between HV and LV windings are almost similar. Indeed, the band (2–20 kHz) shows the interaction between low and high-voltage windings for each and every phase. Therefore, we can conclude that the LV–HV canals between transformer windings are normal. It can also be concluded that there is no radial deformation throughout the windings. Moving from 20 kHz to higher frequencies, the discrepancy between the phases becomes obvious. Figure 5.13 clearly shows that they experienced different resonances and antiresonances. This can be interpreted via structure discrepancy between windings especially the series as well as mutual-inductance. Change in shunt capacitance is not evident here because all spectra appear well matched in the intermediate frequency band. Hence, the focus is on series and mutual-inductance variations. It was discussed earlier that the mutual-inductance is mostly from number of winding turns, height and winding width as well as diameter. Based on the manufacturer report, all these parameters are quite similar for entire LV windings. Therefore, the mutual-inductance cannot be the main cause of mid-frequency discrepancy in the recorded FRA spectra. Thus, the mid-frequency deviation is initiated through different series capacitances rather than the other factors. The series capacitance itself is influenced by many mechanical and insulation factors although the mechanical factor is more significant to affect FRA trace.

One of the factors which can easily affect the winding series capacitance is winding pressure after winding manufacturing or during transformer winding assembling. Improper winding press or lack of pressure on windings during active part assembling time will cause winding interdisk distances become unequal in different windings (phases a, b and c), and consequently, the winding series capacitances become different. This will result in discrepancy in the mid-frequency range of FRA trace for an unused new transformer. It should be noted the argument would be different for FRA interpretation of aged transformers.

Back to low-frequency interpretation, the FRA traces for LV windings show that transformer magnetic core reluctance for phase a is quite different from other phases. Furthermore, series capacitance was changed due to inappropriate winding press during manufacturing or assembling time.

More FRA measurements on the HV side were conducted. Figure 5.15 shows the frequency response traces for each and every HV winding when other windings were left open circuit (end-to-end measurement). It can be seen that the discrepancy in FRA traces of phases A, B and C is quite obvious for very LFs and becomes less when moving to higher frequencies. Discrepancy in the low-frequency region emphasizes the previous hypothesis about different magnetic reluctance for phases A and C. The trace of phase A demonstrates less inductance as compared to phase C. For these types of transformer, the magnetic reluctance should be similar for

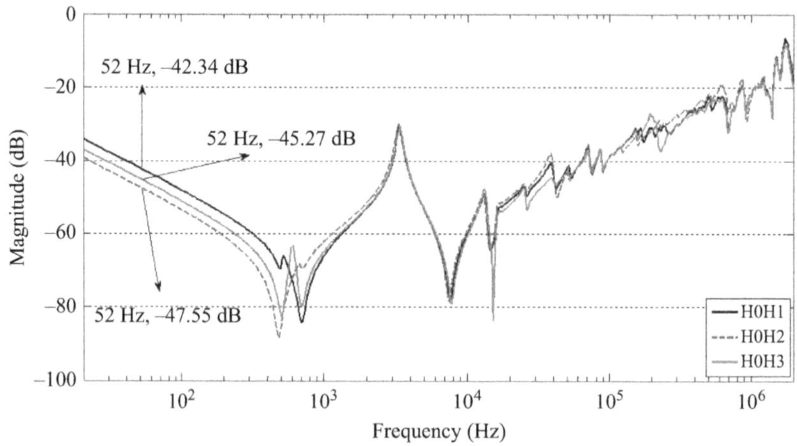

Figure 5.15 Frequency response spectra of HV side (antiresonance frequencies)

phases A and C, and the winding turns have the same construction. Therefore, the inductance values should be almost identical or at least close together. However, based on the measured frequency responses, the discrepancy is around 2.93 dB (or 40% difference in magnetic reluctance) which is significant for a large power transformer.

5.6.2 Statistical assessment of FRA signature

A common method of interpreting FRA data is to use statistical indices (indicators), particularly the correlation coefficient (CC), the standard deviation (SD) and the relative factor. The CC is defined as

$$CC_{(X,Y)} = \frac{\sum_{i=1}^{N_s} X_i Y_i}{\sqrt{\sum_{i=1}^{N_s} [X_i]^2 \sum_{i=1}^{N_s} [Y_i]^2}} \tag{5.9}$$

where X_i and Y_i are the ith elements of the fingerprint and measured FRA traces, respectively, and N_s is the number of elements (or samples). CC is thus a number whose absolute value lies between 0 and 1.

The SD is defined as [11]

$$SD_{(X,Y)} = \sqrt{\frac{\sum_{i=1}^{N_s} [Y_i - X_i]^2}{N - 1}} \tag{5.10}$$

CC and SD have been evaluated for the various subbands within the FRA spectrum [12–14], but the choice of frequency range differs between authors. In [11], several transformer windings were gradually deformed in the radial and axial directions, up to 1% physical deformation, values of CC and SD were calculated at each step. It was concluded that it would be appropriate to use single values of CC and SD as indicators of winding deformation over any frequency band within the range 10 Hz–3 MHz. More specifically, $|CC| < 0.9998$ and $SD > 1$ could each be taken as indicating winding deformation.

The relative factor (R_{XY}) is defined as [15]

$$R_{XY} = \begin{cases} 10 & 1 - P_{XY} < 10^{-10} \\ -\log_{10}(1 - P_{XY}) & \text{otherwise} \end{cases} \tag{5.11}$$

where P_{XY} is given by

$$P_{XY} = \frac{(1/N_s)\sum_{i=1}^{N_s}\left(X_i - (1/N_s)\sum_{i=1}^{N_s} X_i\right)^2 \left(Y_i - (1/N_s)\sum_{i=1}^{N_s} Y_i\right)^2}{\sqrt{D_X D_Y}} \tag{5.12}$$

D_X and D_Y are given by

$$\begin{aligned} D_X &= \frac{1}{N_s}\sum_{i=1}^{N_s}\left(X_i - \frac{1}{N_s}\sum_{i=1}^{N_s} X_i\right)^2 \\ D_Y &= \frac{1}{N_s}\sum_{i=1}^{N_s}\left(Y_i - \frac{1}{N_s}\sum_{i=1}^{N_s} Y_i\right)^2 \end{aligned} \tag{5.13}$$

and are, respectively, the standard variances of the fingerprint (X_i) and measured (Y_i) data [16].

As mentioned before, the frequency response data may be conveniently divided into three bands, namely low-, medium- and high-frequency bands. The frequency bands for R_{XY} and deformation levels related to the R_{XY} values have been defined in the Chinese standard [15] and by some other workers [16,17], see Table 5.4. These definitions have been widely used for FRA trace evaluation using R_{XY} method.

Table 5.4 Deformation levels and the corresponding R_{XY} *values at low frequency (LF), medium frequency (MF) and high frequency (HF) [16]*

Deformation level	**Limits for R_{XY}**
Severe	$R_{LF} < 0.6$
Moderate	$1.0 > R_{LF} \geq 0.6$ or $R_{MF} < 0.6$
Slight	$2.0 > R_{LF} \geq 1.0$ or $0.6 \leq R_{MF} < 1$
Normal winding	$R_{LF} \geq 2.0$, $R_{MF} \geq 1.0$ and $R_{HF} \geq 0.6$

LF: 1–100 kHz, *MF*: 100–600 kHz, *HF*: 600 kHz–1 MHz.

Example 5.2 The main technical specifications of a 400 MVA transformer are given in Table 5.5. This transformer was out of service due to a three-phase short-circuit line fault which occurred quite close to the transformer terminals. Hence, the management of the power plant decided to perform a full investigation before reenergizing the transformer.

Table 5.5 Transformer specification

Type	Power transformer	Type	Power transformer
Manufacture date	May 1993	No-load current (%)	0.37
Rated voltage (kV)	242/20	Number of phases	3
Rated power (MVA)	400	Number of limbs	5
Rated current (A)	954/11,550	Frequency (Hz)	50
No. of coolers	12	Cooling system	OFAF

- ***Assessment***

Different tests were performed on the transformer, and results were analyzed. Most of test results showed desirable values. Hence, it was decided to check the transformer mechanical integrity. FRA measurements were made by applying a 5.66 V signal across each winding, at 801 frequencies in the range 20 Hz–2 MHz. The characteristic impedance of the measurement cables was 50 Ω. The measured FRA traces were compared with fingerprint traces obtained during transformer overhaul (before failure). The results are shown in Figure 5.16. From visual assessment, it can be seen that the measured traces for phases A and C are very similar to their past fingerprints. However, this is not the case for phase B. Figure 5.16(d) shows the measured and fingerprint impedances for phase B. The measured and fingerprint impedances for phases A and C were very nearly identical.

After visual examination of the FRA results, statistical indicators were calculated for transformer condition evaluation. The values of CC and SD for each of the three phases are given in Table 5.6, and the corresponding values of R_{XY} in Table 5.7. It can be seen that, for phase B, CC and SD indicate deformation, and R_{XY} indicates slight deformation at LF and no deformation at medium and HFs. There is no indication of deformation of phase A or phase C.

These findings were validated by subsequent internal inspection after detanking the transformer. Figure 5.17 shows the side and front views of phase B of the HV winding. Clearly, the winding had suffered slight deformation.

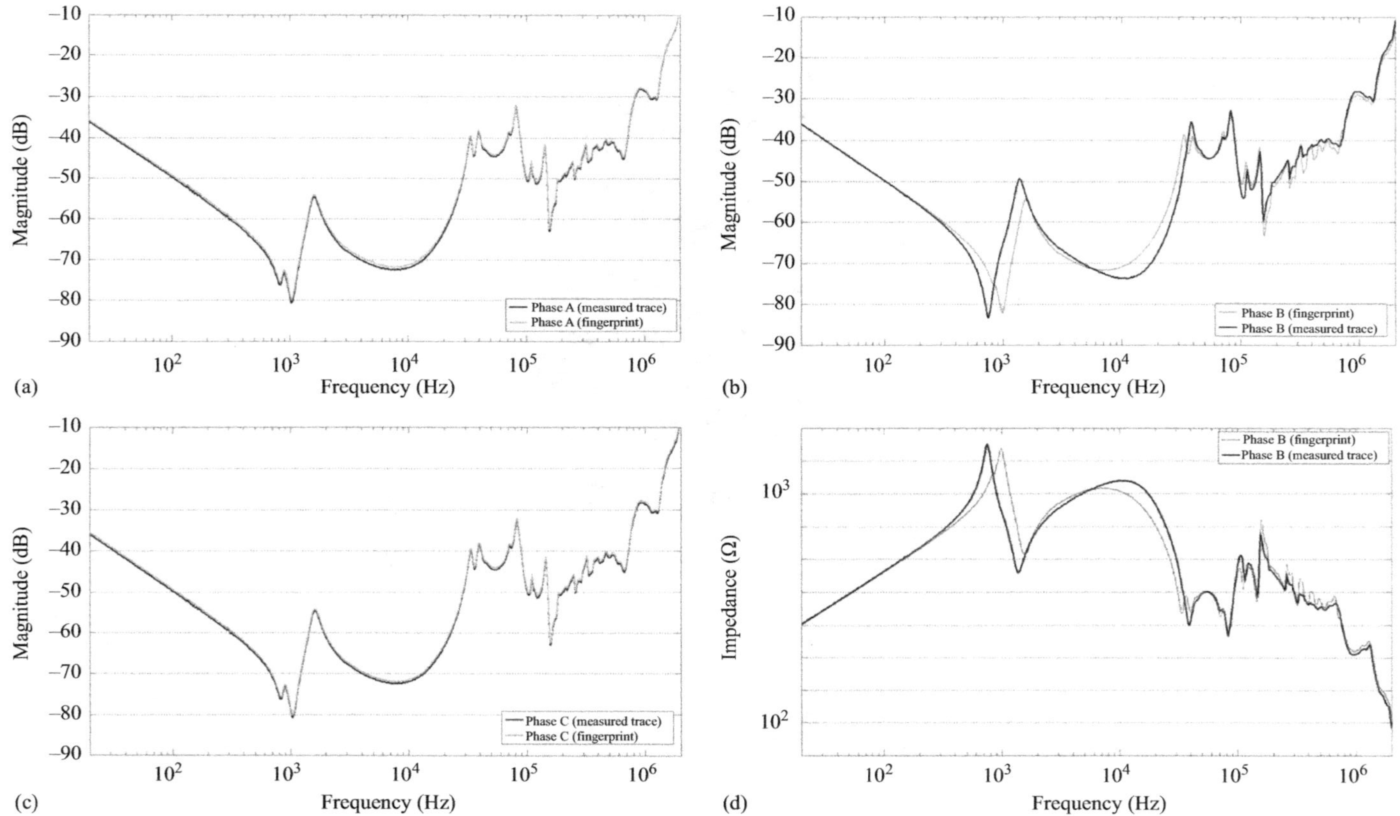

Figure 5.16 (a)–(c) Measured and fingerprint frequency response magnitudes for phases A, B and C, respectively, of the transformer HV side. (d) Measured and fingerprint impedances for phase B

Table 5.6 CC and SD values

Frequency band	CC			SD		
	Phase A	**Phase B**	**Phase C**	**Phase A**	**Phase B**	**Phase C**
20 Hz–2 MHz	0.9999	0.9977	1	0.5859	3.6367	0.5004

Table 5.7 R_{XY} *values*

Frequency band	R_{XY}		
	Phase A	**Phase B**	**Phase C**
Low	10	1.2674	10
Medium	10	1.0724	10
High	10	2.3831	10

Low: 1–100 kHz, medium: 100–600 kHz, high: 600 kHz–1 MHz.

(a) (b)

Figure 5.17 Buckled HV winding of phase B: (a) side view of the middle disks and (b) front view of the upper disks

5.7 Factors affecting frequency response signature

5.7.1 Winding inductance, capacitance

Transformer FRA signature can be deviated from its original spectrum due to variations in the inductance, capacitance, or even transformer winding resistance. The influence of winding resistance appears in the very high-frequency regions of the FRA spectrum. This effect is mixed with the effect of measurement leads and connections. Hence, the influence of winding resistance on FRA trace is not considered. However, winding's inductance and capacitance will be discussed in their normal and deformation conditions hereinafter.

5.7.1.1 Self- and mutual-inductance in circular form

- Biot–Savart law and inductance calculation

One of the most fundamental laws in computation of the magnetic field generated by an electric current is the Biot–Savart law which is given by

$$B = \frac{\mu_0}{4\pi} \int_v \frac{J \times a_R}{R_d^2} dv \tag{5.14}$$

where B denotes the magnetic flux density at some location near the current element, J is the current density vector, v is the volume containing current and R_d is the distance. On the other hand $Jds = I$; therefore, $Jdv = Jdsdl = Idl$. Hence, according to (5.14), dB is given by [18,19]

$$dB = \frac{\mu_0 I}{4\pi R_d^2} dl \times a_R \tag{5.15}$$

According to Figure 5.18, the magnetic field generated through I can be expressed as

$$dB = \frac{\mu_0 Idl}{4\pi R_d^2} \sin\theta' a_n \tag{5.16}$$

where a_n is the unit vector perpendicular to the plane containing dl and a_R, $dl\ a_n = dl \times a_R$

Based on Gauss's law for a circular filament having radius R_a and carrying the current I, the magnetic flux ϕ, is given by

$$\phi = \int_s B \cdot ds = \int_0^{R_a} \int_0^{2\pi} B_z r d\phi' dr \tag{5.17}$$

where $B = Ba_z$ and can be obtained through (5.16), and $rd\phi' = dl$. Thus, for a circular filament having radius R_a, (5.16) can be obtained as

$$dB_z = \frac{\mu_0 I}{4\pi R_d^2} R_a d\phi' \sin\theta' \tag{5.18}$$

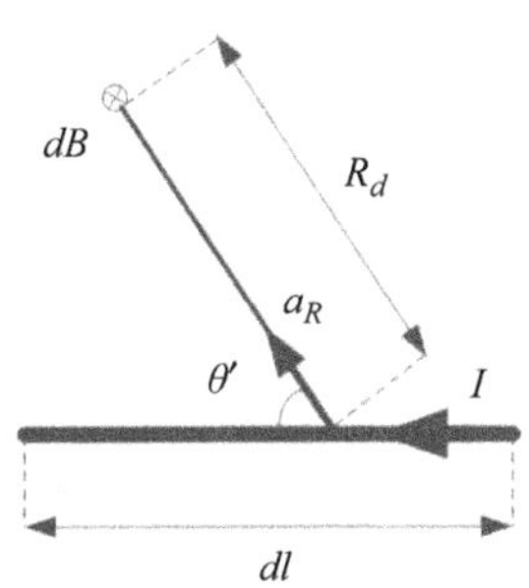

Figure 5.18 Biot–Savart law

where R_d is the distance between the differential current segment and the point of proposed magnetic field and obtained as (5.19) for a specific point r away from the center. θ' would then be the angle between the differential current vector and the vector directed from it to the point as shown in Figure 5.19.

$$R_d^2 = R_a^2 + r^2 - 2R_a r \cos\phi' \tag{5.19}$$

Hence, (5.19) is modified as

$$dB_z = \frac{\mu_0 I R_a}{4\pi \left(R_d^2 = R_a^2 + r^2 - 2R_a r \cos\phi'\right)^2} d\phi' \sin\theta' \tag{5.20}$$

$\theta' = \pi/2 + \alpha'$; thus, $\sin\theta' = \sin(\pi/2 + \alpha') = \cos\alpha'$. Using the law of cosines [19] $r^2 = R_a^2 + R_d^2 - 2R_aR_d \cos\alpha'$, and therefore [19]

$$\sin\theta' = \cos\alpha' = \frac{R_a^2 + R_d^2 - r^2}{2R_aR_d} = \frac{R_a - r\cos\phi'}{R_d} \tag{5.21}$$

Based on this, $B_z(r)$ is given by

$$B_z(r) = \frac{\mu_0 I R_a}{4\pi} \int_0^{2\pi} \frac{R_a - r\cos\phi'}{\left(R_a^2 + r^2 - 2R_a r \cos\phi'\right)^{3/2}} d\phi' \tag{5.22}$$

In [19], it is stated that the second method for B_z determination is using the vector magnetic potential as given by

$$B_z = \nabla \times A_{\phi'} = \frac{\partial(rA_{\phi'})}{r\partial r} \tag{5.23}$$

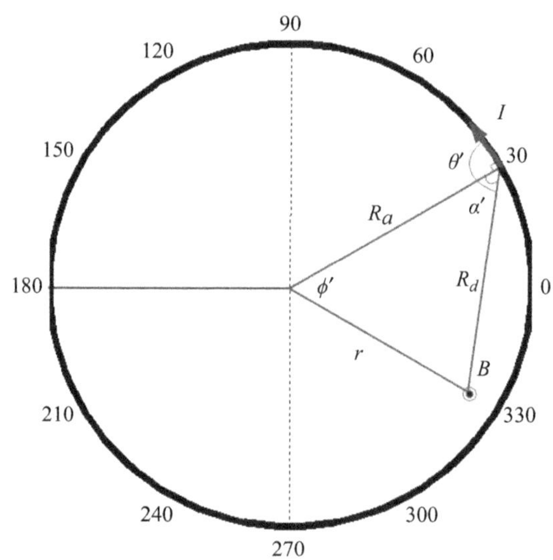

Figure 5.19 Magnetic flux determination for circular filament, taken and modified [19]

Therefore, using (5.18) and (5.14), (5.17) for a filament having radius R_a is given by

$$\phi_a = \int_{S_a} B_z ds = \oint_{C_a} A_{\phi'} dl \tag{5.24}$$

Hence, the mutual-inductance between two circular filaments having radii R_a and R_b is obtained as

$$M_{ab} = \frac{\phi_b}{I_a} \tag{5.25}$$

where ϕ_b is the induced magnetizing flux on the second loop due to the current initiated by the first loop. Hence, M_{ab} is given by (see Figure 5.20)

$$\begin{aligned} M_{ab} &= \oint_{C_b} A_{ab} dl_b \\ &= \frac{\mu_0}{4\pi} \oint_{C_a} \oint_{C_b} \frac{1}{R_{ab}} dl_b \cdot dl_a \\ &= \frac{\mu_0 R_a R_b}{4\pi} \int_0^{2\pi} \int_0^{2\pi} \frac{\cos \vartheta_a}{\left(R_a^2 + R_b^2 + d^2 - 2R_a R_b \cos \vartheta_a\right)^{1/2}} d\vartheta_b d\vartheta_a \end{aligned} \tag{5.26}$$

$\left[\text{Note that: } R_{ab} = \left(R_a^2 + R_b^2 + d^2 - 2R_a R_b \cos \vartheta_a\right)^{1/2}\right]$

It is obvious that the mutual-inductance between two circular filaments is only a function of their shapes as well as orientations. This was stated in [18,19].

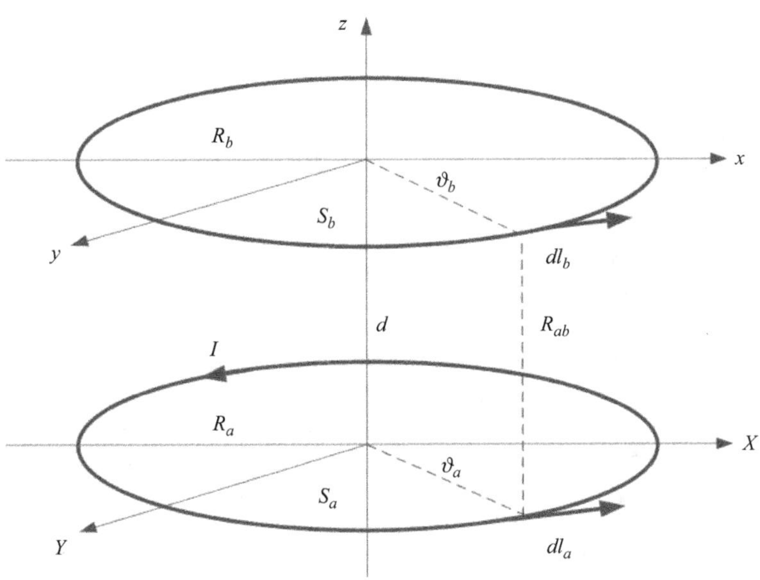

Figure 5.20 Concentric circular filaments

Using Maxwell's advices [18], the solution of (5.26) was discussed in [19]. In fact, changing the variable ϑ_a to $2\theta''$, $\cos\vartheta_a = \cos^2\theta'' = 2\cos^2\theta'' - 1$ and $d\vartheta_a = 2d\theta''$ will simplify the equation. Thus, M_{ab} can be calculated as [19]

$$M_{ab} = \frac{\mu_0 R_a R_b}{2} \int_0^{\pi} \frac{2\cos 2\vartheta_a}{\left((R_a+R_b)^2 + d^2 - 4R_aR_b\cos^2\vartheta_a\right)^{1/2}} d\vartheta_a$$

$$= \frac{\mu_0 (R_aR_b)^{1/2}}{2} \int_0^{\pi} \frac{k\cos 2\vartheta_a}{(1-k^2\cos^2\vartheta_a)^{1/2}} d\vartheta_a, \quad k^2 = \frac{4R_aR_b}{(R_a+R_b)^2 + d^2}$$

$$\left[\text{Note that: } k\cos 2\vartheta_a = (2\cos^2\vartheta_a - 1) = \left(\frac{2}{k} - k\right) - \frac{2}{k}(1-k^2\cos^2\vartheta_a)\right] \tag{5.27}$$

Having information about complete elliptic integrals $K(k)$ and $E(k)$, (5.26) is given by [18,19]

$$M_{ab} = \frac{\mu_0 (R_aR_b)^{1/2}}{2} \int_0^{\pi} \frac{k\cos 2\vartheta_a}{(1-k^2\cos^2\vartheta_a)^{1/2}} d\vartheta_a$$

$$= \mu_0 (R_aR_b)^{1/2} \int_0^{\pi/2} \left[\left(\frac{2}{k} - k\right)\frac{1}{(1-k^2\cos^2\vartheta_a)^{1/2}} - \frac{2}{k}(1-k^2\cos^2\vartheta_a)^{1/2}\right] d\vartheta_a$$

$$= \mu_0 (R_aR_b)^{1/2} \int_0^{\pi/2} \left[\left(\frac{2}{k} - k\right)\frac{1}{(1-k^2\sin^2\vartheta_a)^{1/2}} - \frac{2}{k}(1-k^2\sin^2\vartheta_a)^{1/2}\right] d\vartheta_a$$

$$= \mu_0 (R_aR_b)^{1/2} \left[\left(\frac{2}{k} - k\right)K(k) - \frac{2}{k}E(k)\right]$$

where:

$$K(k) = \int_0^{\pi/2} \frac{1}{(1-k^2\sin^2\vartheta_a)^{1/2}} d\vartheta_a, \quad E(k) = \int_0^{\pi/2} (1-k^2\sin^2\vartheta_a)^{1/2} d\vartheta_a \tag{5.28}$$

Equation (5.28) shows the analytical solution to determine the mutual inductance for concentric circular filaments.

5.7.1.2 Self- and mutual-inductance under buckling

Equation (5.26) is utilized to calculate the impact of buckling on the mutual-inductance of two concentric circular filaments as shown in Figure 5.21.

It should be noted that in the case of inward buckling, the winding radius for the span faced buckling is defined as $R_{bd} = R_b + 0.5r'$ $(\cos\eta\theta - 1)$. For outward buckling, it is mathematically expressed as $R_{bd} = R_b - 0.5r'$ $(\cos\eta\theta - 1)$. η is the ratio of the entire trigonometric circular span (2π) over the deformation span (rad) as illustrated in Figure 5.22, and r' represents the deformation radius.

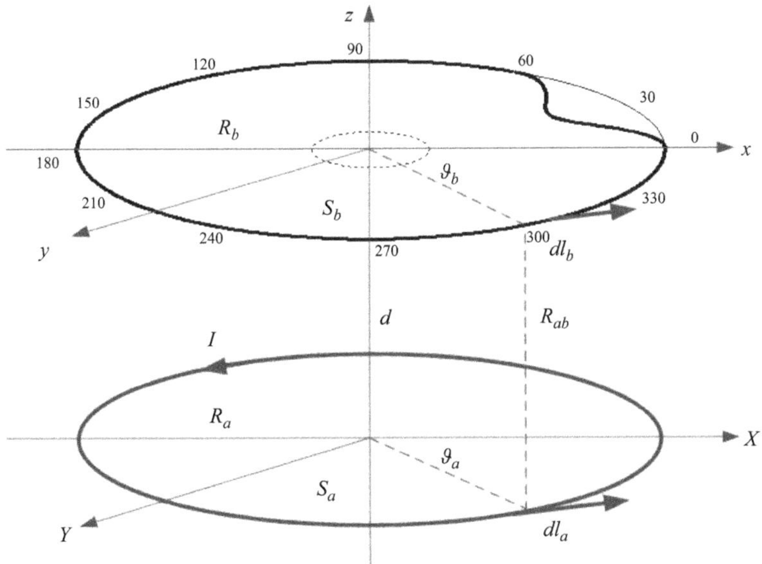

Figure 5.21 Concentric circular filaments, inward buckling demonstration for the second loop

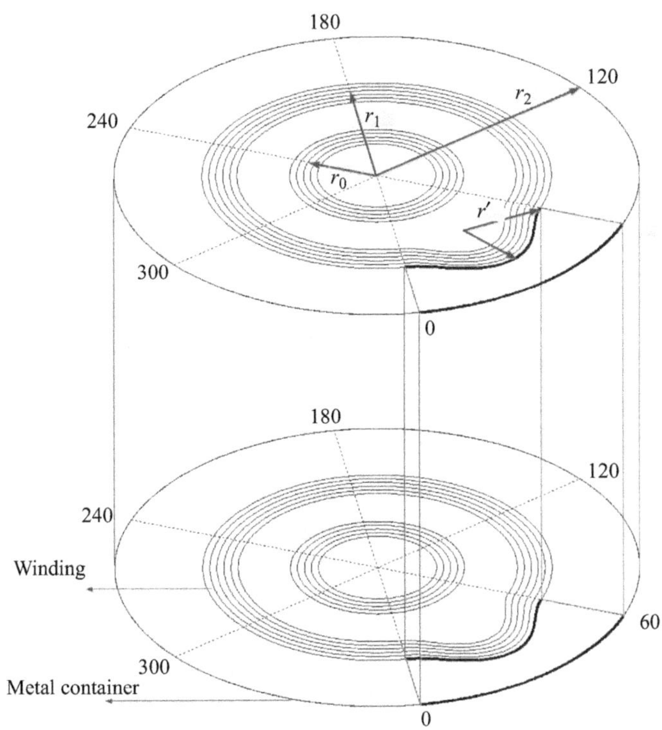

Figure 5.22 Radial deformation, outward buckling, η=6, and r' is the deformation radius

Based on this, the mutual-inductance is given by

$$M'_{ab} = \frac{\mu_0 R_a}{4\pi}\int_0^{2\pi}\int_0^{2\pi/\eta}\frac{\cos\vartheta_1(R_b+0.5r'(\cos\eta\vartheta_2-1))}{\left(R_a^2+(R_b+0.5r'(\cos\eta\vartheta_2-1))^2+d^2-2R_a(R_b+0.5r'(\cos\eta\vartheta_2-1))\cos\vartheta_1\right)^{1/2}}d\vartheta_2 d\vartheta_1$$
$$+\frac{\mu_0 R_a R_b}{4\pi}\int_0^{2\pi}\int_{2\pi/\eta}^{2\pi}\frac{\cos\vartheta_1}{\left(R_a^2+R_b^2+d^2-2R_aR_b\cos\vartheta_1\right)^{1/2}}d\vartheta_2 d\vartheta_1 \quad (5.29)$$

where the first part of (5.29) comes through the span facing deformation and the second part represents the circular part in the second filament. This integral is quite nonlinear and complex to solve analytically. It can be computed numerically using software (such as MATLAB®). However, to continue the equation analytically, we can assume that the influence of R_b on the total value of R_{ab} as the denominator of the first integral is negligible as compared to R_b plays as the numerator. This assumption is reasonable for the filaments which are quite far away, but maybe it is not accurate for close loops. All in all, having this assumption for the equation in (5.29), M'_{ab} is obtained as

$$M'_{ab} = \frac{\mu_0 R_a}{4\pi}\int_0^{2\pi}\int_0^{2\pi/\eta}\frac{(R_b-0.5r')\cos\vartheta_1}{\left(R_a^2+R_b^2+d^2-2R_aR_b\cos\vartheta_1\right)^{1/2}}d\vartheta_2 d\vartheta_1$$
$$+\frac{\mu_0 R_a}{4\pi}\int_0^{2\pi}\int_0^{2\pi/\eta}\frac{0.5r'\cos\vartheta_1\cos\eta\vartheta_2}{\left(R_a^2+R_b^2+d^2-2R_aR_b\cos\vartheta_1\right)^{1/2}}d\vartheta_2 d\vartheta_1$$
$$+\frac{\mu_0 R_a R_b 2\pi(1-\eta^{-1})}{4\pi}\int_0^{2\pi}\frac{\cos\vartheta_1}{\left(R_a^2+R_b^2+d^2-2R_aR_b\cos\vartheta_1\right)^{1/2}}d\vartheta_1 \quad (5.30)$$

where the first integral in (5.29) has split in (5.30). The calculation of the second filament influence on M'_{ab} is given by

$$M'_{ab} = \frac{\mu_0 R_a(R_b-0.5r')}{2\eta}\int_0^{2\pi}\frac{\cos\vartheta_1}{\left(R_a^2+R_b^2+d^2-2R_aR_b\cos\vartheta_1\right)^{1/2}}d\vartheta_1$$
$$+\frac{\mu_0 R_a r'}{8\pi\eta}\int_0^{2\pi}\frac{\cos\vartheta_1}{\left(R_a^2+R_b^2+d^2-2R_aR_b\cos\vartheta_1\right)^{1/2}}d\vartheta_1$$
$$+\frac{\mu_0 R_a R_b(1-\eta^{-1})}{2}\int_0^{2\pi}\frac{\cos\vartheta_1}{\left(R_a^2+R_b^2+d^2-2R_aR_b\cos\vartheta_1\right)^{1/2}}d\vartheta_1 \quad (5.31)$$

After simplification of (5.31), M'_{ab} is then given by

$$M'_{ab} = \left(\frac{\mu_0 R_a r'(1-2\pi)}{8\pi\eta}+\frac{\mu_0 R_a R_b}{2}\right)\int_0^{2\pi}\frac{\cos\vartheta_1}{\left(R_a^2+R_b^2+d^2-2R_aR_b\cos\vartheta_1\right)^{1/2}}d\vartheta_1 \quad (5.32)$$

Comparing (5.32) with (5.27), it is obvious that the term having $(1-2\pi)$ subtly reduces the mutual inductance within inward buckling as it takes a negative value. Accordingly, this coefficient will change to a positive value $(1+2\pi)$ when the outward buckling occurred for the filament (see Figure 5.22).

5.7.1.3 Capacitance

Series-capacitance

Layer winding

Figure 5.23 is a schematic of the layer winding which shows its series and shunt capacitances. There is a capacitance between each conductor and its adjacent conductors. If N_w is the number of turns, the number of series capacitances would be N_w-1, and the series capacitances can be determined through

$$C_s = \frac{C_{tt}}{N_w - 1} \tag{5.33}$$

where C_{tt} is the turn-to-turn capacitance and C_s is the total series capacitance in a layer winding. It should be noted that the value of series capacitances will be reduced when the number of series conductors is increased.

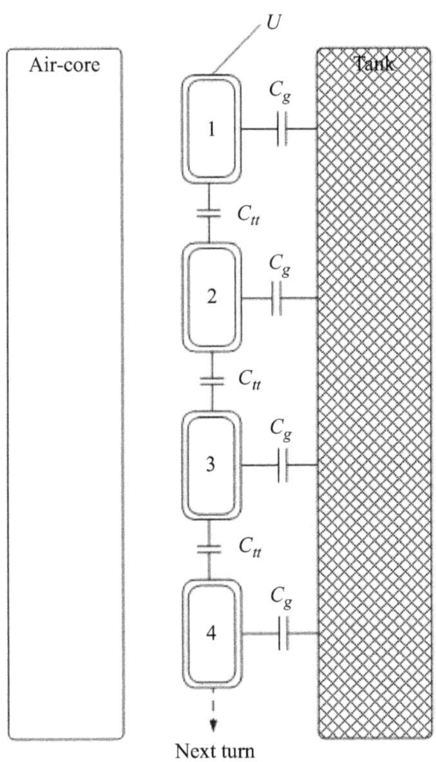

Figure 5.23 The overall layout of a layer winding including equivalent capacitance network

Disk winding

Various types of disk winding are used in power and distribution transformers. Continuous disk winding (also called conventional disk winding) and interleaved winding are the most popular. In some distribution and power transformers, continuous disk winding is used, while interleaved winding is employed in large power transformers at voltage level of 230 kV and above.

Continuous disk winding and related series capacitances

Figure 5.24 shows a continuous disk winding containing two disks, and Figure 5.25 shows its equivalent capacitive network. The series capacitance is divided into two parts:

1. Total series capacitance between the turns, C_t.
2. Total series capacitance between the disks, C_d.

To calculate the equivalent series capacitance in continuous disk winding, the energy summation method is used. According to this method, the summation of

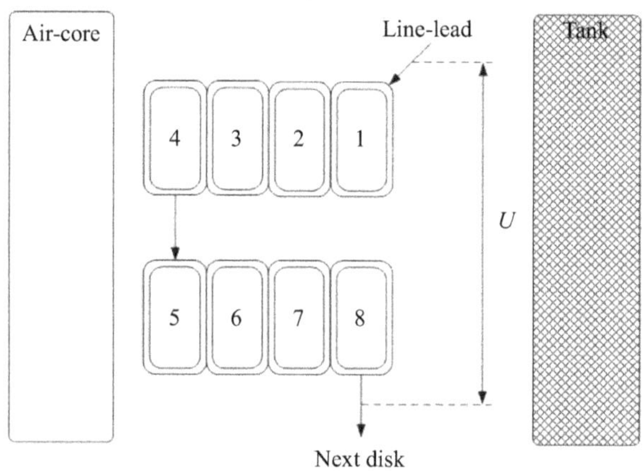

Figure 5.24 Continuous disk winding schematic, taken and modified [20]

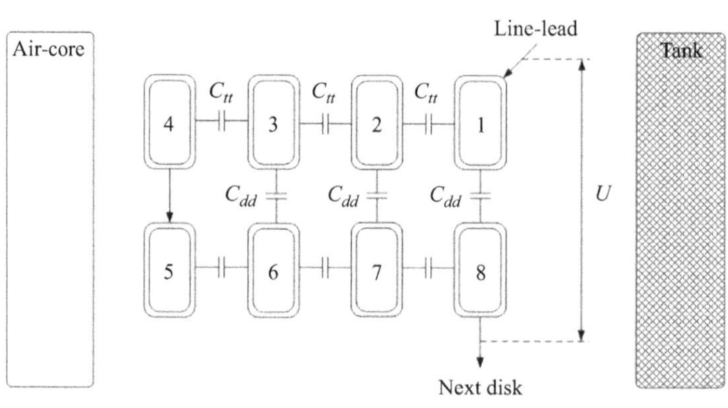

Figure 5.25 Equivalent capacitance network of the continuous disk winding

energies in the capacitances along a pair of disks is equal to the total energy which exists in the winding with those two disks. It is assumed that the number of conductor turns in each disk is N. The number of series capacitors between turns, as shown in Figure 5.25, will be $2N-2$ for a pair of disks.

Therefore, the total equivalent capacitance between the conductors, C_t, is given by [20]

$$\frac{1}{2}C_tU^2 = (2N-2)\frac{1}{2}C_{tt}\left(\frac{U}{2N}\right)^2 \Rightarrow C_t = \frac{1}{2}C_{tt}\left(\frac{N-1}{N^2}\right) \tag{5.34}$$

where U is the voltage drop across the pair of disks (see Figure 5.25), and C_{tt} is given by

$$C_{tt} = \varepsilon_t\varepsilon_0 R\pi\left(\frac{h+2\delta_t}{2\delta_t}\right) \tag{5.35}$$

In the above formula, δ_t is the thickness of interturn insulation, ε_t is the relative permittivity of the paper insulation and ε_0 is the vacuum permittivity.

Calculation of the equivalent capacitance between disks (C_d) is based on the voltage distribution demonstrated in Figure 5.26. When moving from end points starting from conductor number 1 or number 8 toward middle of the winding (conductor number 4 and 5), the voltage on corresponding conductors will change linearly and continually. Hence, the steady-state voltage distributions for the conductor in upper and lower disks are as follows:

$$U_{up}(n) = U\frac{2\lambda' - x}{2\lambda'}, \quad U_{down}(n) = U\frac{x}{2\lambda'} \tag{5.36}$$

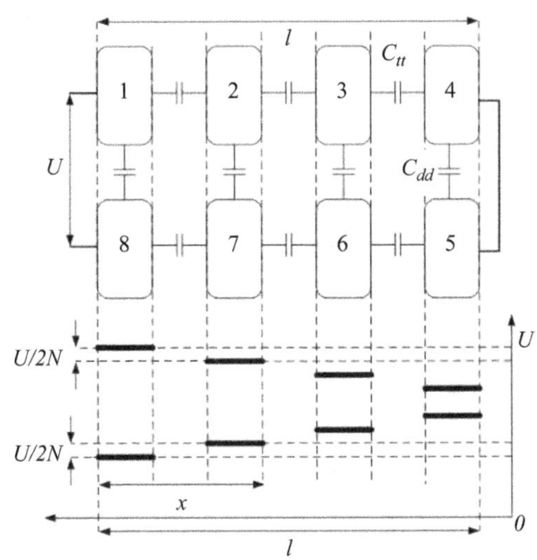

Figure 5.26 Pair of disks, cross-section overview and voltage distribution along disks pair (paper insulation has been ignored)

where n is the turn number, λ' is the total length of conductor in one disk and U denotes the voltage across the disk pair. The equivalent interdisk capacitance between two disks is given by [20]

$$E_d = \frac{1}{2}C_d U^2 = \frac{1}{2}C_{dd}\left(U_{up}(n) - U_{down}(n)\right)^2$$
$$= \frac{1}{2}C_{dd}U^2 \int_0^{\lambda'} \left(1 - \frac{x}{\lambda'}\right)^2 dx \quad (5.37)$$

$$\therefore C_d = \frac{C_{dd}}{3}\lambda' \quad (5.38)$$

The summation of C_t and C_d gives the equivalent series capacitance $C_{s\text{-}pair}$ for a pair of disks in a continuous disk winding:

$$C_{s-pair} = C_d + C_t \quad (5.39)$$

The first part in (5.39) corresponds to the capacitance between the disks, which is obtained with stored energy and the second part corresponds to the capacitance between the conductors. The equivalent capacitance for the entire winding (C_s) is then obtained by

$$C_s = \frac{1}{N_d}\left(4\frac{N_d - 1}{N_d}C_d + \frac{(N_w/N_d) - 1}{(N_w/N_d)^2}C_{tt}\right) \quad (5.40)$$

where N_d is the number of transformer winding disks and N_w is the number of winding turns.

Interleaved winding and related series capacitances

In an interleaved winding, C_s increases considerably as compared to continuous disk winding; therefore, it is utilized to improve the electric stress distribution. Various methods to interleave the disk windings are available. One of the simplest techniques is shown in Figure 5.27.

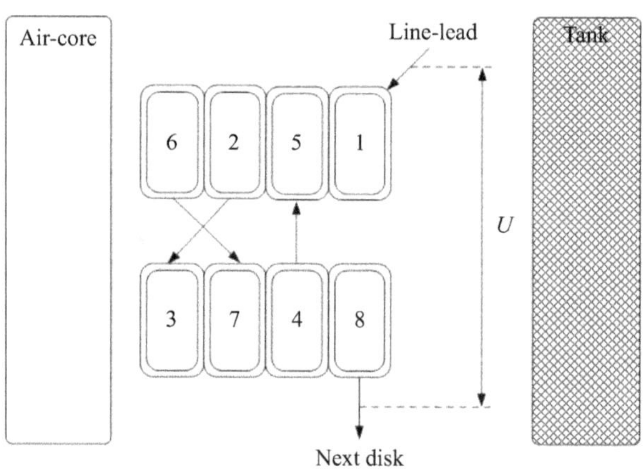

Figure 5.27 The interleaved disk winding

The series capacitance of this interleaved winding is given by [20]

$$C_s = E_{int} \frac{C_{tt}(N-1)}{4} \tag{5.41}$$

where E_{int} is the number of disks used for interleaving and equal to 2 for Figure 5.27. Increasing E_{int} will considerably increase the manual welding work required within the interleaved winding. The magnitude of series capacitance is increased in interleaved windings, and this leads to a more uniform voltage distribution throughout the winding. However, the turn-to-turn potential difference is greatly increased for steady-state operation, and thus the turn-to-turn insulation should be made properly to withstand against this potential.

Intershield winding

In Figure 5.28, the configuration of a disk winding with electrostatic shields in each disk is shown. In this winding, the shield turns, which can be made from copper or aluminum conductor, are placed between the winding main conductors at predetermined places, while the shield or shield turns of each disk are insulated from the conductors. Electrostatic shield conductors from the upper disk (in a pair of disks) are connected to the electrostatic shield conductors of the lower disk at the outermost shield turn as shown in Figure 5.28. For instance, the shield conductor between main conductors 1 and 2 of the upper disk is connected to the shield conductor between conductors 7 and 8 from the lower disk and both are isolated from the main conductors.

The series capacitance of the intershield winding is given by

$$C_s = \frac{1}{2} C_{tt} \left(\frac{N-1}{N^2} + \frac{2N_{sh}(N-1)}{N} \right) + \frac{C_{dd}}{3} \lambda' \tag{5.42}$$

where N_{sh} is the number of shield turns per disk.

Shunt capacitance

The shunt capacitance between the winding and the cylindrical metal housing tank is given by

$$C_g = \frac{2\pi H_w \varepsilon}{\ln(r_2/r_1)} \tag{5.43}$$

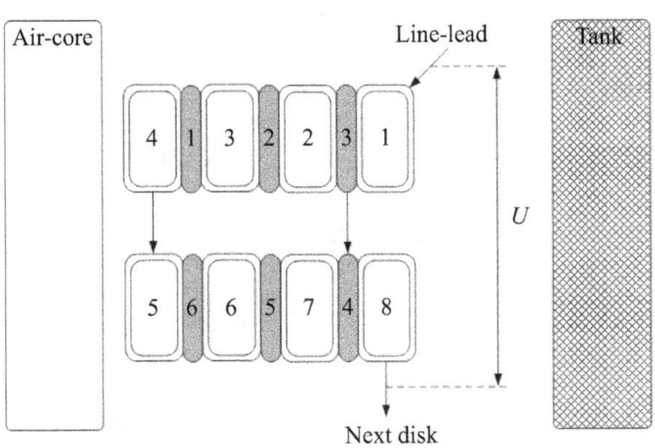

Figure 5.28 The intershield disk winding

5.7.2 Series capacitance under buckling

According to (5.34) through (5.40), it can be concluded that the series capacitance including turn-to-turn and interdisk capacitances of the buckled winding shown in Figure 5.22 is not changed significantly as the dimensions still remained unchanged for these parameters.

5.7.3 Shunt capacitance under buckling

The electric field generated by the transformer winding is not uniform across the deformed turns. Therefore, calculation of the shunt capacitance requires the electric field determined by finite element method.

In the case of analytical approach, if assuming a uniform electric field across the deformed section (highlighted in Figure 5.22), the equivalent shunt capacitance can be obtained through the summation of the shunt capacitance of the normal section, C_{normal}, as given by (5.44) paralleled with the shunt capacitance of the deformed section, C_{deform}, as given by (5.45)

$$C_{normal} = \int_{(2\pi-(2\pi/\eta))}^{2\pi} \frac{\varepsilon_0 \varepsilon_r H}{\ln(r_2/r_1)} d\varphi \tag{5.44}$$

$$C_{deform} = \int_{0}^{2\pi/\eta} \frac{\varepsilon_0 \varepsilon_r H}{\ln(r_2/(r_1 - 0.5r'(\cos(\eta\varphi) - 1)))} d\varphi \tag{5.45}$$

where η is the ratio of the entire trigonometric circular span (2π) over the deformation span (rad) as illustrated in Figure 5.17, H is the winding height and r' represents the deformation radius.

The total shunt capacitance of the buckled winding is eventually obtained as

$$C'_g = C_{normal} + C_{deform} \tag{5.46}$$

where H_w is the height of the winding, ε is the dielectric permittivity, r_1 is the radial dimension of the winding and r_2 denotes the radial dimension of the tank, and δ_t as compared to r_2 is small and thus ignored.

5.7.4 Tap-changer

Transformer failure is usually due to the tap-changer problem. This could be either in the transformer tap-changer diverter switch or the selector switch. Therefore, it is recommended to record frequency response signatures for all transformer tap terminals during transformer commissioning. These recorded data will become valuable later when the transformer needs diagnosis. In transformer FRA standard, it is only suggested to perform the FRA test on the (1) tap-position with the highest number of effective turns in circuit and (2) on the tap-position with the tap winding out of circuit [9].

Transformer frequency response signature is obviously influenced by the tap-position setting. Low-, mid- and even high-frequency regions in the FRA spectrum can be deviated from the reference values due to tap-position changes. The number of transformer winding turns will be changed. Therefore, winding self- and mutual-inductance as well as series capacitance can be changed substantially, and ultimately, the transformer frequency response is changed. Furthermore, the winding resistance is also altered from its nominal value in nominal tap-position. Therefore, the influence of winding resistance changes may become detectable in the very high-frequency region of the FRA spectrum. Detailed discussion on this issue is provided through the following example.

Example 5.3 Consider a 40-MVA, 33,000/7,070-V two-winding three-phase transformer with linear tap-changer arrangement in the HV side. Table 5.8 provides the transformer specifications, and Figure 5.29 shows the tap-terminal diagrams. The frequency response has been taken for both HV and LV sides with different tap-changer positions to study tap-changer position influence on the FRA spectrum.

Figure 5.30(a) shows the full FRA spectra for the HV side, phase V for tap-position on terminals 1, 3 (nominal position) and 5. Figure 5.30(b) shows the close-up view of the low-frequency region. From Table 5.8, the voltages provided at tap

Table 5.8 Transformer specifications with tap-changer terminals information

Tap-position no.	Volts	Amps	Power rate (kVA)	VG
1	34,650	666/500	40,000/30,000	Y_Ny0
2	33,825	683/512		
3	33,000	700/525		
4	32,175	718/530		
5	31,350	737,552		

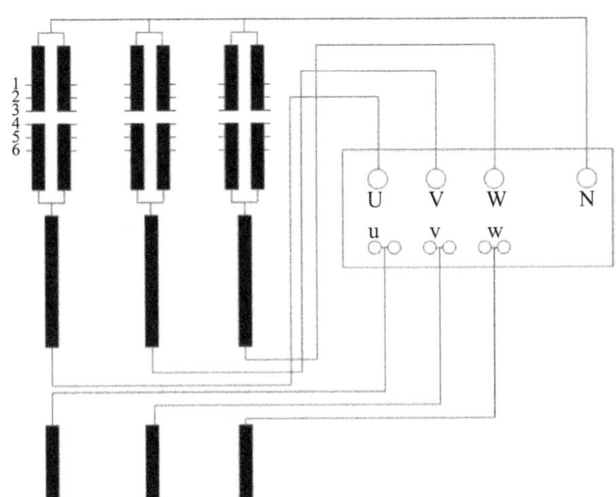

Figure 5.29 40-MVA transformer connections

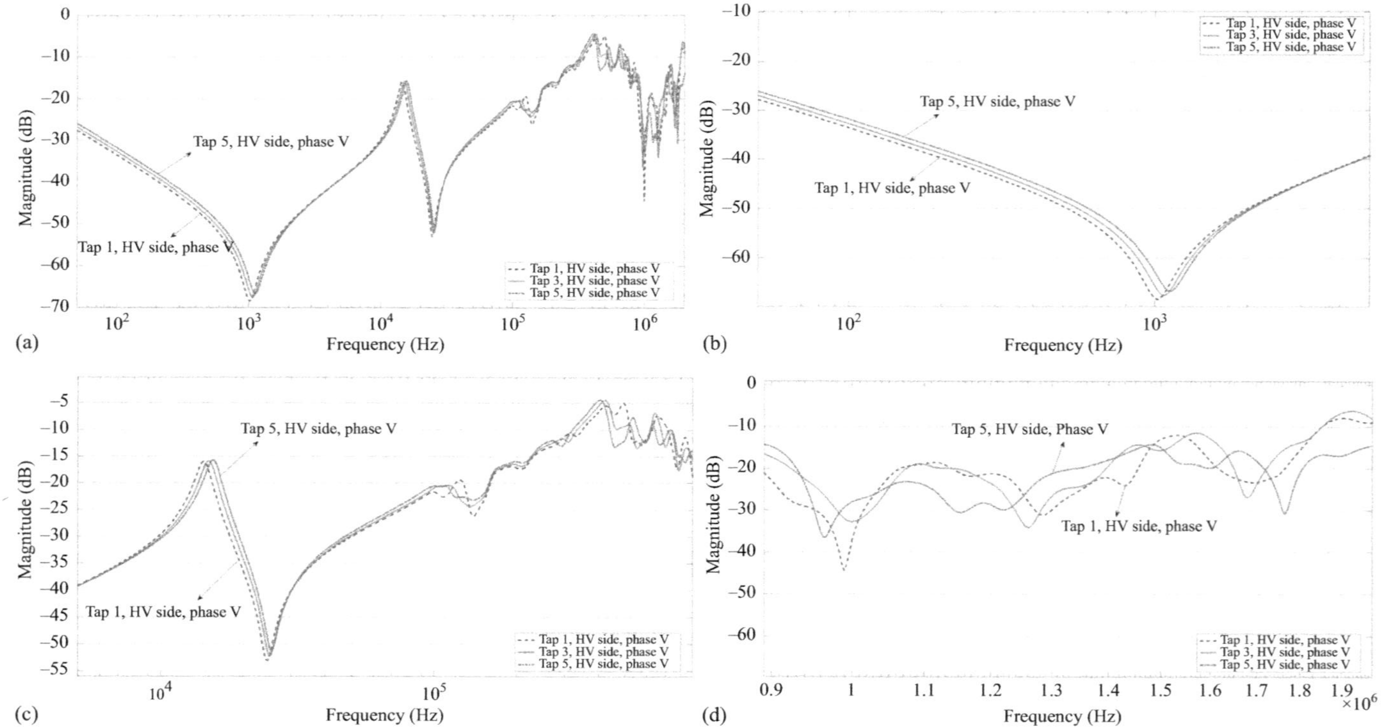

Figure 5.30 Influence of transformer tap-changer on FRA spectrum, transformer HV side, phase V (end-to-end measurement): (a) full spectrum 50 Hz–2 MHz, (b) 50 Hz–5 kHz, (c) 5–900 kHz and (d) 900 kHz–2 MHz

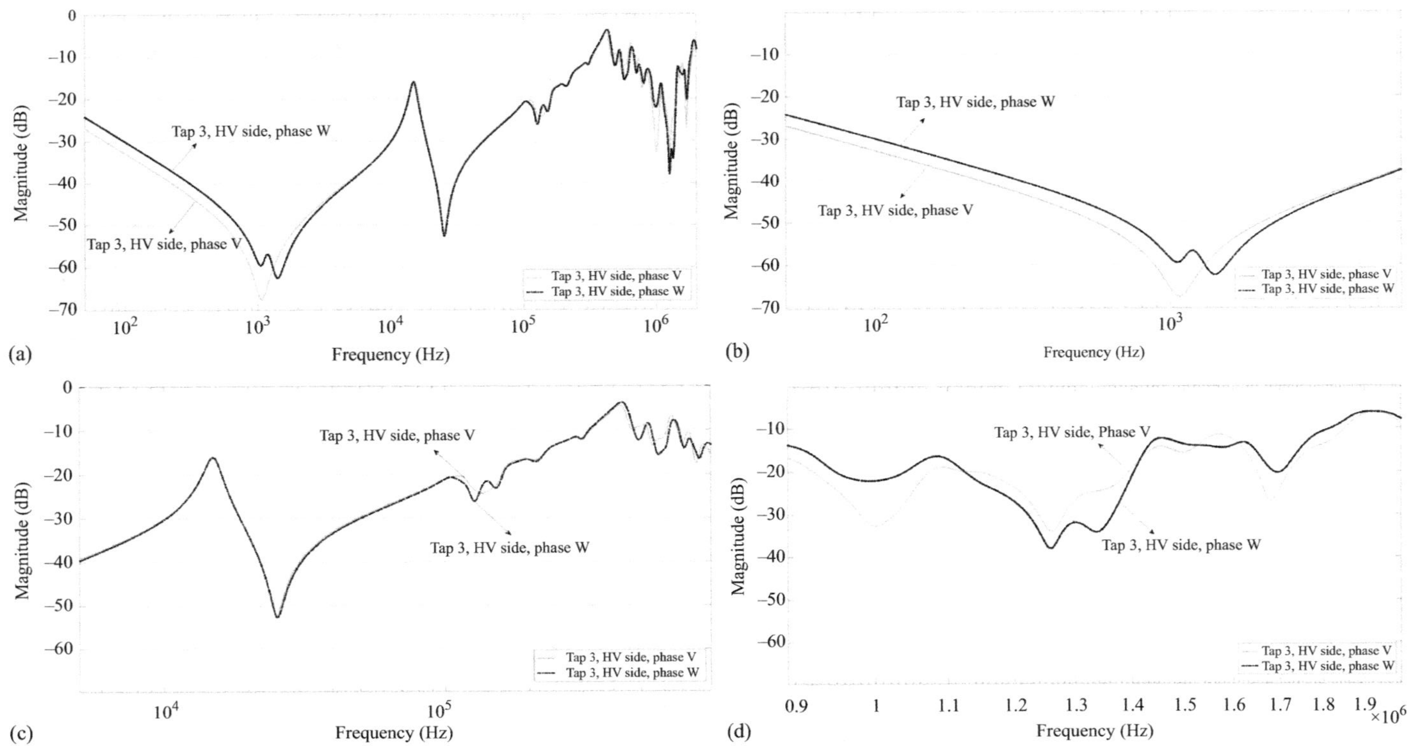

Figure 5.31 HV side FRA spectra, phase V and phase W, tap-changer terminal 3 (end-to-end measurement): (a) full spectrum 50 Hz–2 MHz, (b) 50 Hz–5 kHz, (c) 5–900 kHz and (d) 900 kHz–2 MHz

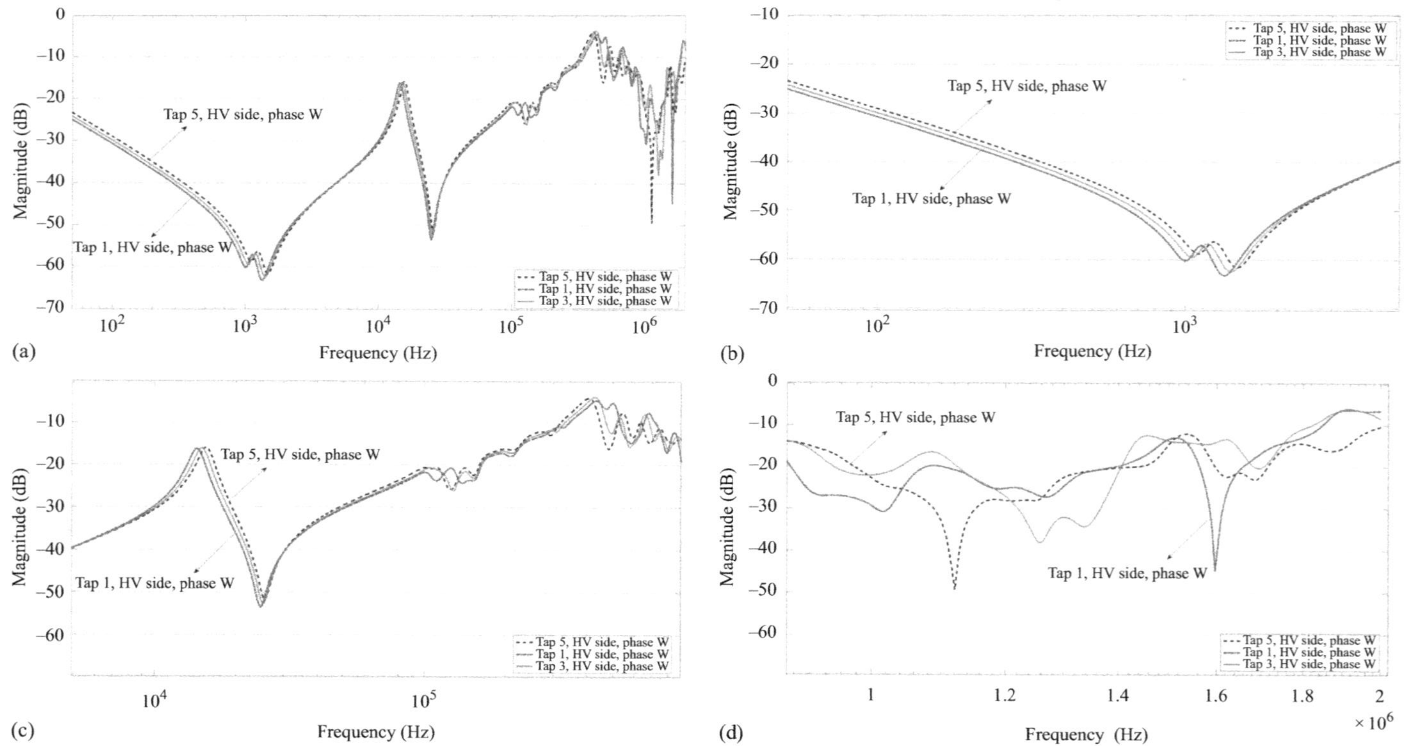

Figure 5.32 Influence of transformer tap-changer on FRA spectrum, transformer HV side, phase W, tap-changer terminal 3 (end-to-end measurement): (a) full spectrum 50 Hz–2 MHz, (b) 50 Hz–5 kHz, (c) 5–900 kHz and (d) 900 kHz–2 MHz

terminals 5, 3 and 1 are in increasing order, respectively; therefore, the number of transformer winding turns in tap-position 5 is less than that in tap terminals 3, and 1. Hence, tap-position 5 has less inductance than tap-position 3, and tap-position 3 has less inductance as compared to tap-position 1. It is anticipated that different inductive reactances would affect the FRA spectrum in the low-frequency region as revealed in Figure 5.30(b). In addition, the lower inductive reactance of tap-position 5 causes the first antiresonance in the low-frequency region to shift slightly to a higher frequency.

Moving to higher frequencies in the intermediate frequency range, at first, all spectra in Figure 5.30(c) are aligned which means that the winding shunt capacitance has not been altered significantly from its original value due to tap-position changes. However, moving forward to reach the first resonance point shows that this resonance point follows similar habit as the first antiresonance point in the low-frequency range. This might be due to variation of mutual inductive reactance or even changes in winding series capacitance. Moving to higher frequencies within the mid-frequency range, slight discrepancy between spectra can be interpreted through changes of winding series capacitance. In Figure 5.30(d), discrepancy in the high-frequency band could be due to different winding resistance values for different tap-positions. Figure 5.31 provides the FRA spectra for HV winding, phases V and W. Study on frequency response of HV side, phase W shows similar habit to phase V, see Figure 5.32.

According to the transformer specifications, the tap-changer was designed for the HV side. Hence, to find its influence on the LV winding FRA spectra, the frequency responses of phases uw (concentric to phase V) and uv (concentric to phase W) were recorded and shown in Figure 5.33(a) and (b). It can be seen that they are well matched over the very low-frequency region. This is predictable as the winding inductive reactance on the LV side is not affected by tap-position changes on the HV side. However, moving from low- to mid- and high-frequency range, Figure 5.33 shows discrepancy between the traces, specifically in the intermediate frequency range. This in turn could be due to change of mutual-inductance between concentric HV and LV windings. It increased when the transformer tap-position changed from 5 to 1, and frequency response resonant points are shifted slightly to low-frequency in Figure 5.33 for both central and lateral windings.

5.7.5 Paper insulation deterioration

Insulation deterioration is quite common in distribution and power transformers. Usually, the cylindrical shape pressboards and also the paper insulation wrapped around the winding conductors have the most significant aging rate as compared to other transformer insulation media. Specifically, high level of voltage and thermal stresses on transformer winding turn-to-turn insulation medium results in a fast aging rate. Transformer overload as well as water content can accelerate the aging.

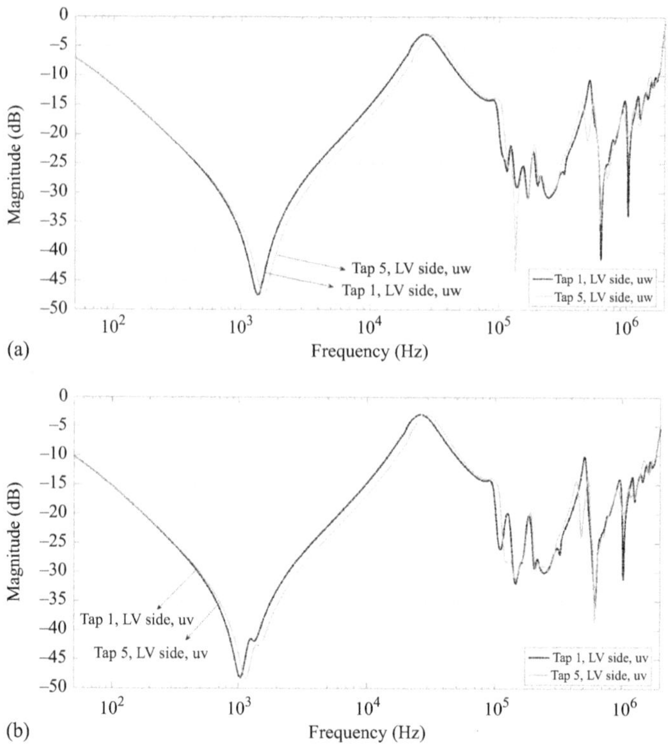

Figure 5.33 Influence of transformer tap-changer on FRA spectrum, transformer LV side, tap-changer terminals 1 and 5 (end-to-end measurement): (a) full spectrum, phases uw 50 Hz–2 MHz and (b) full spectrum, phases uv 50 Hz–2 MHz

Transformer insulation media, in particular turn-to-turn paper insulation, will influence the transformer winding series capacitance directly. Thus, the transformer frequency response can be changed due to changes of its winding series capacitance. In addition, if the transformer turn-to-turn insulation is partially broken or failed, partial discharge would be initiated due to the creepage current initiation between the winding turns. If the transformer winding is not changed or properly repaired, the partial discharge will get worse and develop into a continuous leakage current between winding turns. This leakage current is not large enough to activate the transformer protection system. However, it will lead to turn-to-turn circulating current which generates heat in transformer winding and also impedes the main flux flow in transformer core. Flux flow changes in transformer winding will affect the magnetic reluctance seen by the resistive, inductive, capacitive (RLC) network, and ultimately the transformer inductance becomes different to that calculated for normal transformer. This phenomenon will influence the transformer frequency response trace.

For better understanding, a practical example is provided and turn-to-turn transformer insulation deterioration is emulated on a 2-kVA transformer.

Example 5.4 An open-wound three-phase 2-kVA transformer was used as the test object and a turn-to-turn fault was emulated to model high leakage current. The frequency response measurement was performed when the transformer was out of service. A variable resistor was connected to phase B, HV side, between turns 3 and 4, as depicted in Figure 5.34 and a switch (SW) was used to connect or disconnect it from the circuit. This variable resistor helps to model insulation deterioration as well as turn-to-turn short-circuit on phase B. The switch is open for modeling the transformer normal condition.

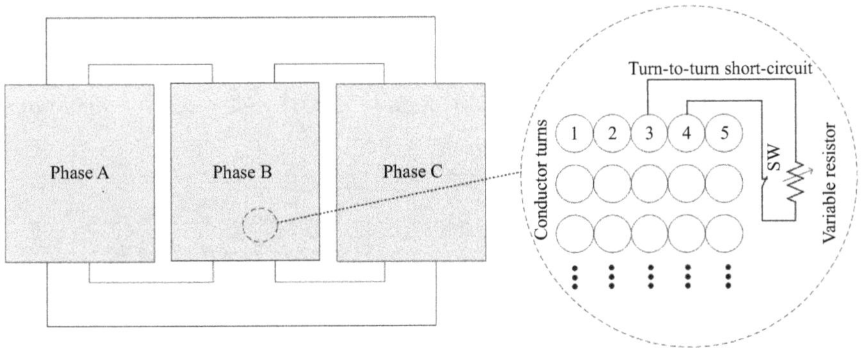

Figure 5.34 Schematic of turn-to-turn short-circuit model, HV side, phase B

The reason to select such a small size transformer was to perform a real turn-to-turn fault on transformer. Here, the fault current level is small and can be easily controlled through a variable resistor; this is not simple in large power transformers.

- ***Assessment***

Using end-to-end FRA measurement setup, the frequency response was first performed and baseline signatures recorded for different phases, see Figure 5.35. Afterwards, the switch was closed to effect a fault between turns 3 and 4. The short-circuit circulating current was 100 mA when the variable resistor was set at its maximum value. The resistor was gradually adjusted to simulate turn-to-turn insulation deterioration, and different short-circuit currents were obtained (200, 500, 1,000 and 2,000 mA). The FRA spectra for all conditions were recorded and shown in Figure 5.36.

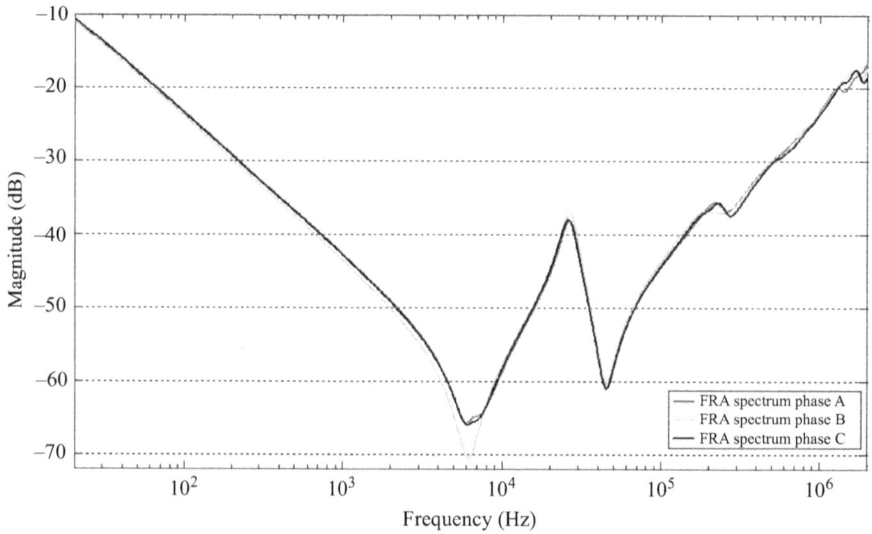

Figure 5.35 FRA spectra for phases A, B and C on HV side (baseline signatures)

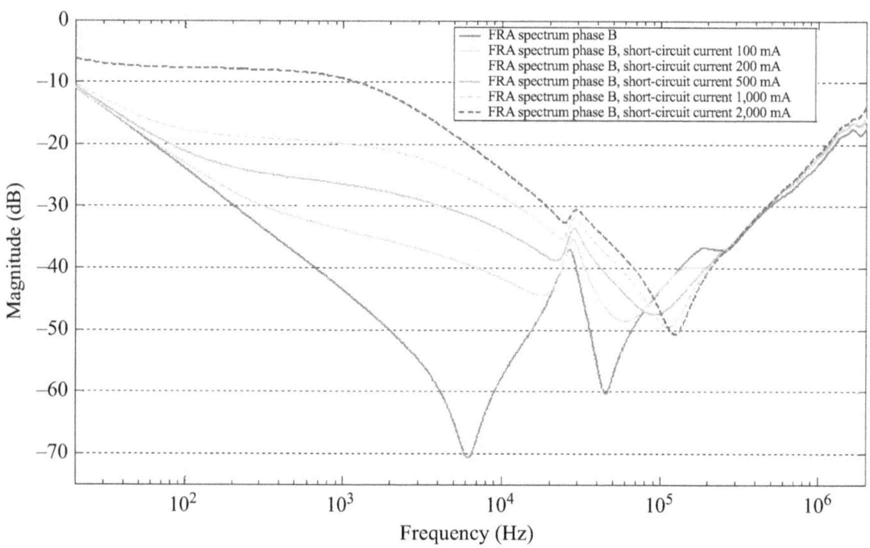

Figure 5.36 FRA spectra for different short-circuit conditions

---

FRA measurement results for normal and faulty conditions in Figure 5.36 show that even the small short-circuit circulating current (100 mA) can cause significant change in the spectrum. The first and second antiresonance points are both shifted to higher frequencies and their magnitude significantly reduced. In the meantime, the falling trend in very LFs is also changed and reduced. As the short-circuit circulating current is increased to higher values, the traces follow the same habit. However, it is interesting to note that the resonance point around 20 kHz is not significantly shifted. The technical reason is explained hereinafter.

The antiresonances in the very low- and low-frequency ranges are initiated and influenced mostly by the transformer inductive- and shunt capacitive-reactance. The inductive-reactance is influenced by transformer core reluctance, the latter is given by

$$\Re = \frac{l}{\mu A} \tag{5.47}$$

where l denotes the magnetic circuit length (m), $\mu = \mu_0 \mu_r$ is the permeability of the material, μ_0 is the permeability of vacuum ($4\pi \times 10^{-7}$ H/m), μ_r is the relative magnetic permeability of the material and A is the effective cross-sectional area of the circuit (m^2). If any of these parameters is changed, the magnetic reluctance and ultimately the inductive-reactance would be changed.

In this example, when the turn-to-turn insulation deterioration or short-circuit occurred, the circulating current in the short-circuit loop opposes the flow of magnetic flux in the transformer core, see Figure 5.37. Hence, from the flux division point of view, a transformer core leg can be blocked magnetically due to the turn-to-turn or disk-to-disk short-circuit, or, its cross-section area is effectively reduced. If A is decreased in (5.47), the magnetic core reluctance will increase. Consequently, the inductive-reactance of phase B transformer leg will be reduced. Lower inductive-reactance will cause the antiresonances shifting to higher frequencies.

Similarly, the resonance points are also influenced by the inductive and shunt capacitive-reactance in the very low- and low-frequency ranges. However, in this case, the shunt capacitive-reactance influences more than the inductive-reactance. That is the reason, in Figure 5.36, the resonance point around 20 kHz remained essentially unchanged as the shunt capacitive-reactance did not change in all the experiments. The very slight shifting of this resonance point is due to minor influence of the inductive reactance.

5.7.6 Temperature and moisture content

To discuss about temperature and moisture influence on the FRA spectrum, transformer water dynamic is first briefly explained.

5.7.6.1 Transformer water dynamic

Residual moisture in a transformer due to the water ingress through atmosphere, insulation aging, cellulose decomposition or even after dry-out process will transfer from the oil into the paper insulation and from the paper toward oil insulation when at low and high temperatures, respectively. Indeed moisture migration from one phase (liquid/solid) to the other phase (solid/liquid) could be due to the moisture concentration, temperature and pressure gradients [21]. Nevertheless, water

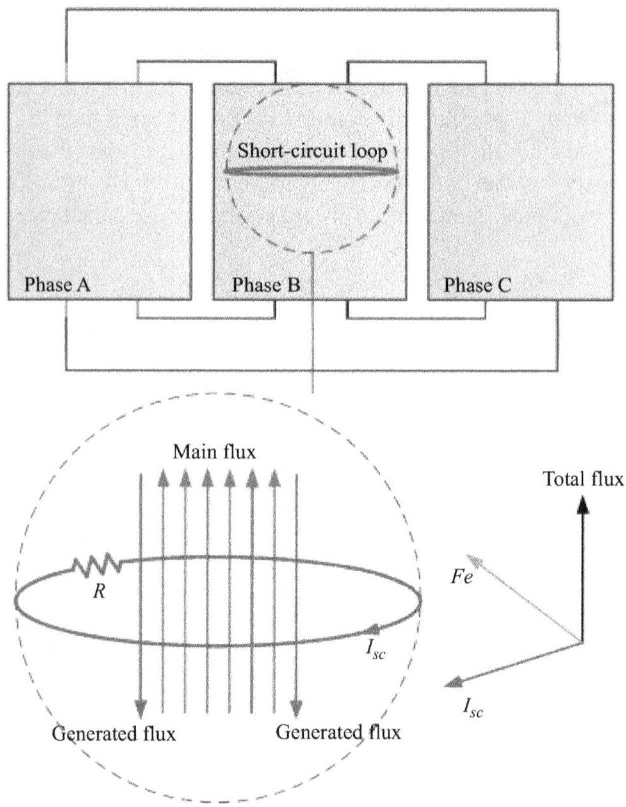

Figure 5.37 Opposite flux initiated by short-circuit current

dynamics in transformer can be classified into transient and steady-state periods [22]. The transient period involves moisture diffusion activity, whereas water equilibrium between paper and oil insulations is attained in the steady state. Figure 5.38 shows the water dynamic in paper and oil insulations for different temperatures.

Transient

In the case of moisture diffusion, Fick's second law as one of the basis equations can be expressed as

$$\frac{\partial C_c}{\partial t} = D\frac{\partial^2 C_c}{\partial x^2} \tag{5.48}$$

where D is the diffusion coefficient (m^2/s), C_c is the concentration of substance (moisture) (mol/m^3) and x is substance movement position (m).

Temperature changes can lead to different activation energy for the molecules in adjacent regions and ultimately moisture migration. Guidi and Fullerton employed a diffusion model to estimate moisture migration from the transformer

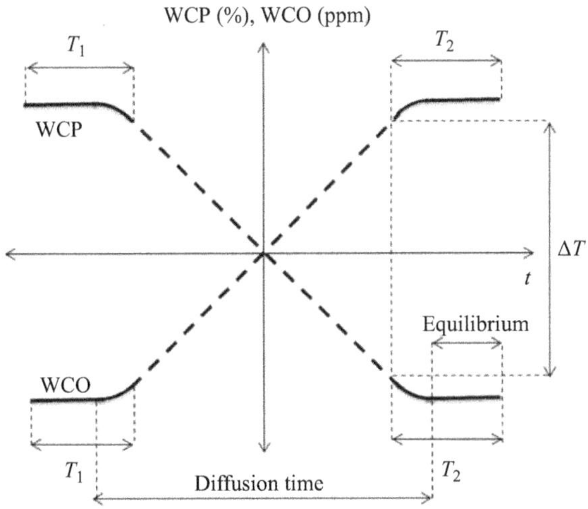

Figure 5.38 Water dynamic in paper and oil insulation for different temperatures T_1 and T_2 ($T_1 < T_2$), WCO (water content in oil) and WCP (water content in paper); t *denotes the time*

paper insulation [21,23,24]. They specifically focused on power transformer drying time, temperature and moisture adsorption rates for insulation when exposed to the atmosphere [21]. Their estimation is expressed by

$$D = D_0 e^{(k'c + E_0((1/T_0) - (1/T_k)))} \tag{5.49}$$

where c is the local moisture concentration (H_2O/kg), T_k is the current temperature (K), T_0 is the reference temperature (298 K), k' is a dimensionless parameter and is equal to 0.5 [25,26], D_0 is a preexponential factor (m^2/s) and E_0 is the activation energy of the diffusion process (kJ/mol). D_0 and E_0 have been estimated for oil-free as 2.62×10^{-11} and 8,140 and for oil-impregnated paper as 1.34×10^{-13} and 8,074 by [25,26], respectively. Substitution of (5.49) into (5.48) will lead to (5.50) which can explain moisture migration as a function of temperature. The solution of (5.50) is detailed in [27]:

$$\frac{\partial C_c}{\partial t} = 2.62 \times 10^{-11} e^{(0.5c + 8{,}140((1/T_0) - (1/T_k)))} \frac{\partial^2 C_c}{\partial x^2} \tag{5.50}$$

Equilibrium

Water content equilibrium between paper and oil insulation has been widely discussed in [22–28]. Equilibrium curves showing paper–water content versus oil–water content for different temperatures have been achieved; hence, it is possible to determine the value of one of them once knowing the other [22]. Most of the abovementioned literatures have provided the equilibrium curves up to at most 100 ppm, a wider range of equilibrium curve from 0 to 100 °C and moisture in oil up to 800 ppm was presented in a work by Du *et al.* at MIT university and named MIT oil–paper equilibrium curve [25]. Due to the wide range available, the MIT equilibrium curves were employed in the current study to derive the moisture content.

The diffusion time constant for moisture diffusing from one side of insulation is calculated as [25]

$$\tau = 4\frac{d_p^2}{D\pi^2} \tag{5.51}$$

while the equation for double-side diffusion through paper insulation is given by [25]

$$\tau = \frac{d_p^2}{D\pi^2} \tag{5.52}$$

where τ is the diffusion time constant, and d represents the thickness of paper insulation or pressboard.

Example 5.5 To conduct a practical study on temperature and moisture influences on FRA spectrum of winding, a model transformer was fabricated according to transformer manufacturing standards and used in this example.

The test object is shown in Figure 5.39. Oil is required to be injected into the transformer glassy tank. Hence, a drain valve was fitted onto the top plate to enable oil injection and also taking oil sample.

Three different methods can be used for heating the test object: LF heating, oil circulation through an oil circulator, and using an electric oven. The third approach was chosen for this study as it would be more accurate for controlling the temperature. The test object was placed inside the oven and wiring connections for FRA measurement were brought out through a bushing mounted on a small opening on the oven top. The oven was equipped with a sensitive thermostat and a digitally controlled heater to govern the internal temperature. Two thermocouples were used to monitor the internal temperature of the oven as well as the test object. The FRA test setup was then kept unchanged for the entire experiments.

Figure 5.39 Manufactured glassy air-core transformer (setup preparation to study temperature and moisture impact)

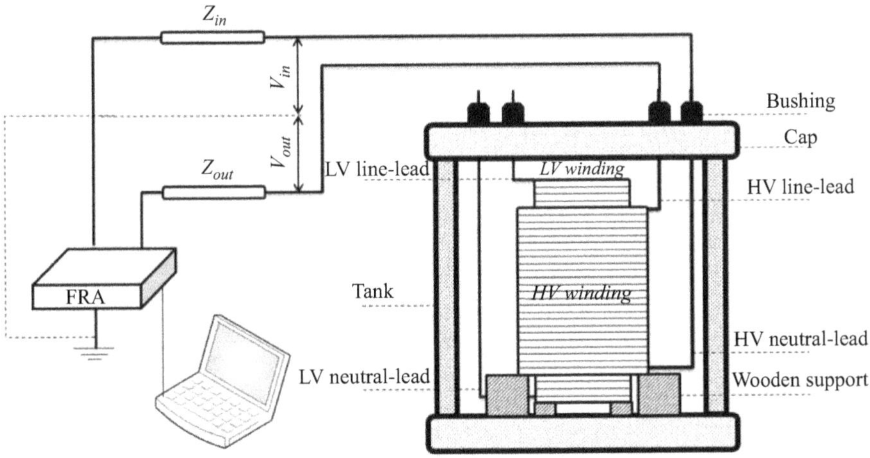

Figure 5.40 FRA test setup to examine temperature and moisture variation

- ***Assessment***

FRA measurement was performed on a winding by injecting a swept sinusoidal signal V_{in} at the line-lead, and detecting the response V_{out} at the neutral-lead (end-to-end measurement) as shown in Figure 5.40.

In this example, the significant value of the air-core magnetic reluctance of the test object results in a small self-inductance for the windings. Small self-inductances will lead to small inductive reactances. Therefore, the resonance frequencies in FRA traces for HV and LV windings would be shifted considerably to higher frequencies. Hence, the upper band limits for FRA measurements were extended from 2 to 20 MHz to display entire oscillations similar to what conducted for FRA measurement.

To study the effects of temperature and moisture variation on the FRA trace, at first the "wet" test object was examined. The transformer drain valve was opened, and the test object was deliberately left exposed to the laboratory ambient for 2 weeks so that the paper insulation was saturated with moisture. The average readings of ambient temperature and relative humidity were 23 °C and 26%, respectively. Hence, the initial moisture content for paper insulations was 4.1%, calculated using the following data in Table 5.9 and also (5.53) provided by Du *et al.* [25] on air relative humidity method:

$$WCP = a_6(RH)^6 + a_5(RH)^5 + a_4(RH)^4 + a_3(RH)^3 + a_2(RH)^2 + a_1(RH) + a_0 \quad (5.53)$$

where *WCP* is the moisture in paper in percent by weight and *RH* is the air relative humidity in percent.

Table 5.9 Ambient air relative humidity and moisture in paper (polynomial fitting parameters for various temperatures), taken and modified [26]

	30 °C	40 °C	50 °C	60 °C	70 °C
a_0 ($\times 10^{-1}$)	2.4131270	1.6954583	1.0483257	1.3978572	0.7441865
a_1 ($\times 10^{-1}$)	3.2828657	2.9079147	2.4316118	2.1359436	1.7762623
a_2 ($\times 10^{-3}$)	−14.929696	−11.950117	−7.2850779	−6.2300223	−2.7797731
a_3 ($\times 10^{-4}$)	4.3831525	3.2448905	1.2731316	1.1731076	−0.27101029
a_4 ($\times 10^{-6}$)	−6.3395879	−4.2926236	−0.37397578	−0.57129397	2.2473555
a_5 ($\times 10^{-8}$)	4.2446633	2.5228351	−1.1755019	−0.75286519	−3.3218692
a_6 ($\times 10^{-11}$)	−9.3468655	−3.8729882	9.5319144	7.4380470	16.252499

Afterwards, the transformer oil was dried out, and the tank was filled with dry transformer oil (<5 ppm, at 70 °C) then left standing until the paper was fully impregnated by oil, and eventually both of them reached equilibrium. The equilibrium time was calculated as 244 h using (5.51) and (5.53) at 23 °C. Thus, the oil sample was taken from the container using a syringe after 11 days (>244 h). This is long enough to ascertain that oil and paper insulation are in equilibrium.

The moisture content of the oil was measured using Karl Fischer Titration (KFT) method at 23 °C and moisture content of the paper insulation was derived through MIT oil–paper equilibrium curves [25], accordingly. These values were 11 ppm and 4%, respectively. Derived moisture contents of the paper insulation through air relative humidity and KFT methods gave similar results (4.1% and 4%).

Next, the test object was heated up to 30 °C to be ready for the initial stage of FRA measurement. FRA measurements were performed over the range of 30–90 °C in 10 °C increments, but the spectra were recorded for HV and LV windings at 30, 50, 70 and 90 °C. The frequency response trace for HV winding was recorded over the range 20 Hz–20 MHz when the LV winding terminals were left open circuit (end-to-end measurement). Likewise, FRA measurement was also performed for the LV winding. On completion, the oven temperature was increased gradually to reach 50 °C. The test object was left standing for 25 h (equilibrium time at 50 °C) to complete heat exchange between the test object insulations and oven environment and also to pass the equilibrium time.

Then, frequency response traces for HV and LV windings were remeasured. Similar experiments were done for 70 and 90 °C. FRA traces for HV and LV windings for different temperatures are shown in Figures 5.41 and 5.42 over the frequency range of 5 kHz–20 MHz. The frequency response magnitude for the frequency range 20 Hz–5 kHz was 0 dB and thus is not shown in Figures 5.41 and 5.42. Figures 5.41(b), (c) and 5.42(b), (c) show the close-up view of the measurement results.

The majority of the insulation system for manufactured test object was between HV and LV windings including Kraft paper, oil canal, spacers and pressboard. Hence, parallel to the other measurements, the dielectric dissipation factor (DDF) values between HV and LV windings were measured at the power

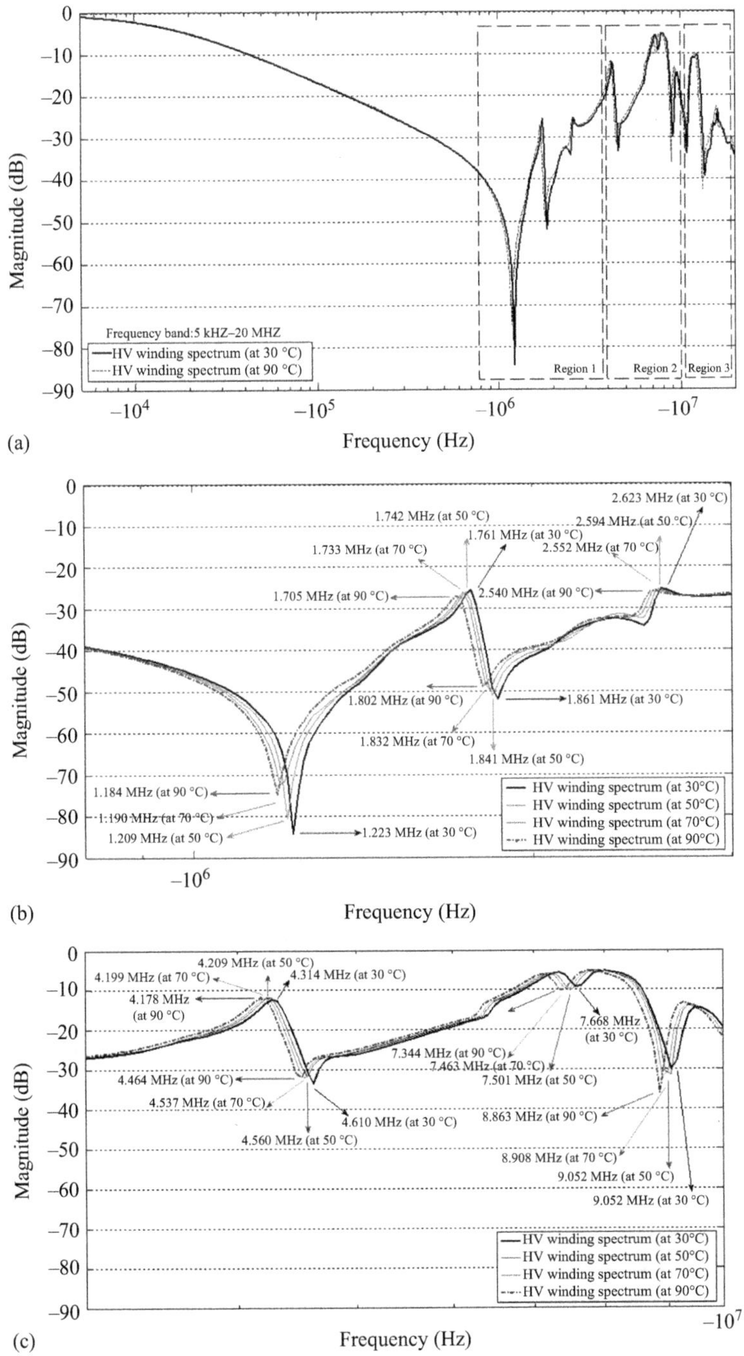

Figure 5.41 FRA spectra for "wet" model transformer, HV side: (a) entire trace for 30 and 90 °C, (b) close-up view of region 1, frequency band 800 kHz–3 MHz, (c) close-up view of region 2, frequency band 3–10 MHz

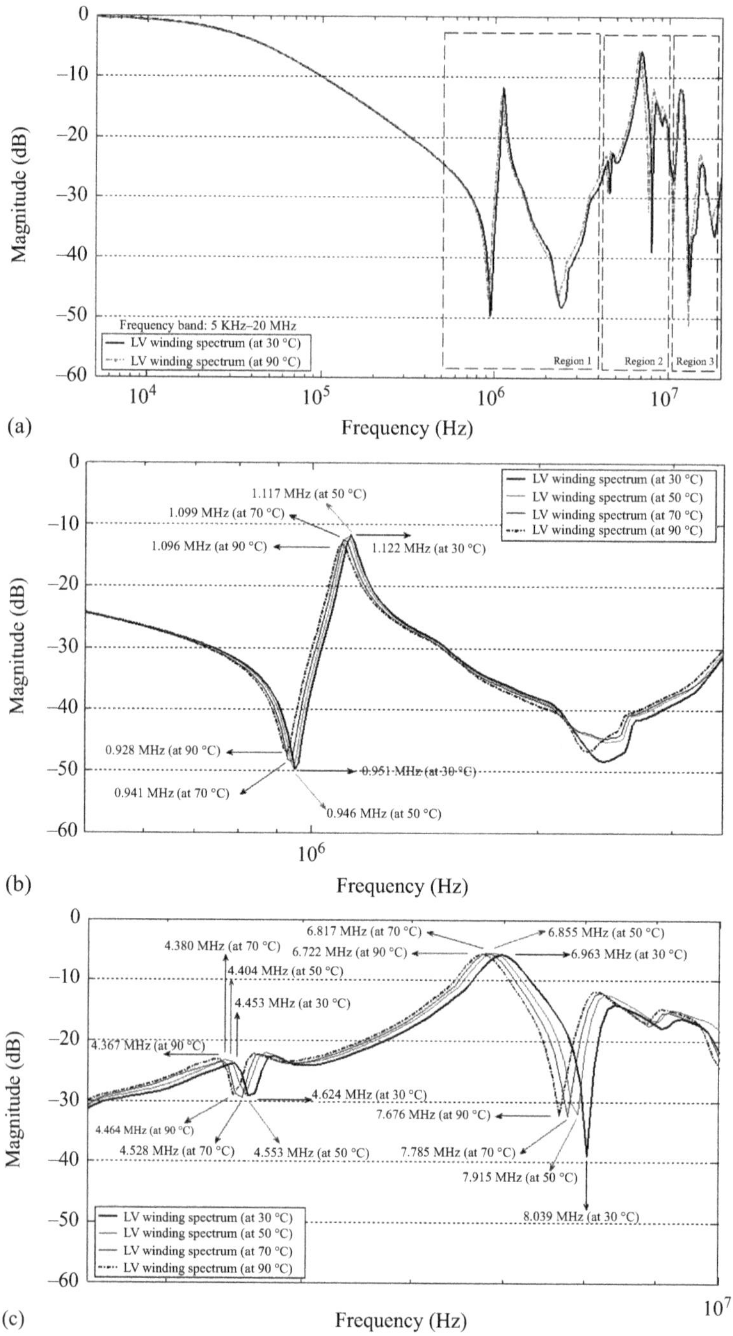

Figure 5.42 FRA spectra for "wet" model transformer, LV side: (a) entire trace for 30 and 90 °C, (b) close-up view of region 1, frequency band 500 kHz–3.5 MHz, (c) close-up view of region 2, frequency band 3.5–10 MHz

frequency (50 Hz) for different temperatures to determine the insulation quality. To avoid flash-over in the test object, the maximum applied voltage for DDF measurement was kept low at 5 kV.

According to Figures 5.41 and 5.42, when examining from LFs to 800 kHz all of the traces are perfectly matched. Moving from 800 kHz to higher frequencies, the discrepancy between the traces becomes obvious.

As the test object temperature changed from 30 to 90 °C, the FRA traces have slightly shifted to lower frequencies. In the case of 90 °C, this movement seems to be more significant. Some of the resonances and antiresonances have been highlighted in Figures 5.41(b), (c) and 5.42(b), (c).

Detailed investigation of Figures 5.41 and 5.42 shows that the oscillations trend of the spectra seems to be similar, while the resonance frequencies have shifted to lower frequencies as the temperature increased and equilibrium time passed.

In the meantime, some of the resonant magnitudes are reduced. This deviation comes certainly through changes in the windings' inductance, total capacitance, resistance or insulation conductance. From a mathematical point of view, resonances and antiresonances in the FRA trace can be generated due to interaction between the inductive and capacitive reactances. As frequency increases, each resonance point indicates the changing from capacitive toward inductive behavior, while the antiresonance point shows the turning from inductive to capacitive behavior of the winding impedance. Hence, every resonance or antiresonance can be explained through (5.54) if it is considered independently:

$$f_i = \frac{1}{2\pi\sqrt{L_i C_i}} \tag{5.54}$$

where f_i denotes the ith resonance frequency, L_i and C_i are the inductance and capacitance (involving series and shunt capacitances) at that resonance frequency. According to (5.54), resonance frequencies would be changed if the inductance or total capacitance is changed. Thus, to clarify whether inductance variation due to the temperature changes can influence FRA trace, the windings' inductances were measured for some frequencies and given in Tables 5.10 and 5.11 at 30 and 90 °C

Table 5.10 HV winding electrical parameters for 30 and 90 °C

Parameter	***R***	***X***	***Z***	**θ∡**	***L***
100 Hz (30 °C)	0.087 Ω	0.444 Ω	0.453 Ω	78.91	707.90 μH
100 Hz (90 °C)	0.104 Ω	0.449 Ω	0.461 Ω	76.95	715.30 μH
Change	19.54%	1.12%	1.76%	2.48	1.04%
1 kHz (30 °C)	0.333 Ω	4.13 Ω	4.14 Ω	85.39	657.43 μH
1 kHz (90 °C)	0.347 Ω	4.17 Ω	4.19 Ω	85.24	664.57 μH
Change	4.20%	0.96%	1.20%	0.17	1.08%
100 kHz (30 °C)	11.65 Ω	337.76 Ω	337.95 Ω	88.02	537.55 μH
100 kHz (90 °C)	13.12 Ω	338.57 Ω	338.82 Ω	87.78	538.85 μH
Change	13.49%	0.23%	0.25%	0.27	0.24%

Table 5.11 LV winding electrical parameters for 30 and 90 °C

Parameter	R	X	Z	$\theta\measuredangle$	L
100 Hz (30 °C)	0.056 Ω	0.174 Ω	0.183 Ω	72.15	278.3 μH
100 Hz (90 °C)	0.067 Ω	0.176 Ω	0.188 Ω	69.15	280.3 μH
Change	19.64%	1.14%	2.73%	4.15	0.71%
1 kHz (30 °C)	0.114 Ω	1.67 Ω	1.68 Ω	86.09	267.25 μH
1 kHz (90 °C)	0.123 Ω	1.69 Ω	1.69 Ω	85.84	268.97 μH
Change	7.89%	1.19%	0.59%	0.29	0.64%
100 kHz (30 °C)	2.94 Ω	148.55 Ω	148.57 Ω	88.86	236.41 μH
100 kHz (90 °C)	3.33 Ω	148.93 Ω	148.97 Ω	88.72	237.03 μH
Change	13.26%	0.25%	0.27%	0.15	0.26%

Table 5.12 HV winding capacitance ratio, antiresonance and resonance frequencies for quoted points in Figure 5.41(b) and (c)

Frequency	f_1 (30 °C)	f_2 (90 °C)
First minimum	1.223 MHz	1.184 MHz
First maximum	1.761 MHz	1.705 MHz
Second minimum	1.861 MHz	1.802 MHz
Second maximum	2.623 MHz	2.540 MHz
Third maximum	4.314 MHz	4.178 MHz
Third minimum	4.610 MHz	4.464 MHz
Fourth minimum	7.668 MHz	7.344 MHz
Fifth minimum	9.052 MHz	8.863 MHz
Sixth minimum	10.810 MHz	10.580 MHz
Seventh minimum	13.630 MHz	13.200 MHz
Fourth maximum	16.090 MHz	15.750 MHz

for HV and LV windings, respectively. These tables also show the measured values for windings' resistances, inductive reactances and impedances.

According to Tables 5.10 and 5.11, the maximum inductance deviation due to temperature changes is less than 1.1% for the entire measurement. Hence, it has insignificant impact on FRA resonant peaks; therefore, deviation of resonant points in FRA trace seems to be coming through changes in the total capacitance.

Total capacitance variation could be influenced by series and/or shunt capacitances alteration. Technically speaking, series and shunt capacitance can be altered due to the moisture migration. As a hypothesis, the shunt capacitance is more influenced by transformer oil insulation, and thus it could be more significant in changing the total capacitance when temperature and moisture change.

Tables 5.12 and 5.13 provide detailed information of deviated frequencies at 30 and 90 °C where f_1 and f_2 represent the same frequencies of the antiresonances and resonances for HV and LV windings at 30 and 90 °C, respectively.

Table 5.13 LV winding capacitance ratio, antiresonance and resonance frequencies for quoted points in Figure 5.42(b) and (c)

Frequency	f_1 **(30 °C)**	f_2 **(90 °C)**
First minimum	0.951 MHz	0.928 MHz
First maximum	1.122 MHz	1.096 MHz
Second maximum	4.543 MHz	4.367 MHz
Second minimum	4.624 MHz	4.464 MHz
Third maximum	6.963 MHz	6.722 MHz
Third minimum	8.039 MHz	7.676 MHz
Fourth minimum	13.240 MHz	13.170 MHz

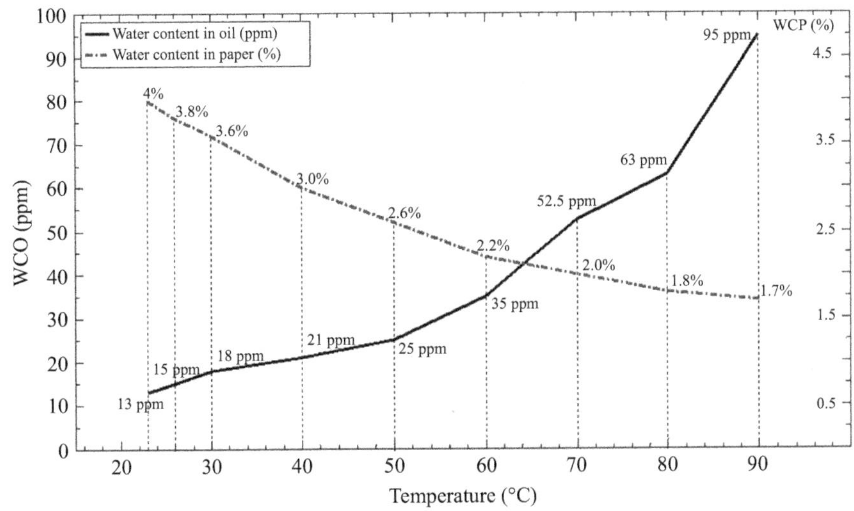

Figure 5.43 Moisture content of oil and paper (wet model transformer)

The moisture content of the oil and paper insulations for each and every temperature was recorded and shown in Figure 5.43. Moisture migration from the paper into the oil is clearly obvious in this figure. The water content of the oil has increased from 18 ppm at 30 °C to 95 ppm at 90 °C. In addition, the water content in paper (WCP) insulation has decreased from 3.6% at 30 °C to 1.7% at 90 °C.

Increasing the temperature will certainly cause the dielectric loss in paper and oil insulations to increase. Dielectric loss variations will marginally influence the magnitude of resonant peaks. Hence, some of the resonant peaks in Figures 5.41 and 5.42 are only slightly damped due to the temperature increase. Considerable dielectric loss will increase attenuation of the peaks.

Study on frequency response spectra when the temperature decreases from 90 to 30 °C led to similar results. Although moisture absorption for paper insulation is different with desorption, the equilibria should be the same and hence similar spectra were observed in the reverse procedure.

Example 5.6 A three-phase two-winding core type 20/0.4-kV, 1.6-MVA transformer was chosen as another test object to examine the effect of temperature and moisture content variation on the FRA spectrum. It was a spare transformer which was never in service and had been kept in stock as reserve. The windings are homogenous conventional disk type.

- ***Assessment***

At first, an oil sample was taken from the transformer sampling valve, and water content in oil (WCO) and WCP were derived using KFT method and MIT equilibrium curves, respectively (the oil temperature was 10 °C). These values were 3 ppm and 3%, respectively.

Frequency response traces for the HV windings were then recorded. To heat up the transformer in order to study the temperature and moisture impacts, the secondary side of the test object was short-circuited and the voltage was increased through the primary side to achieve the nominal current on the secondary side. It took 8 h to reach to 60 °C for the test object. After that, the test object was left for 48 h under this condition to reach moisture equilibrium between oil and paper insulations. The oil sample was then taken and FRA spectra were rerecorded. WCO achieved was 26 ppm and WCP was derived as 1.9%. Figure 5.44 shows the FRA spectra for phase U on HV side at 10 and 60 °C. In addition, Table 5.14 provides detailed information of the deviated frequencies in Figure 5.44.

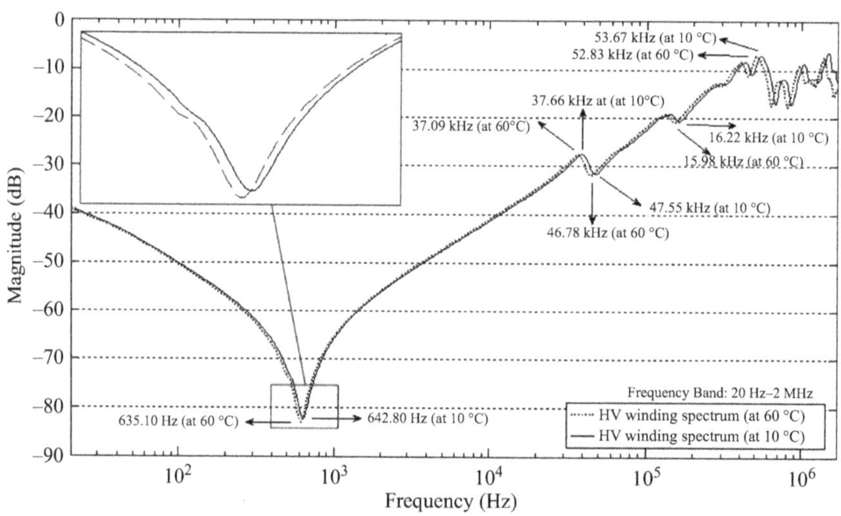

Figure 5.44 HV winding spectra at 10 and 60 °C (1.6-MVA transformer)

Table 5.14 HV winding capacitance ratio, antiresonance and resonance frequencies for quoted points in Figure 5.44

Frequency	f_1 (30 °C)	f_2 (90 °C)	f_1/f_2
First minimum	642.80 Hz	635.10 Hz	1.0121
First maximum	37.66 kHz	37.09 kHz	1.0153
Second minimum	47.55 kHz	46.78 kHz	1.0164
Third maximum	16.22 kHz	15.98 kHz	1.0150
Second maximum	53.67 kHz	52.83 kHz	1.0159

Based on the results, the total moisture variation for the paper insulation of the transformer is calculated as 1.1% from 10 to 60 °C. This variation has caused almost 1.5% alteration in the FRA trace.

5.8 Online transformer winding deformation diagnosis

Online monitoring and diagnosis of transformers have been investigated in the last decade. Obtaining real-time information about the transformer condition is crucial to ensure its reliable operation with minimum down-time and maintenance cost. In addition, having access to transformer real-time data will help to perform prognosis rather than diagnosis. This has been done for transformer oil insulation analysis; however, real-time assessment of transformer active part is still under development. In this section, different techniques for online transformer winding deformation diagnosis are introduced, and OFRA application is discussed in detail.

5.8.1 Methods for online transformer active part assessment

5.8.1.1 Vibration analysis

Transformer tank vibration has been recommended as an online transformer winding deformation diagnosis method. Transformer vibration can be considered as repetitive movement of transformer inner parts that are covered by the transformer tank. This movement oscillates around a reference position, i.e., the position when the transformer is at rest (deenergized). Vibration can be characterized using parameters such as winding displacement, velocity and acceleration.

Studies have shown that transformer tank vibration depends on voltage square and current square [29–31]. Furthermore, the main vibration frequency is twice (100 Hz) the fundamental power frequency (50 Hz). The core vibration, caused by magnetostriction and magnetic forces, is

$$v_{tank-100\text{ Hz}} = (\alpha + \beta\theta_{to})i_{100}^2 + (\gamma + \delta\theta_{to})u_{100}^2 \tag{5.55}$$

where $v_{tank\text{-}100\text{ Hz}}$ is 100 Hz frequency of tank vibration, i_{100}^2 is 100 Hz harmonic of the current square, u_{100}^2 is 100 Hz harmonic of the voltage square, θ_{to} is oil temperature measured at the top of the tank, and α, β, γ and δ are proposed coefficient in [29].

5.8.1.2 Communication method

Communication method [32] is based on scattering parameters. The magnitude and phase of scattering parameters for normal transformer winding are measured by several antennas as finger print. Proposed antennas could be placed outside or inside the transformer tank. In this method, the mean absolute magnitude distance and mean absolute phase distance are introduced as displacement indices, calculated according to (5.56) and (5.57), respectively.

$$MAMD = \frac{\sum_{i=1}^{n} \left||S_i| - |S_{ref}|\right|}{n} \tag{5.56}$$

$$MAPD = \frac{\sum_{i=1}^{n} \left|\measuredangle S_i - \measuredangle S_{ref}\right|}{n} \tag{5.57}$$

where S_i is the measured scatter parameters and S_{ref} is the reference scatter parameters.

As mentioned in [32], any kind of transformer winding deformation can cause changes to the abovementioned indices and so deformation can be detected.

5.8.1.3 Current deformation coefficient (CDC) method

The CDC method was introduced by [33,34]. A HF low-voltage signal is applied to live power system line. The line-end and neutral-end HF currents are continuously measured using isolated precision current probes and digital filtering technique [34]. Any change in the capacitive reactance due to transformer winding deformation is reflected in deviations of HF terminal currents from the fingerprint. When these deviations are measured, the ratio of deviations at the two ends is calculated. Hence, the CDC is introduced and calculated as follows [34]:

$$CDC = \log_{10}\left(\frac{I_{1H} - I'_{1H}}{I_{2H} - I'_{2H}}\right) \tag{5.58}$$

where I_{1H} and I_{2H} is the fingerprint values of measured terminal currents at the selected HFs, and I'_{1H} and I'_{2H} are the Terminal current values at the winding terminals after deformation.

5.8.1.4 Ultrasonic method

Ultrasound refers to sound with a frequency greater than the upper limit of human hearing (about 20 kHz). In this method [35], an ultrasonic signal is used as the reference signal. The basis of this method concentrates on ultrasound reflection due to the nonmatching acoustic impedance between the oil and the winding.

5.8.1.5 Short-circuit impedance (SCI) and winding stray reactance methods

This method relies on time-based comparison. The newly measured transformer SCI is compared to the value previously recorded [36–44]. Considering (5.5), the SCI of a transformer is related to the windings configuration and distance between windings. Offline transformer SCI measurement is performed when the secondary

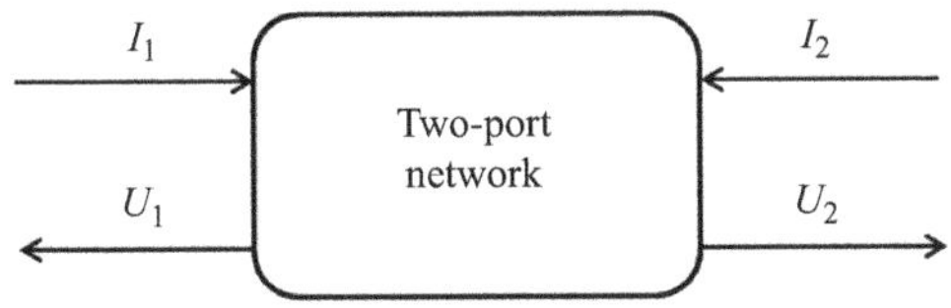

Figure 5.45 Two-port network

side of transformer is shorted and voltage excitation is applied at the primary side. Since secondary short-circuit setup is not possible to be implemented for an energized transformer, the following approach for online SCI calculation is utilized. A two-port network as a model for transformer is considered, see Figure 5.45.

The relationship between input and output signals is given by

$$\begin{pmatrix} U_1 \\ U_2 \end{pmatrix} = \begin{pmatrix} Z_{11} & Z_{12} \\ Z_{21} & Z_{22} \end{pmatrix} \begin{pmatrix} I_1 \\ I_2 \end{pmatrix} \tag{5.59}$$

where Z_{11} is the open circuit input impedance, Z_{12} is the open circuit reverse transfer impedance, Z_{22} is the open circuit output impedance and Z_{21} is the open circuit forward transfer impedance.

When short-circuit happens, U_2=0 and (5.59) will be recalculated as follows:

$$\begin{aligned} U_1 &= Z_{11}I_1 + Z_{12}I_2 \\ 0 &= Z_{21}I_1 + Z_{22}I_2 \end{aligned} \tag{5.60}$$

Hence the SCI (Z_{sc}) is given by

$$Z_{sc} = Z_{11} - \frac{Z_{12}Z_{21}}{Z_{22}} \tag{5.61}$$

Any deviation in Z_{11}, Z_{22}, Z_{12} and Z_{21} will affect Z_{sc}. Apply (5.59) at two different time instants:

$$\begin{cases} U_{1_{t1}} = Z_{11}I_{1_{t1}} + Z_{12}I_{2_{t1}} \\ U_{2_{t1}} = Z_{21}I_{1_{t1}} + Z_{22}I_{2_{t1}} \end{cases} \tag{5.62}$$

$$\begin{cases} U_{1_{t2}} = Z_{11}I_{1_{t2}} + Z_{12}I_{2_{t2}} \\ U_{2_{t2}} = Z_{21}I_{1_{t2}} + Z_{22}I_{2_{t2}} \end{cases} \tag{5.63}$$

where $U_{1_{t1}}$ and $U_{1_{t2}}$ are the primary voltages in Figure 5.45 corresponding to t_1 and t_2, respectively; $I_{1_{t1}}$ and $I_{1_{t2}}$ are the primary currents corresponding to t_1 and t_2 and $U_{2_{t1}}$ and $U_{2_{t2}}$ are the secondary voltages corresponding to t_1 and t_2 and $I_{2_{t1}}$ and $I_{2_{t2}}$ are the secondary currents corresponding to t_1 and t_2; t is the time of measurement.

From (5.62) to (5.63), all impedances can be calculated by using

$$\begin{pmatrix} Z_{11} & Z_{12} \\ Z_{21} & Z_{22} \end{pmatrix} = \begin{pmatrix} U_{1_{t1}} & U_{1_{t2}} \\ U_{2_{t1}} & U_{2_{t2}} \end{pmatrix} \begin{pmatrix} I_{2_{t2}} & -I_{1_{t2}} \\ -I_{2_{t1}} & I_{1_{t1}} \end{pmatrix} \cdot \left(\frac{1}{I_{1_{t1}}I_{2_{t2}} - I_{2_{t1}}I_{1_{t2}}} \right) \tag{5.64}$$

When the matrix on the left hand side in (5.64) changes, Z_{sc} will change. Online SCI measurement has been utilized as an economical method for transformer winding deformation diagnosis. However, some researchers have raised concern about the accuracy of this method.

5.8.1.6 Voltage–current locus diagram

The *V–I* locus diagram to detect internal transformer fault online was introduced in [45]. In contrast to the FRA method covering a wide frequency range (20 Hz–2 MHz), this technique relies on the power frequency (50/60 Hz) signals from which the mathematical concept is formulated to analyze transformer mechanical integrity. The voltage and current measured at transformer bushing terminals are used to construct the *V–I* ellipse. In [45], extensive simulation results on disk space variation, axial displacement, buckling stress, interdisk fault, leakage (disk to ground) fault were presented and analyzed with the *V–I* locus technique.

5.8.1.7 Transfer function method

Whilst offline transfer function measurement is already established as a common method for transformer winding deformation diagnosis, online transfer function measurement is still very much under development. Since most power transformers utilize capacitive bushings, the bushing tap can be exploited as an input point for low-voltage signal injection during online transfer function (TF) measurement [40,41,46]. Figure 5.46 shows the side cutoff of a capacitive bushing. The bushing tap is connected to the last layer of capacitive grading which is brought out insulated at the bushing flange via a small auxiliary bushing. This tap provides a much

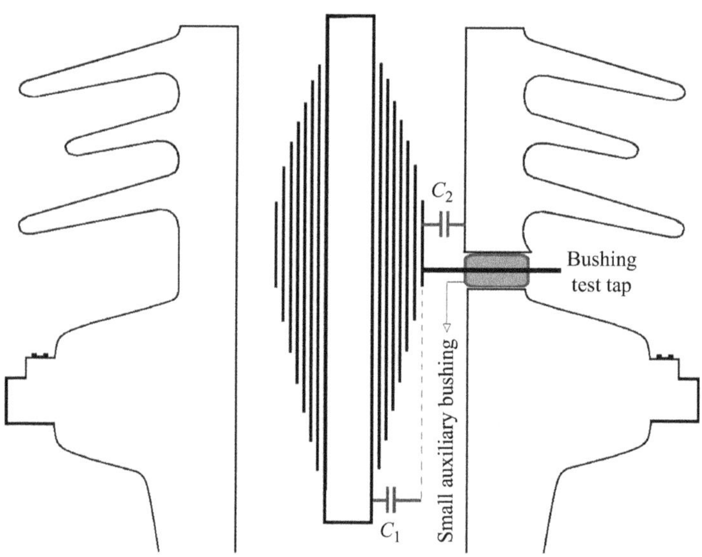

Figure 5.46 Side cutoff of a capacitive bushing

Table 5.15 Typical values for bushing capacitances

Rating (kV)	C_1 (pF)	C_2 (pF)	Ratio (C_2/C_1)	Tap voltage (kV)
230	608	6,192	10.18	20.56
500	498	10,021	20.12	23.67

reduced terminal voltage due to the capacitive divider. In fact during operation, the transformer bushing tap is grounded through the cover cap.

Online TF measurement requires opening the grounded bushing tap so to be able to inject signal into the transformer without requiring a direct connection to the main feeder. Hence, C_1 and C_2 values as a capacitive divider must be taken into consideration. Table 5.15 shows typical capacitance values of 230 and 400 kV bushings.

Measured values of C_1 and C_2 by bushing manufacturers show that the division ratio results in high voltage at the bushing tap when it is not grounded. Obviously, the creepage distance and air clearance between the bushing tap and grounded flange is not coordinated to withstand against the estimated voltage as calculated using (5.65). Also, the instrument to generate the injection signal cannot tolerate the high voltage either.

$$V_{tap} = \frac{C_1}{C_2 + C_1} V_{phase} \tag{5.65}$$

where V_{tap} is the bushing tap voltage and V_{phase} is the phase voltage.

Therefore, during online measurement when the bushing tap is not grounded, it is compulsory to shunt appropriate impedance in parallel with C_2 as illustrated in Figure 5.47. With this arrangement, the voltage at the bushing tap is

$$V_{tap} = \frac{Z_p||1/j\omega C_2}{Z_p||1/j\omega C_2 + 1/j\omega C_1} V_{phase} \tag{5.66}$$

where Z_p impedance in parallel with bushing tap and ω is the angular frequency.

Furthermore, the capacitance values of the bushing remain relatively constant over a wide frequency range, which should limit the effect of masking the actual transformer signature due to the bushings' own frequency response [41,46,47].

Wye and delta circuits for online TF measurement are shown in Figures 5.48 and 5.49, respectively. For wye connection shown in Figure 5.48, online TF can be measured by signal injection at the phase bushing tap, and response can be recorded through the neutral bushing tap. For delta-type connection shown in Figure 5.49, it can be measured between two phases.

5.8.2 Online FRA setup

In the previous section, it was supposed that an impulse signal is injected into the transformer bushing tap and the response is recorded in time-domain. Alternatively,

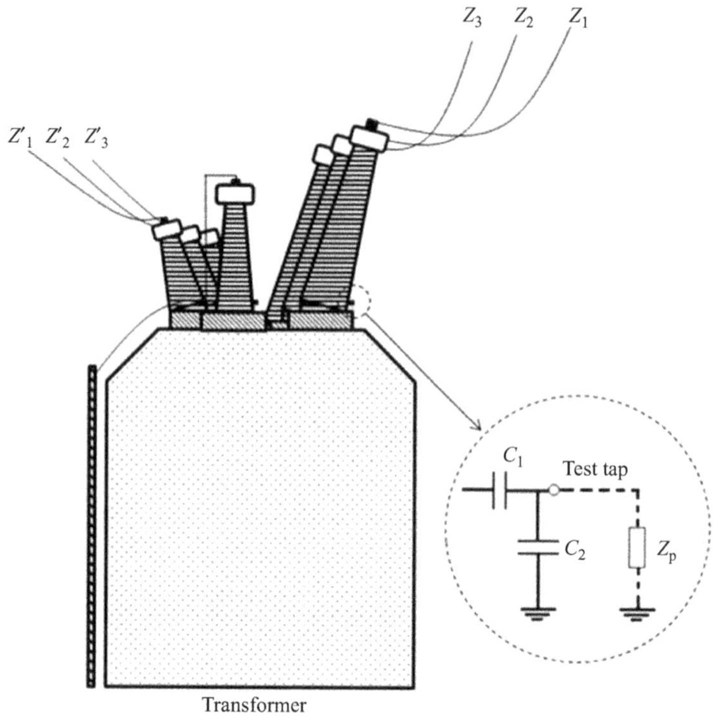

Figure 5.47 Paralleled impedance with bushing tap (test tap) on phase U

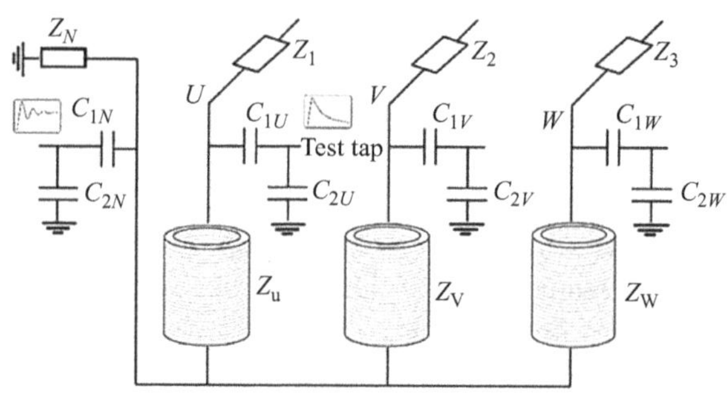

Figure 5.48 Wye circuits for online TF measurement

a sinusoidal signal source can be used for injection to obtain the response at one discrete frequency and then repeat at other frequencies to obtain the full FRA spectrum. The setup for online frequency response measurement is illustrated in Figure 5.50.

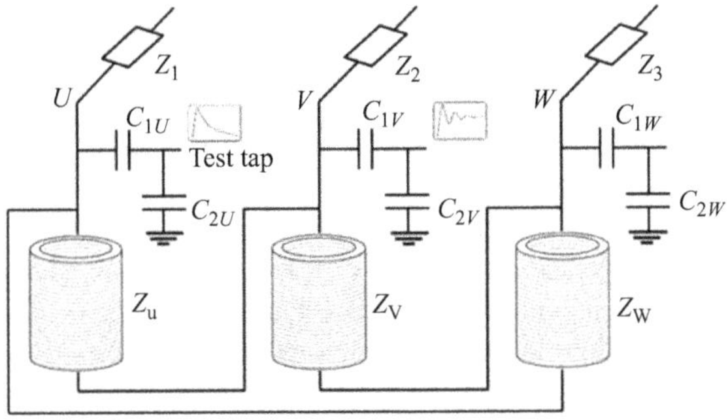

Figure 5.49 Delta circuits for online TF measurement

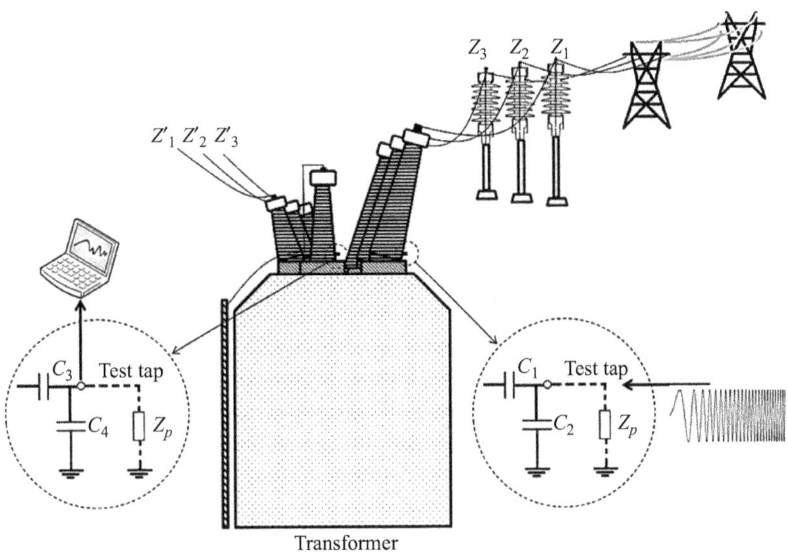

Figure 5.50 Online frequency response analysis (OFRA) setup, end-to-end measurement

5.8.3 Online FRA (OFRA) progress and influence of bushing tap

OFRA has its own practical challenges to overcome for proper implementation. Some of the challenges are listed as follows:

- The crucial one is validity of OFRA measurement results. Transformer bushing HV terminals are connected to the network. Here, the problem is how to separate the response of external systems such as connections, busbars, feeders from the winding frequency response.
- With offline FRA measurements, it has already been explored that the transfer function of the primary side of a transformer is different for open circuit and

short-circuit configuration of the secondary in LFs. With OFRA measurement, the transformer is energized and secondary side is normally loaded. Loading variation during operation is inevitable which can be considered as a parallel variable impedance on the transformer secondary side. This in turn will result in variations of OFRA data.

- OFRA needs to be performed separately for each winding. In fact, signal injections for three-phase transformers must not be exercised simultaneously to avoid the overlap and superposition disturbance. This can be achieved by using a multiplexer for signal injection to each phase from a single signal generator.
- The effects of the input signal the on protection system while OFRA is conducted should be investigated in detail. As the voltage of the injected signal could be about 230 V [40], its influence would be negligible.
- A number of studies have suggested that high frequency current transformer (HF CT) should be used as an output probe at the neutral point if available. In this case, the CT frequency response would cause additional error in results and increase the uncertainty of the main winding's transfer function. The error and uncertainty formulas are

$$e(TF) = e(winding) + e(CT) \tag{5.67}$$

$$u(TF) = \sqrt{u(Winding)^2 + u(CT)^2} \tag{5.68}$$

where e is the measurement error and u is the estimated uncertainty.

- As discussed, OFRA measurement would be applicable for HV side of transformer through the bushing tap. As for the LV side, oil bushing is often used which does not have bushing tap. A Rogowski coil can be used as the sensor [47], and the output voltage is as follows:

$$V = -\frac{N\mu_0 A}{L}\frac{dI}{dt} \tag{5.69}$$

where N is the number of coil turns for Rogowski coil, L is the length of the winding and A is the cross-section of each small loop in Rogowski coil.

This formula assumes the turns are evenly spaced, and that these turns are small relative to the radius of the coil itself. In addition, since a Rogowski coil has an air core rather than an iron core, it has a low inductance and can respond to fast-changing currents. Also, because it has no iron core to saturate, it is highly linear even when subjected to large currents. This coil would be appropriate regarding to immunity against electromagnetic interference because of uniformly spaced winding. However, its frequency response would interfere with the transformer winding transfer function. This interference should be minimized as much as possible.

- Since the OFRA data of HV and LV transformer windings are affected by self and mutual inductances as well as series and ground capacitances, any changes in these elements will lead to OFRA data alteration. Offline FRA is mostly affected by self-inductance, series and ground capacitances. In offline measurement, the test voltage is often applied to primary winding (HV or LV), whilst the secondary winding (LV or HV) is open circuit, and without carrying any alternating current, mutual inductances do not contribute to FRA spectrum formation.

References

[1] E. P. Dick and C. C. Erven, 'Transformer diagnostic testing by frequency response analysis', IEEE Trans. Power App. Syst., 1978, 97(6), pp. 2144–2150.

[2] M. Bagheri, M. S. Naderi, T. Blackburn and B. T. Phung, 'Frequency response analysis and short circuit impedance measurement in detection of winding deformation within power transformers', IEEE Electr. Insul. Mag., 2013, 29(3), pp. 33–40.

[3] A. Abu-Siada, N. Hashemnia, S. Islam and M. A. S. Masoum, 'Understanding power transformer frequency response analysis signatures', IEEE Electr. Insul. Mag., 2013, 29(3), pp. 48–56.

[4] S. V. Kulkarni and S. A. Khaparde, Transformer Engineering Design and Practice, Marcel Dekker, Inc., New York, USA, 2004.

[5] R. Doebbelin and A. Lindemann, 'Leakage inductance determination for transformers with Interleaving of Windings', Progress in Electromagnetics Research Symposium, Cambridge, USA, July 5–8, 2010, pp. 439–443.

[6] M. Jablonski and E. Napieralska-Juszczak, 'Internal faults in power transformers', IET Electr. Power Appl., 2007, 1(1), pp. 105–111.

[7] IEEE Guide for Diagnostic Field Testing of Electric Power Apparatus – Part 1: Oil Filled Power Transformers, Regulators, and Reactors. In *IEEE Std 62-1995*, pp. 1–68, December 1995, doi: 10.1109/IEEESTD.1995.122640.

[8] Power transformers – Part 5: Ability to withstand short circuit. In *IEC 60076-5:2006*, pp. 1–71, February 2006.

[9] Power Transformers – Part 18: Measurement of Frequency Response. In *IEC 60076-18:2012*, pp. 1–90, July 2012.

[10] IEEE Guide for the Application and Interpretation of Frequency Response Analysis for Oil-Immersed Transformers. In *IEEE Std C57.149-2012*, pp. 1–72, March 2013. doi: 10.1109/IEEESTD.2013.6475950.

[11] P. M. Nirgude, D. Ashokraju, A. D. Rajkumar and B. P. Singh, 'Application of numerical evaluation techniques for interpreting frequency response measurements in power transformers', IET Sci. Meas. Technol., 2008, 2(5), pp. 275–285.

[12] J. R. Secue and E. Mombello, 'Sweep frequency response analysis (SFRA) for the assessment of winding displacements and deformation in power transformers', Electr. Power Syst. Res., 2008, 78(6), pp. 1119–1128.

[13] S. A. Ryder, 'Methods for comparing frequency response analysis measurements', IEEE Int'l Sympos. Electr. Insul., Boston, MA, USA, 2002, pp. 187–190.

[14] G. M. Kennedy, A. J. McGrail and J. A. Lapworth, 'Using cross-correlation coefficients to analyze transformer sweep frequency response analysis

(SFRA) traces', IEEE PES Power Africa, Johannesburg, South Africa, July 16–20, 2007, pp. 1–6.

[15] Frequency response analysis on winding deformation of power transformers. In *Professional Standard of the People's Republic of China, DL/T 911–2004*. ICS27.100, F24, Document No. 15182-2005, June 1, 2005. Available: https://www.chinesestandard.net/PDF-Excerpt/ShowPDFexcerpt.aspx?ExcerptID=DL/T%20911-2004.

[16] A. Kraetge, M. Kruger and P. Fong, 'Frequency response analysis – Status of the worldwide standardization activities', Int'l Conf. Condition Monitoring and Diagnosis, Beijing, China, April 21–24, 2008, pp. 651–654.

[17] R. Wimmer, S. Tenbohlen, M. Heindl, A. Kraetge, M. Kruger and J. Christian, 'Development of algorithms to assess the FRA', 15th Int. Symp. High Voltage Engineering, Paper No. T7-523, Ljubljana, Slovenia, 2007.

[18] J. C. Maxwell, A Treatise on Electricity and Magnetism. Vol. 2, 3rd Ed., Clarendon Press, Oxford, 1892, Section 696–701, p. 334, p. 336, p. 339.

[19] C. R. Paul, Inductance Loops and Partial, 1st Ed., Published by John Wiley, Hoboken, NJ, 2009.

[20] K. Karsai and D. Kerenyi, Large Power Transformer, Elsevier, New York, 1987, pp. 187–214.

[21] D. F. Garcia, B. Garcia and J. C. Burgos, 'A review of moisture diffusion coefficients in transformer solid insulation – Part 1: Coefficients for paper and pressboard', IEEE Electr. Insul. Mag., 2013, 29(1), pp. 46–54.

[22] B. Garcia, J. C. Burgos, A. M. Alonso and J. Sanz, 'A moisture-in-oil model for power transformer monitoring – Part I: Theoretical foundation', IEEE Trans. Power Delivery, 2005, 2(20), pp. 1417–1422.

[23] M. Bagheri, B. T. Phung, T. Blackburn and A. Naderian, 'Influence of temperature on frequency response analysis of transformer winding', IEEE International Electrical Insulation Conference (EIC'13), Ottawa, Canada, June 2–5, 2013, pp. 40–44.

[24] W. Guidi and H. Fullerton 'Mathematical methods for prediction of moisture take-up and removal in large power transformers', Proceedings of IEEE Winter Power Meeting, No. C-74, 1794, pp. 242–244.

[25] Y. Du, A. V. Mamishev, B. C. Lesieutre and M. Zahn, 'Measurement of moisture diffusion as a function of temperature and moisture concentration in transformer pressboard', IEEE Conf. Electr. Insul. Dielectr. Phenomena, Annual Report, Vol. 1, 1998, pp. 341–344.

[26] Y. Du, M. Zahn, B. C. Lesieutre, A. V. Mamishev and S. R. Lindgren, 'Moisture equilibrium in transformer paper-oil systems', IEEE Electr. Insul. Mag., 1999, 15(1), pp. 11–20.

[27] L. Jian, Z. Zhaotao, S. Grzybowski and L. Yu, 'Characteristics of moisture diffusion in vegetable oil–paper insulation', IEEE Trans. Dielectr. Electr. Insul., 2012, 19(5), pp. 1650–1656.

[28] P. J. Griffin, 'Water in transformers-so what!', National Grid Condition Monitoring Conf., 1996, pp. 1–16.

[29] B. García, J. C. Burgos and A. M. Alonso, 'Winding deformations detection in power transformers by tank vibrations monitoring', Electr. Power Syst. Res., 2004, 74(1), pp. 129–138.

[30] B. García, J. C. Burgos and A. M. Alonso, 'Transformer tank vibration modeling as a method of detecting winding deformations – Part I: Theoretical foundation', IEEE Trans. Power Delivery, 2006, 21(1), pp. 157–163.

[31] B. García, J. C. Burgos and A. M. Alonso, 'Transformer tank vibration modeling as a method of detecting winding deformations – Part II: Experimental verification', IEEE Trans. Power Delivery, 2006, 21(1), pp. 164–169.

[32] M. A Akhavanhejazi, G. B. Gharehpetian, R. Faraji-Dana, G. R. Moradi, M. Mohammadi and H. A. Alehoseini, 'A new on-line monitoring method of transformer winding axial displacement based on measurement of scattering parameters and decision tree', Expert Syst. Appl., 2011, 38(7), pp. 8886–8893.

[33] P. M. Joshi and S. V. Kulkarni, 'Transformer winding diagnostics using deformation coefficient', IEEE Power and Energy Soc., General Meeting, Conversion and Delivery of Electrical Energy in the 21st Century, 2008, pp. 1–4.

[34] P. M. Joshi and S. V. Kulkarni, 'A novel approach for online deformation diagnostics of transformer windings',IEEE Power and Energy Society, General Meeting, 2010, pp. 1–6.

[35] N. Shu, C. Zhou, F. Hu, Q. Liu and L. Zheng, 'Study on ultrasonic measurement device for transformer winding deformation',IEEE Int'l. Conf. Power Syst. Techn., (POWERCON), Vol. 3, 2002, pp. 1401–1404.

[36] M. Alpatov, 'On-line detection of winding deformation', IEEE Int'l. Sympos. Electr. Insul., 2004, pp. 113–116.

[37] S.-B. Gao and G. Wang, 'Study on on-line monitoring of windings deformation of power transformer',IEE 8th Int'l. Conf. Developments in Power System Protection, Vol. 1, 2004, pp. 335–338.

[38] G. Hu, L. Zhang, X. Wu, D. Correia and W. He, 'Detecting the capacity of distribution transformer based on an on-line method', Asia-Pacific Power and Energy Eng. Conf. (APPEEC), 2011, pp. 1–4.

[39] A. Palani, S. Santhi, S. Gopalakrishna and V. Jayashankar, 'Real-time techniques to measure winding displacement in transformers during short-circuit tests', IEEE Trans. Power Delivery, 2008, 23(2), pp. 726–732.

[40] T. De Rybel, A. Singh, J. A. Vandermaar, M. Wang, J. R. Marti and K. D. Srivastava, 'Apparatus for online power transformer winding monitoring using bushing tap injection', IEEE Trans. Power Delivery, 2009, 24(3), pp. 996–1003.

[41] M. Bagheri, M. S. Naderi and T. Blackburn, 'Advanced transformer winding deformation diagnosis: Moving from off-line to on-line', IEEE Trans. Dielectr. Electr. Insul., 2012, 19(6), pp. 1860–1870.

[42] N. Pinhas, S. Islam and J. Hullett, 'On the development of transfer function method for fault identification in large power transformers on load', IEEE Conf. Electr. Insul. Dielectr. Phenomena (CEIDP), Vol. 2, 2000, pp. 747–751.

[43] D. K. Xu and Y. M. Li, 'A simulating research on monitoring of winding deformation of power transformer by on-line measurement of short-circuit reactance', IEEE Int'l. Conf. Power System Technology (POWERCON), Vol. 1, 1998, pp. 167–171.

[44] Z. Berler, P. Butyrin and M. Alpatov, 'On-line diagnostics of mechanical deformations of winding', Proceedings of TechCon'05, 2005, pp. 1–4.

[45] A. Abu-Siada and S. Islam, 'A novel online technique to detect power transformer winding faults', IEEE Trans. Power Delivery, 2012, 27(2), pp. 849–857.

[46] A. Setayeshmehr, H. Borsi, E. Gockenbach and I. Fofana, 'Online monitoring of transformer via transfer function', IEEE Electr. Insul. Conf. (EIC), 2009, pp. 278–282.

[47] A. Setayeshmehr, A. Akbari, H. Borsi and E. Gockenbach, 'On-line monitoring and diagnoses of power transformer bushings', IEEE Trans. Dielectr. Electr. Insul., Vol. 13, 2006, pp. 608–615.

Chapter 6

Monitoring of power transformers by mechanical oscillations

Michael Beltle[1]

6.1 Introduction

Condition-based asset management is gaining importance, which leads to a prospering of power transformer monitoring [1,2]. Different methods, for example the dissolved-gas-analysis and oil-aging evaluation [3], partial discharge measurements [4–6] and also frequency response analysis [7] have been established to determine the status of the transformer's electric insulation and windings.

Mechanical oscillations of transformers during operation are mostly recognised for being the cause of transformer noises [8]. They are taken into consideration during the design process in order to avoid core geometries which would enforce mechanical resonances at excitation frequencies and hence could lead to critical mechanical forces to the structure or undesirable high-noise levels [9]. This chapter is aimed at providing an overview of transformer's vibration including origins and sources, signal propagation, practical measurements techniques and the diagnostic interpretation in terms of continuous monitoring. In most investigations, accelerometers on the tank surface are used, but some research also considered local measurements directly on winding or core using optical strain sensors [10,11].

Concerning asset management, the question arises if they can supplement transformer diagnosis in terms of transformers' mechanical status during service. In theory, vibrations during operation allow to survey mechanics, since they are either originated by the structure (windings and core) or lie within the signal path (like the mounting of the active part on the tank top). This chapter evaluates the practical value by a laboratory setup and long-term measurements in the field on a transformer in service in accordance with different publications showing the correlation between mechanical status and oscillations [12–15].

The last part of this chapter correlates vibrations to direct current (DC) components running through transformers' grounded star points impacting the magnetic operational status. In addition, vibration measurements are used to estimate the

[1]Institute of Power Transmission and High Voltage Technology, Universität Stuttgart, Germany

resulting changes of transformer's noise due to DC which gains importance for power transformers and in particular low-noise transformers [16].

6.2 Physics of mechanical oscillations

The mechanical oscillations of a transformer originate from the periodical movement of windings and the core plates under the influence of the magnetic field. The signals couple through the transformer as structure- and fluid-borne sound (longitudinal and transversal waves) into the tank. The tank wall, the transformer's baseplate, etc., act as speakers and emit audible noise.

6.2.1 Oscillations of the core

The core movement is driven by magnetostriction [17]. The crystal structure of the core's electric steel plates consists of Weiss' domains. Each domain has its own magnetic polarisation I_{mag}. In a non-magnetised ferromagnetic material all polarisations superimpose destructively and the macroscopic behaviour is neutral, see Figure 6.1(a). In the presence of an external applied magnetic field H, the domains align themselves along this field. Every domain claims a specific volume and size in the material. Depending on the steel type and treatment during production, domain sizes can differ. If the externally applied magnetic field H alternates, the domains follow the field. This is obtained by two different effects; the movement of

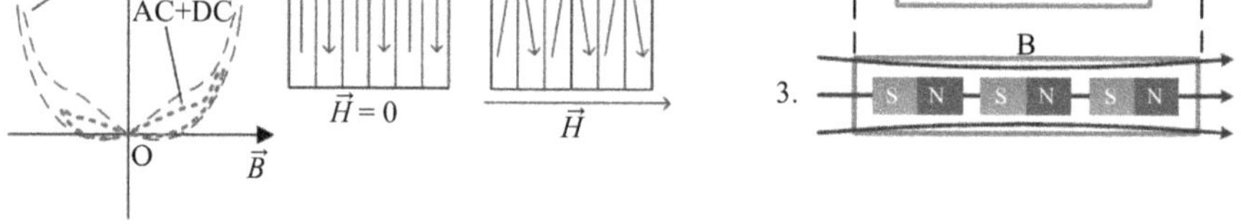

Figure 6.1 Left: (a) Schematic of ideal ferromagnetic cube with Weiss' Domains orientated in closed flux circles without stray flux. (b) Bloch–Wall displacement caused by external applied magnetic field H. (c) General resulting alternation of length in one direction at AC only and superimposed AC + DC condition. Right: Rotation of elementary magnets caused by external applied magnetic flux during one period. (1) and (2) represent maxima in each half period. (d) Rotation of Weiss' Domains by external field H

their boundaries and the Bloch–Walls, see Figure 6.1(b), or by a rotation of the domain, see Figure 6.1(d). The movement of Bloch–Walls causes domains aligned along the external field to gain volume. Rotation does not change the volume of a single domain. The summarised volume of all domains remains constant, thus wall movement and rotation lead to a mechanical deformation. Both effects require energy, which is represented as part of the core losses. In a Gedankenexperiment, Bloch–Wall movement and domain rotation can be merged into rotating elementary magnets.

If the applied external field alternates sinusoidal, so do the elementary magnets align themselves. Hence, in each period of the magnetic flux, each elementary magnet has a certain orientation twice, which leads to a mechanical oscillation with doubled electric frequency. For example an electrical frequency of $f_{n,\text{electric}} =$ 50 Hz leads to an equal magnetic frequency $f_{n,\text{magnetic flux}} = 50$ Hz but to a mechanical oscillation at $f_{n,\text{core}} = 100$ Hz, see Figure 6.1(b), right side which involves the following three steps:

- Step 1: Due to the external magnetic flied, magnetic polarisation to the right will take place. The magnetic polarisation forces the domains to align themselves as shown in accordance with the field.
- Step 2 shows the transition during the change of the external field. In this process, the domains re-orientate themselves which means they start mechanically rotating. This leads to a reduction in length Δl.
- Step 3: The magnetic flux density is oriented anti-parallel to step 1. The original length is retained and the mechanic conditions of steps 1 and 2 are the same, although magnetic conditions are different.

The one-dimensional mechanic deformation or strain depending on the flux density can be visualised using the butterfly plot shown in Figure 6.1(c) [18]. The y axis (strain) represents the change of length in one direction. For only AC conditions, the butterfly plot is symmetric. The two maxima during one electrical period indicate the frequency doubling of the effect. If a constant magnetic field is superimposed, e.g. by a DC component through a transformer winding, the plot gets asymmetrical which also leads to a different frequency spectrum. Details of this effect are discussed in Section 6.7.

6.2.2 *Oscillations of the windings*

The winding movement is driven by the Lorentz force F occurring between the windings. It depends on the load current $i(t)$ as shown in (6.1), where F is the resulting Lorentz force, l the length of the considered part of winding, ρ the distance between the conducting windings and $\omega = 2\pi f$ is the angular frequency.

$$|F| = \frac{\mu_0 \mu_r i^2(t)}{2\pi\rho} l, \quad i(t) = \hat{\imath}\sin(\omega t) \tag{6.1}$$

$$|F| = \frac{\mu_0 \mu_r}{2\pi\rho} l \left(\hat{\imath}^2 \frac{1 - \cos(2\omega t)}{2} \right) \tag{6.2}$$

Given a sinusoidal current, the quadratic term can be written as in (6.2), where the angular frequency is doubled. Therefore, the Lorentz force has two maxima during one electrical period, which means current driven mechanical oscillations do also have the doubled electrical frequency as basic frequency.

The sources of winding and core vibrations provide a very similar frequency spectrum which are superimposed when measured e.g. on the tank surface. With rising load current, winding vibrations tend to be the dominating source. The entire spectrum consists of the basic mechanical frequency and its harmonics which occur due to non-linear effects of both sources and structure-dependent influences.

6.3 Measurement of vibrations

A common way to measure mechanical oscillations is performed by a piezoelectric accelerometer as shown in the schematic plot of Figure 6.2. The sensor is fixed on the reference surface where its lower part moves with the surface, if a mechanical force F is applied. The seismic mass element is an inertial mass and does not follow the movement. Hence, the piezoelectric material in between is compressed which leads to an electric potential difference between the contacts. The sensor shown in the figure is one dimensional and is only sensitive to forces perpendicular to the reference surface.

The generated electric potential between the contacts depends on the used material properties and the compressed volume which are included into (6.3) by a constant proportional factor k_s. The measured voltage can be derived from the generated charge and the overall capacity of the sensor which includes the plate capacity between the electrodes C_0 and all occurring stray capacities C_s.

$$u = \frac{q}{C} = \frac{k_s F}{(C_0 + C_s)} \tag{6.3}$$

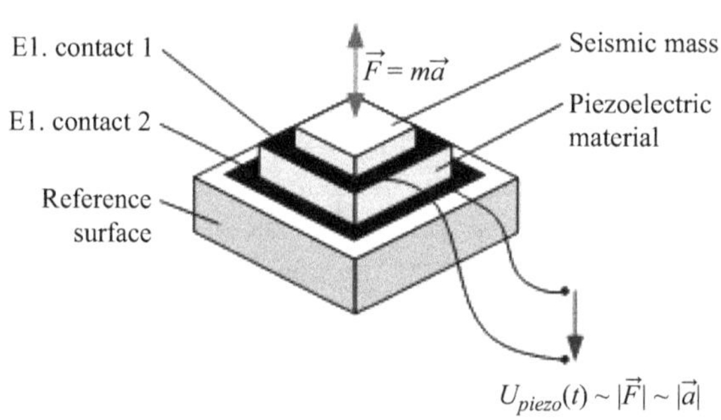

Figure 6.2 Working principle and different layers of a one-dimensional piezoelectric sensor used to measure acceleration orthogonal to the reference surface

A common sensor frequency range is similar to the hearable audio spectrum with a 3-dB corner frequency at approx. $f_{3\,\mathrm{dB}} = 20$ kHz with an average sensitivity of $k_s \approx 10$ mV/(m/s^2). As Figure 6.3 shows, mechanical oscillations of significant amplitude can be observed in a frequency range up to 1,000 Hz. The shown measurement is taken from a generator step-up transformer during service. Usually, mechanical frequencies $f_{\mathrm{mech}} < 1$ kHz cover about 95% of the entire signal power.

In the presented cases, vibrations are measured in the field using the introduced one-dimensional acceleration sensor attached on the transformer tank's wall. The mechanical displacement normal component of the wall is recorded in time domain. Figure 6.4 shows a typical measurement setup for onsite, online operation

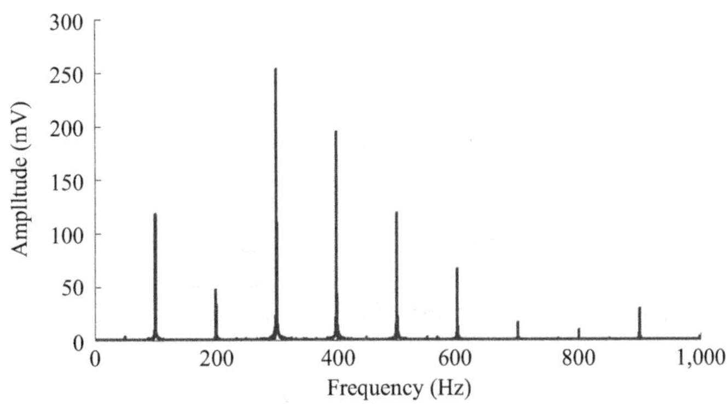

Figure 6.3 Frequency domain signal of raw data output of a vibrations measurement system

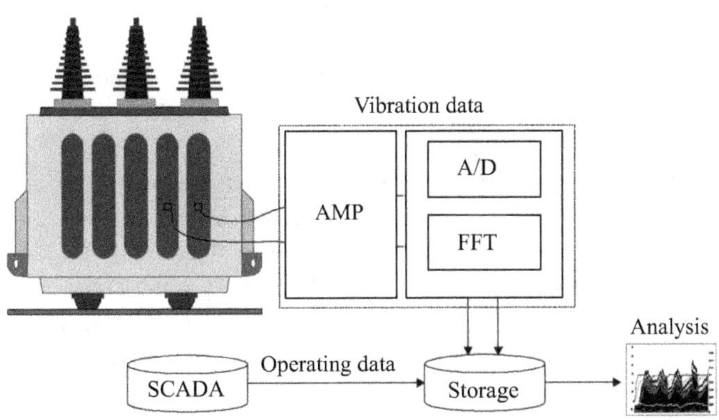

Figure 6.4 Typical measurement setup for tank wall vibration measurements. Signals from the accelerometers are amplified and analogue to digital concerted into the frequency spectrum. Post-processing allows correlation to SCADA data

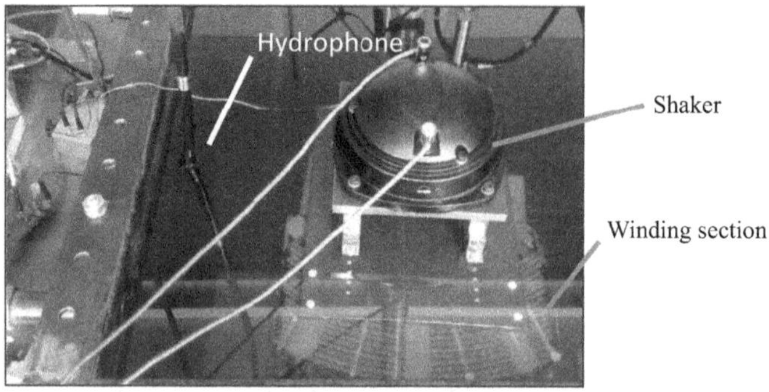

Figure 6.5 Measurement setup with excited winding part, in-oil sensor (sensor)

with two sensors. Laboratory measurements use an identical setup, without the supervisory control and data acquisition (SCADA) system. The low voltage sensor signals generated by the piezo-effect require low-noise amplification. Afterwards, signals can be digitised and converted into frequency domain using Fast Fourier Transformation (FFT).

The influence of the transformer's load and the position of the tap changer will be discussed in the following sections. If the measurement is performed regularly, the introduced system is a monitoring device allowing vibration behaviour surveillance and change-detection on long-time basis.

6.3.1 Comparison of tank wall and in-oil measurement

Advantages of outside tank wall measurement include the utilisation of low equipment costs as well as easy and fast applicability. Yet, the influence of the tank on the signal has to be taken into consideration. Therefore, an alternative method is briefly introduced. Sonic waves are measured directly in oil using a hydrophone sensor. The hydrophone is also a piezoelectric sensor, but it detects changes in the oil-pressure instead of acceleration. A comparison of both methods is performed in a test tank using a small winding section which is excited with a shaker at a designated mechanical frequency, see Figure 6.5.

Figure 6.6 illustrates that the amplitudes measured by accelerometer and hydrophone are not identical, which is attributed to the sensor sensitivities at identical signal amplification. In this experiment the tank wall does not significantly change vibrations in terms of frequency distortion. Nevertheless, practical experiences show that the sensor position on the outside-tank wall and hence tank structure of power transformers also effect signal strength (see Section 6.4.2).

In principle, in-oil measurements provide a better signal-to-noise ratio which is independent on the tank-wall structure. However, the main disadvantage of this technique lies in the difficulty to use sensors in operating power transformers, because only limited access by the number of available oil vales is given. Hence, they are not used in practice, yet.

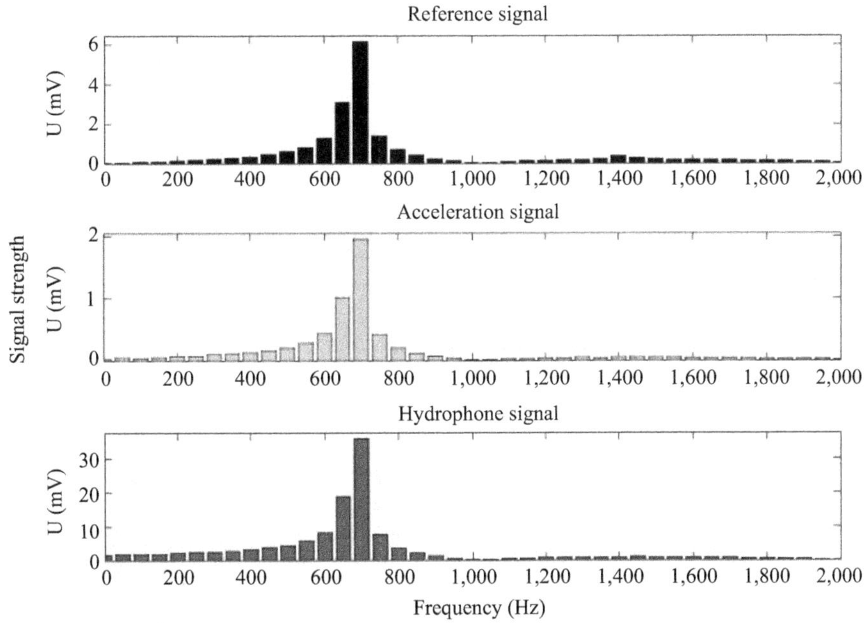

Figure 6.6 Frequency spectrum of excitation, outside tank wall acceleration sensor and in-oil hydrophone

6.4 Sensitivity of surface tank measurements

The correlation between the mechanical status of the transformer active parts and the measured vibration signal are determined in a lab setup which allows the manipulation of clamping torque of the core. The resulting influence on the mechanical oscillations is evaluated at the basic mechanical frequency $f_{n,\mathrm{mech}} =$ 100 Hz. Core magnetisation is provided by an adjustable three-phase voltage source.

6.4.1 Laboratory setup

The measurements are conducted on a 7.5 kV/400 V Yy0 modified distribution transformer. Two acceleration sensors are installed on the yoke of the core, as shown in Figure 6.7. Sensor S_V is attached on the middle of the yoke, above the limb of phase labelled '*V*'. Sensor S_W is positioned above the limb of phase labelled '*W*'. The third sensor S_{ext} is installed on the tank wall which represents a typical position for onsite measurements. The transformer is energised on its high voltage (HV) side using an adjustable three phase 7 kV AC source at nominal frequency f_n =50 Hz. The transformer's 400 V side is left open circuited. The core is fixed with several threaded bolts on the active part's structure, see Figure 6.8. The torque of the bolts is adjustable for the analysis. The active part is reinstalled

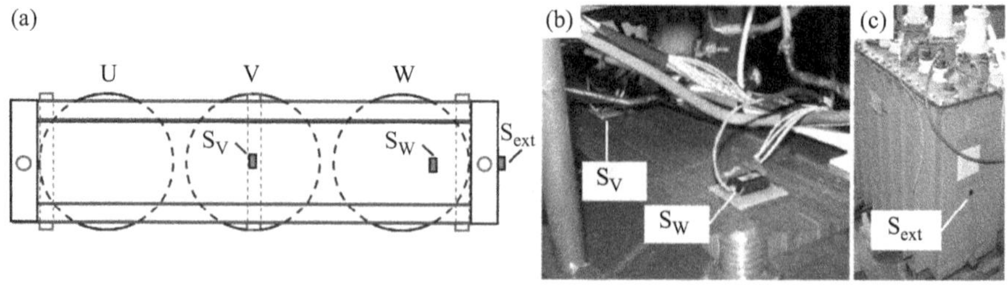

Figure 6.7 *Measurement setup using two acceleration sensors on the core yoke (sensors S_V and S_W) and an external sensor S_{ext} on the tank wall. (a) Top view, (b) sensors on active part, (c) sensor on outside tank wall*

Figure 6.8 *Side view of test transformer's active part. Threaded bolts (grey circles) fixate the core to the structure with defined torque*

into the tank after torque adjustments at each test of the series. The entire setup is filled with oil after extracting the air from the tank using a vacuum pump.

Three different torque values are considered: the original (14 N m), a loosened (0 N m) and a tightened (45 N m) setup. The basic mechanical frequency $f_{n,\mathrm{mech}} =$ 100 Hz for each measurement position at different torques is shown in Figure 6.9. Regarding both, the original core torque and the loosened setup, the outer limb (phase *W*) shows highest oscillations. Additionally, oscillations at phase *W* show a non-linear dependency to the supply voltage, which results into a decaying signal power for excitation levels above 5 kV. Mechanical oscillation is lowest at high torque, because a threaded bolt is near to the position of sensor S_W which tightens the core and therefore limits the mechanical movement. The measurements obtained from the geometric centre of the setup (sensor S_V) show only small

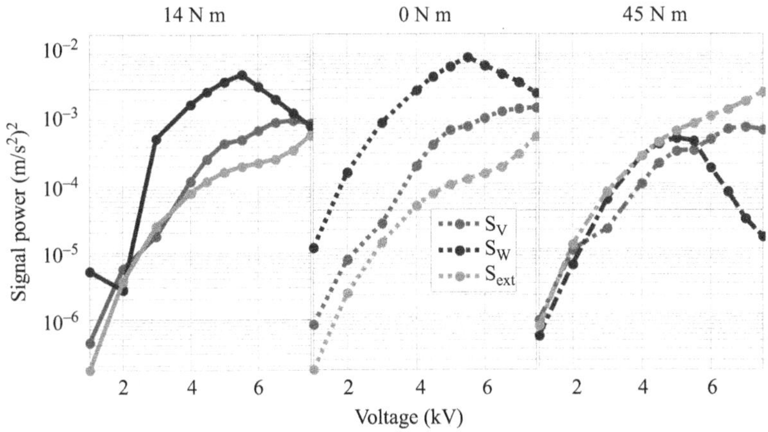

Figure 6.9 Signal power of acceleration signals at $f_{n,mech} = 100$ Hz over AC supply voltage (RMS) at different torques applied to the core measured at different positions

changes between the original and the loosened setup because the distance between threaded bolts and sensor is large compared to the other sensor position. Changes can only be detected at high torque, where the signal power decreases.

The signals measured on the outside tank wall using S_{ext} show a different behaviour. Signal power increases at rising torques due to the rising mechanical coupling which improves transportation of waves not through the oil but through the solid structure. As the setup shows, every change of torque leads to a change of the structure-borne mechanical oscillations measured directly on the active part or the tank wall. The changes within the signal power strongly depend on the individual sensor position, and they do not show a uniform behaviour on the core itself or in comparison to the tank wall measurement. This very basic evaluation indicates that vibration measurement tends to be used for long-term measurement and monitoring comparing measurements form a fixed sensor position over time because mechanical changes are detectable at every tested measurement position but are not comparable against each other.

6.4.2 Field test: sensor positions

Whereas Section 6.4.1 covered the signal propagation of vibrations from core to tank, the influence of the tank itself is determined in this section. Figure 6.10(a) shows the side view of a five-limb power transformer which provides ten wall segments. Acceleration measurements are performed at the same height and width within each segment. The same boundary conditions are applied for all measurements: constant temperature and load factor of the transformer as well as a fixed tap-changer position (TCP). The oil pumps are active but not the vents of the air cooling system.

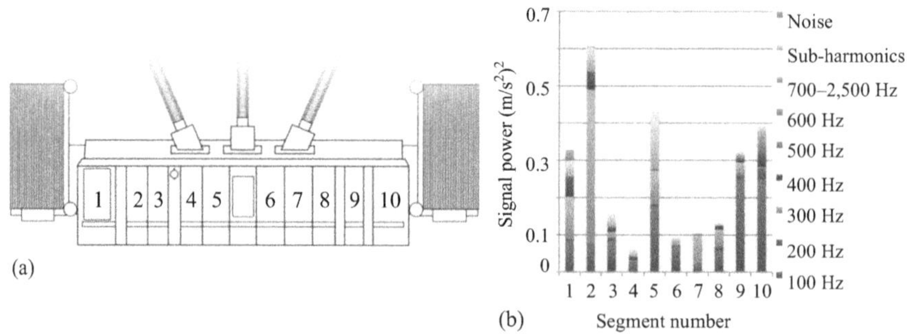

Figure 6.10 (a) Side view of a 400 kV/100 kV power transformer. Tank wall divided into ten segments for sensor positions. (b) Summarised signal power (envelope) and frequency components for each segment

The signal power is divided into basic frequency (100 Hz), harmonic contents, sub-harmonics and noise as shown in Figure 6.10(b). Harmonics are shown up to the fifth degree. Due to the very small signal power at higher orders, the 7th–25th harmonics are only shown as summarised value. Odd sub-harmonics (50, 150, ..., 550 Hz) are also plotted as summarised value. The entire signal power, as well as single frequency components, show high variations between the different measurement positions. There is a tendency to rising signal power at the edges of the transformer (compare segments 1, 2 and 9, 10). In conclusion, no general similarities between different punctual tank wall measurements can be made which supports the assumption made in Section 6.4.1: a comparison of different measurement position seems to be not possible.

6.5 Superimposing effects on tank wall measurements

6.5.1 Effects of on-load tap-changer position

The influence of the TCP on the vibration behaviour is determined in this section. Different TCPs add or remove windings from the current path and hence influences the overall Lorentz force, the magnetic flux in the core and hence magnetostriction. A sensor is attached on the previously introduced power transformer in segment 7, see Figure 6.10(a), and a long-term, 4-month measurement is performed. Acceleration is recorded every 15 min and then correlated with the TCP. Figure 6.11 shows the basic frequency (100 Hz) drawn against the TCPs; harmonics show similar behaviour. Each light grey dot represents one measurement. The lines show the standard deviation and the mean value at each TCP. The high variations at TCPs 5–8 do not indicate any obvious correlation. Variation is caused by changing load factors and oil temperatures over the measurement period (as will be elaborated below).

Variation at TCPs 4 and 9 is smaller, but statistically less reliable, because only few measurements exist at this TCP: the available data of TCPs 4 and 9 only

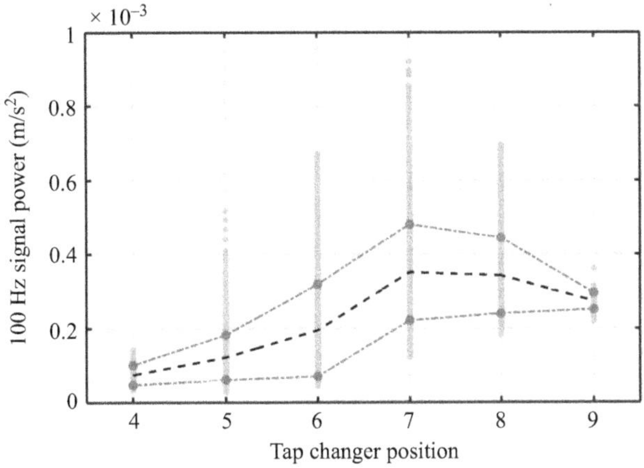

Figure 6.11 Light grey dots: signal power of 100 Hz acceleration at different tap changer positions (light grey). Black: mean values, dark grey: standard deviations of measurements at each position

represent 7% of the entire database. Positions 5–8 have comparable shares; each represents about 23% of the entire database (85,500 measurements). The comparison of the mean values for each position does not indicate correlation, too (see black line in Figure 6.11). Taking the tap changer control mechanism into account, the missing correlation is plausible. In order to keep the 100 kV voltage level constant, the tap changer control tries to compensate for the voltage changes by adding or removing windings (reactive power compensation). Assuming the control is successful and the output voltage is maintained at 100 kV, the magnetic flux has to be constant due to Faraday's law of induction. So core vibrations are regulated to remain constant. The high standard deviation is because other load and temperature effects superimpose independently from the TCP.

6.5.2 Effects of transformer load and operating temperature

Mechanical oscillations show correlation with load current and oil temperature, depending on individual transformer. The load effect is due to the load–current-driven Lorentz force (see Section 6.2.2). The oil temperature which depends on the load and ambient conditions also effects on the measured vibration signal. Transformer operating temperature influences mechanical oscillations through the expansion temperature coefficient of the core plates and structure materials. In addition to load, temperature-driven dependencies also change with the used cooling system, e.g. oil with natural convection and air with natural convection, oil forced by pumps and air forced by ventilators or oil directed in specific cooling channels and air water cooled, etc.

The evaluation of both effects is performed by two long-term measurements on two step-up transformers at different coal power plants: a 120 MVA three-limb transformer and a 525 MVA five-limb generator step-up transformer.

Figure 6.12(a) shows the 100 Hz signal power of both transformers against the transformers' load currents. Each dot represents the 100 Hz vibration of a single measurement. Measurements were performed every 3 min for approx. 3 months. The 525 MVA transformer shows a quadratic dependency on the load current (see quadratic fit, shown as a line), whereas the mechanical oscillations of the 120 MVA transformer do not correlate to load and its signal power is distributed over the entire load current range. Figure 6.12(b) shows the same acceleration measurements plotted against the top oil temperature of each transformer. In this comparison, the 120 MVA transformer can be assumed to show a linear dependency to the temperature (with high standard deviation) starting from approximately 10 to 55 °C [see linear fit in Figure 6.12(b), shown as a line]. On the other hand, the water cooled 525-MVA transformer does not show this correlation but a cluster from 35

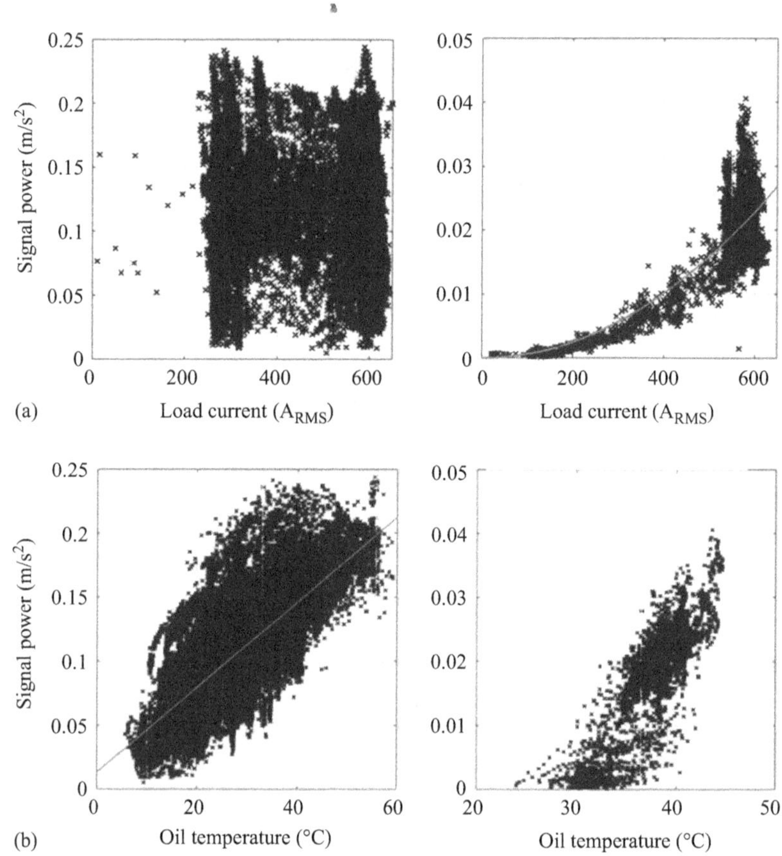

Figure 6.12 (a) Influence of load current on mechanical oscillations (100 Hz) at two step-up transformers. (b) Influence of oil temperature on mechanical oscillations (100 Hz) at two step-up transformers. Left: 120 MVA OFAF, right: 525 MVA ODWF

to 40 °C (due to the regulated oil directed water forced (ODWF) cooling system). Hence, a simple, direct comparison between transformers' vibration behaviour is not possible.

6.6 Practical case studies

Until this point, it can be stated that mechanical oscillations of different transformers cannot generally be compared. Nevertheless, a time-based surveillance of one transformer might be a possibility by allowing detection of changes using trend analysis. Therefore, determination of a 4-years experiment is evaluated through the 120 MVA step-up transformer.

6.6.1 Mechanical oscillations over time

Over a 4-year period, tank accelerations measurements were recorded every 3 min using 44.1 kHz sampling rate. Figure 6.13 shows the behaviour of basic frequency, harmonics, load current and oil temperature at start-up and during 4 days of continuous service.

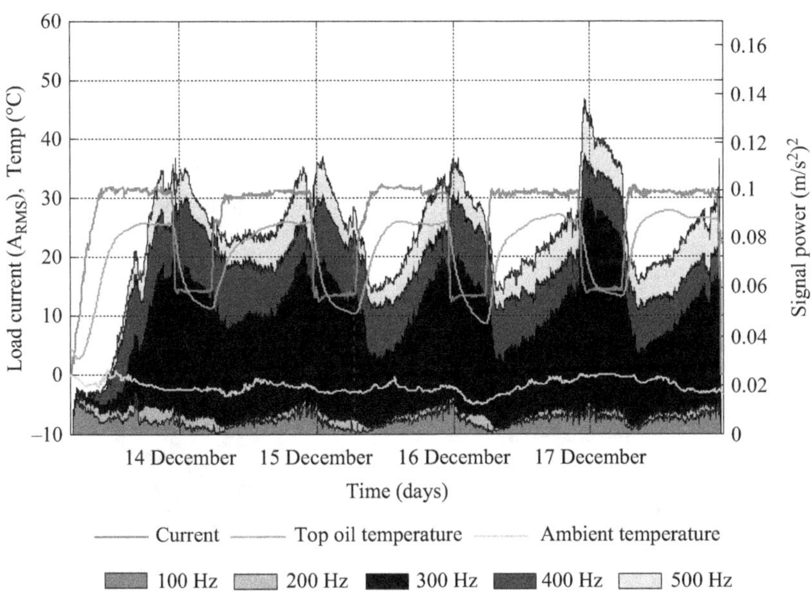

Figure 6.13 Exemplary mechanical oscillation behaviour at start-up and during 4 days of continuous service at changing load conditions 'stacked chart of mechanical oscillations' signal power. The envelope signal power represents the summarised signal power of all considered frequencies

The strong correlation between load current and oil temperature is evident where the temperature follows the load with its time constant τ_{temp}. For a long-term trend analysis, only mechanical oscillations of measurements during continuous service are taken into account. Therefore, only measurements performed 10 h after start-up or later are evaluated. Using this boundary condition, additional effects such as magnetic core remanence or transient heating effects can be excluded from the evaluation. Only the basic frequency (100 Hz) is considered. At this transformer, the basic frequency shows strong correlation to the temperature, which dominates the vibration behaviour; see Figure 6.12(b). In order to evaluate the load dependency, the influence of temperature must be compensated first as explained below.

Figure 6.14 shows the oil-temperature vs. mechanical oscillation comparison separated into single years. Measurements taken during the period 2010–11 do show similar behaviour compared to each other. During the year 2012, a rise of signal power can be observed (at approx. 20 °C), which continues in 2013 (at lower oil temperatures from 10 to 28 °C due to winter season). The linear correlation known from the previous section is used to derive a fitted function, based on (6.4). The factors of the linear approximation are derived from the practical measurements during the 4-year period. The presented approach is considered valid because the annual approximations do show only minor derivations.

$$Sig_{100\text{Hz}} = k\, t_{\ddot{o}l} + d \tag{6.4}$$

where $Sig_{100\text{Hz}}$ is the signal power of 100 Hz component, k is the linear factor in $(\text{m/s}^2)/\text{K}$, d is the offset-correction of linear interpolation.

For each measurement, the 100 Hz acceleration signal power component driven by the temperature is calculated according to (6.4) and then subtracted from the measurement to compensate the influence of the temperature.

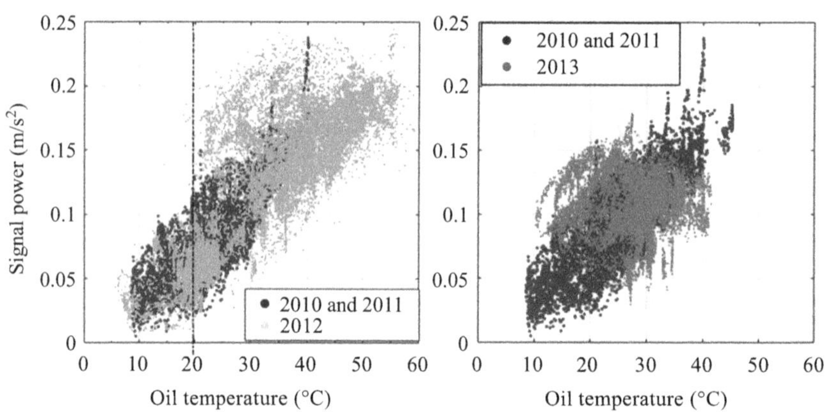

Figure 6.14 Annual comparison of basic frequency (100 Hz) against oil temperature

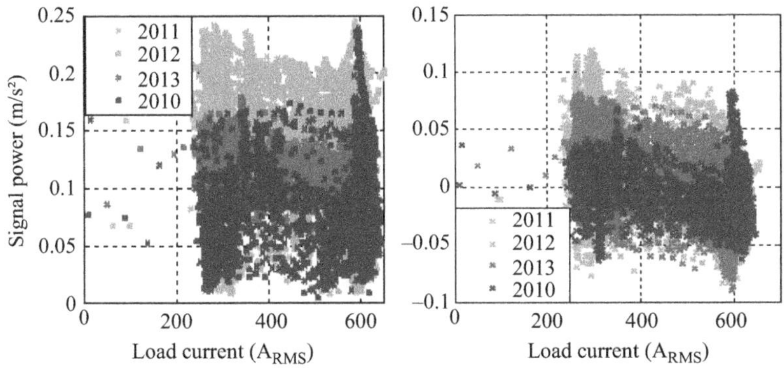

Figure 6.15 Annual comparison of basic frequency (100 Hz) against load current. Left: original measurements, right: after temperature compensation

Figure 6.15 shows the measured signal power against load with (left) and without (right) temperature compensation for annual comparison. Considering the non-compensated plot in Figure 6.15, the signal power in 2012 exceeds those of previous years because of the higher temperatures. Using temperature compensation in Figure 6.15, the overall deviation between different years diminishes with one exception; the deviation at lower load currents (approx. at $I_{\text{load}} = 250–400$ A) does remain. In both years 2012 and 2013, there are higher signal powers than in the previous years. At nominal load $I_{\text{load,nom}} = 600$ A, the annual deviation is not present. Hence, the mechanical oscillation behaviour of the considered step-up transformer has changed during the 4 years of service at certain operational states in partial load which does not originate from external influences, namely the ambient temperature. A surveyed scrapping of the transformer after it has been decommissioned will possibly provide further knowledge of the status of its mechanical structure and is always recommended in such cases.

6.7 Behaviour of mechanical oscillations at DC superimposition

Usually, DC are not associated with the conventional AC transmission grid. Nevertheless, different effects can cause DC currents in the AC grid as shown below. The undesirable influences of DC on transformers have been determined using both, practical measurements and simulation models [19–21]. Known effects of power transformers include high transient alternating currents, rising losses and high harmonic contents of the non-active power [22].

6.7.1 DC-coupling path into power transformers

DC-like currents can be driven by geomagnetically induced currents (GIC), which is shown in [19,23,24] and other publications. Eddy currents of ionised particles in the ionosphere cause a magnetic field which induces currents in the MHz-range according to the Law of Induction into a loop consisting of overhead lines and the ground. The induced currents can be considered as quasi DC compared to the grid's nominal frequency.

An ohmic coupling driven by corona effects can be caused by high voltage direct current (HVDC) systems which are in close proximity to conventional AC systems, typically in the range below 15 m. This constellation can occur if a single system on an existing AC rod is changed to a HVDC system [25]. A field test on a short transmission test line shows a weather-dependent, corona-driven DC coupling from an HVDC line to a grounded AC line. The coupled DC can rise to several mA/km [26] in hybrid overhead line systems.

DC currents can also occur as potential equalising currents which couple into the grid through transformer's grounded star points. Depending on the grid infrastructure, they can couple into the electrical grid by grounded transformer star points and partially flow over the grid to another grounded star point in order to close the current loop using the path with minimal ohmic resistance (both transmission lines and transformer windings of the transmission grid represent a low ohmic path).

Mainly transmission grids and its assets are affected by these effects. In all cases transformer star points must be grounded in order to provide a closed DC loop. Practically, this is always given in HV-transmission grids: a certain amount of star points has to be grounded in order to detect single- and multi-phase errors by the resulting short circuit currents (relevant for power-system protection). If a specific transformer is not affected by these protective circumstances, its star point might be kept open to avoid DC running through its winding causing undesirable interactions.

6.7.2 *Saturation and its effect on magnetostriction*

The superimposition of AC with DC components in a transformer winding leads to an additional magnetic flux offset Φ_{DC} in the core. Hence, the original operating point is shifted in the B-H diagram, see Figure 6.16(left). If the flux amplitude exceeds the knee point of the magnetising curve, the core material cannot be magnetised any further and saturation occurs. This leads to a sudden rise of the current Δi through the winding (because the winding inductance drops). The asymmetric magnetic flux density causes asymmetric effects in magnetostriction, which is shown in Figure 6.16(right). As a result, the mechanical strain has two local maxima with different amplitudes during one electrical period. If the strain is evaluated in frequency domain, it becomes apparent that the spectrum of vibrations during DC consists of two frequency components (idealised): the already known doubled electrical frequency and a new component at the line frequency. For example in the European grid, the basic mechanical frequency 100 Hz is superimposed with 50 Hz.

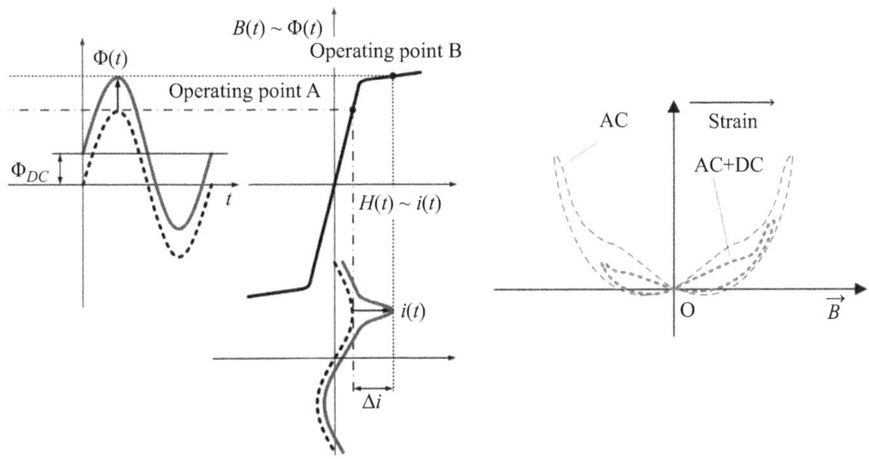

Figure 6.16 Left: influence of DC on the magnetisation curve and resulting (half wave) saturation. Right: resulting butterfly plot of strain within the electrical steel sheets of the core with asymmetric distribution (dark grey)

6.7.3 Test setup for DC superimposed effects

The evaluation of DC influences on power transformers is performed by using two identical 30 MVA Yy0 110/10 kV power transformers. The transformers are connected in back-to-back (on the 110 kV side). The first transformer in the series connection is connected to a three-phase voltage source on its 30 kV side (a three-phase inverter which includes internal reactive power compensation). The first transformer only provides the magnetising current for the second transformer which is used in open loop. The entire setup is shown in Figure 6.17. Both star points are connected by a dedicated cable, where a current source is inserted to drive a DC current through both transformers. The DC source is isolated from the AC components by a bypass capacitor. Shunt resistors are used in the interconnections between the 110 kV bushings to force the DC distribution in the 110 kV phases. Variable power resistors can be inserted into each phase in addition to the shunts to form a current divider which directs the DC imbalanced thorough the single phases. The current measurement is isolated to ground by an optical fibre data transmission system. The phase voltages are measured using capacitive dividers.

Two different scenarios are considered: balanced DC scenarios, where all phases (and hence all magnetic fluxes) are faced with the same amount of DC component, and imbalanced DC scenarios, where the DC distribution of each phase depends on the additional shunt resistors. Effects driven by GIC and potential equalising currents will, potentially, both cause symmetric DC distributions. Hybrid grids might cause imbalanced DC because the interaction between an AC phase and the HVDC system depends on the individual distance between them.

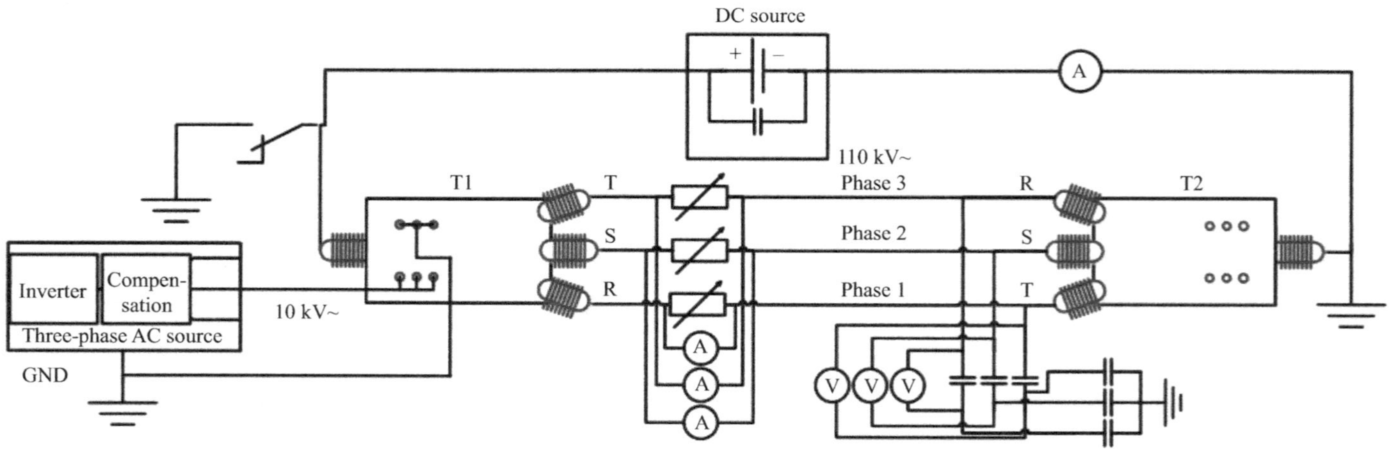

Figure 6.17 Test setup consisting of two 30 MVA transformers (110 kV/10 kV) connected in back-to-back. AC supply provided by a mobile inverter including reactive power compensation. DC injection through star points on 110 kV side

Table 6.1 DC for balanced scenarios. Total DC, DC on each phase and DC referred to the magnetising current

$I_{\mathrm{DC,total}}$ (mA)	$I_{\mathrm{DC,phase}}$ (mA)	Relative to $I_{\mathrm{mag,110\ kV}}$ (%)
630	210	90
2,400	800	343

Table 6.2 DC for imbalanced scenarios. Total DC and DC per phase

No.	$I_{\mathrm{DC,total}}$ (mA)	$I_{\mathrm{DC,phase1}}$ (mA)	$I_{\mathrm{DC,phase2}}$ (mA)	$I_{\mathrm{DC,phase3}}$ (mA)
1	219	22	22	175
2	600	60	60	480
3	369	37	295	37
4	370	22	174	174
5	600	36	282	282

Table 6.1 shows the overall DC current of the star point $I_{\mathrm{DC,total}}$ for balanced scenarios, the resulting phase current $I_{\mathrm{DC,phase}}$ and for reference the ratio between $I_{\mathrm{DC,phase}}$ and the magnetising current ($I_{\mathrm{mag,total}} = 700$ mA at 110 kV).

Table 6.2 shows the overall DC current of the star point $I_{\mathrm{DC,total}}$ for imbalanced scenarios and the resulting current $I_{\mathrm{DC,phase1,2,3}}$ for each phase. The relative DC distribution of scenarios 1 and 2 is the same which provides a higher absolute current on one outer phase. Scenario 3 evaluates a setup injecting a high current into the middle phase. Scenarios 4 and 5 consider the case if the middle and one outer phase have the same DC.

6.7.4 DC-detection using vibration measurement

Accelerometers are installed to measure tank oscillations on both transformers at comparable positions. Acceleration is measured in time domain and converted into frequency domain using FFT.

Two types of frequencies (basic mechanical frequencies and harmonics along with the additional frequencies due to DC superimposition) are considered. For both types of frequencies, the sum of their signal power is calculated and compared against each other. Frequencies are considered up to 1 kHz. Higher frequencies show low signal power (<2% of the overall signal power). The frequency behaviour of both transformers is comparable because they share the same design and accelerometer positions on the tank are also identical. Therefore, only the signal power of transformer T1 is elaborated below.

Figure 6.18(a) shows the mechanical spectrum components of the balanced DC scenarios. No significant changes in the spectrum can be observed. The relative

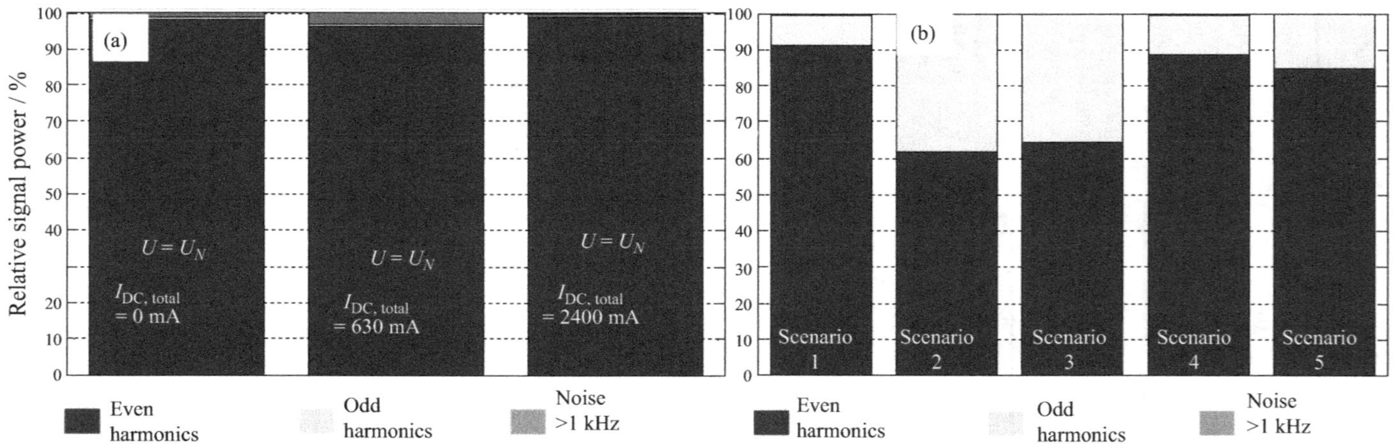

Figure 6.18 Relative signal power of even harmonics (100, 200 Hz, etc.), odd harmonics (50, 150 Hz, etc.) and noise at scenarios with (a) balanced DC and (b) imbalanced DC

signal power of even harmonics is typically above 95% and therefore comparable to vibrations without DC. DC do not affect the transformer at balanced scenarios: no interferences with the core material occur and hence no changes in the mechanical oscillations arise. The reason is to be found in the magnetic equivalent circuit. The direct magnetic flux components of all limbs are directed equally, e.g. they are all directed into the upper yoke. Hence, the direct magnetic flux component is mainly a stray flux outside the core. It has to be mentioned that this behaviour applies to three-limb transformers, not to the five-limb transformer. Five-limb transformers always provide a path for direct magnetic fluxes inside the core through the outer limbs. Therefore, any DC superimposition on five-limb transformers causes interactions with the core [27].

The mechanical oscillations change significantly if imbalanced DC scenarios are applied, see Figure 6.18(b). The resulting changes of the operating point lead to high signal power of 50 Hz and odd harmonics. The new frequency spectrum covers a range between 10% (scenario 1) and approx. 40% (scenario 2) of the entire mechanical oscillation signal power. The relative signal power mainly depends on the imbalanced DC distribution on the phases, where higher imbalance leads to rising odd harmonics, see worst case scenario 2 compared to scenario 5. The influence of the overall $I_{DC,total}$ is determined using scenarios 1 and 2 (for a DC distribution mainly on one outer phase) and scenarios 4 and 5 (for higher DC components on the middle and one outer phase). If only the outer phase exhibits DC, the rising current leads to a significant increase in the relative signal power distribution for sub-harmonics of approx. 30%. Considering scenarios 4 and 5, the changes in the relative signal power distribution are small; an increased $I_{DC,total}$ by 62% only increases odd harmonics signal power by 5%.

The empirical determination in this experiment shows that both the total amount of DC $I_{DC,total}$ and its distribution on the phases influence the resulting mechanical oscillations. The severity of $I_{DC,total}$ depends on the DC distribution on the phases. Considering the worst cases, already relatively small DC below the range of the HV magnetising current $I_{DC,total} \leq I_{mag,HV}$ leads to significant changes of mechanical oscillations. Usually, $I_{mag,HV}$ is smaller than 1 A per phase. Further research will have to provide information about the influence of different core materials and the core geometry itself (three-limb or five-limb design).

It can be concluded that acceleration measurement is an easy applicable method to detect problematic DC in transformers if DC leads to changes in the transformer's magnetic operating point. Tank wall sensors can be installed during service, which allows a fast way to decide if a transformer is impacted by DC or not.

6.7.5 Dependency of DC-driven vibration and transformer noise

Figure 6.19 compares the frequency spectrum of vibrations and transformer noise with and without DC. For both vibration and noise measurement, the reference measurement without DC shows typical peaks at doubled electrical frequency and harmonics (100, 200 Hz, etc.) as introduced in Section 6.2.1. Applying DC to the

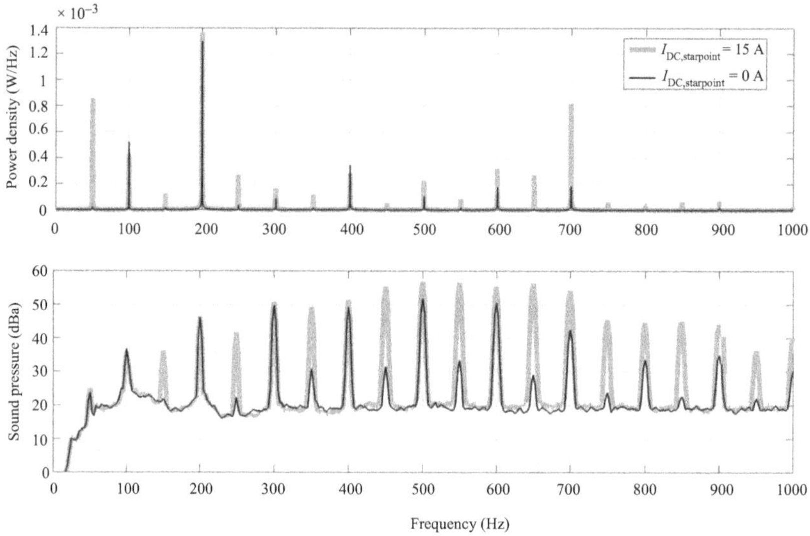

Figure 6.19 Comparison between vibrations and transformer noise measured according to [8]. Black: no DC applied, grey: $I_{DC,total} = 15$ A

setup causes several changes. Odd harmonic components of 50 Hz become apparent (50, 150, 250 Hz, etc.). In addition, the amplitudes of AC harmonics partly increase.

In order to quantify the increase of vibrations and audible noise at even and odd harmonics, both are measured at different DC levels. The noise measurement is performed according to [18]. Measurements are performed on a five-limb core, 350 MVA grid coupling transformer with symmetric DC. Figure 6.20 A shows the increase of noise (left y axis) and vibrations (right y axis) at increasing DC currents. The x axis shows the applied DC in relation to the magnetising current of the transformer's 380 kV windings. Vibrations are plotted in relation to the reference measurement without DC. Hence, only the increase of vibrations is plotted. Vibrations and audible noise have the same gradients and show a very similar trend; at first with a high gradient at low DC values and a flattened trend at higher DC. The plot shows similarities to the magnetising curve of the core sheets. At high DC levels the half wave saturation is approx. complete, meaning no further material effects can occur which limits the further increase of vibrations and noise. Figure 6.20(b) compares the vibration components with the audible noise. Dispersion can be seen at higher DC levels, when the saturation kicks in. If the DC is in the range of the magnetising current, even and odd harmonics behave similarly and provide equally to the signal power.

As Figure 6.20 illustrates, changes of the locally measured tank wall vibrations can be compared to the measured noise signal. Concluding, vibration analysis provides two benefits: surveillance of DC-driven influences on transformers in general and to estimate the resulting increase of transformer noise.

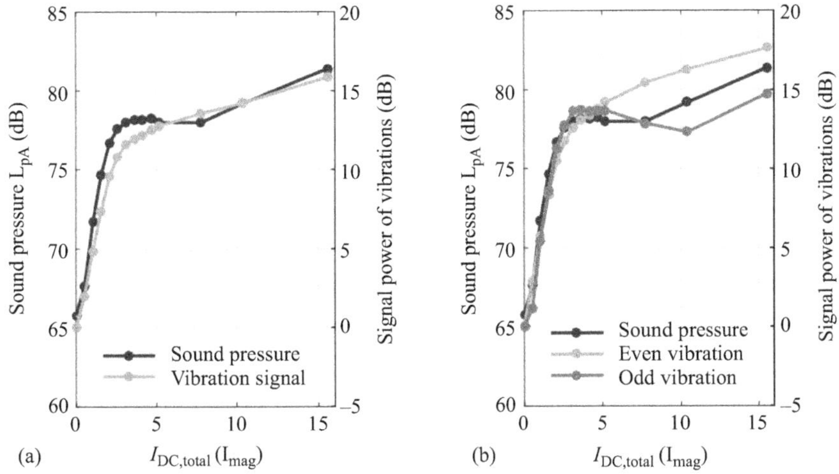

Figure 6.20 Increase of noise and vibrations at rising DC. DC is plotted in relation to the AC magnetising current of the transformer: (a) sound pressure (dark grey, left y axis) and increase of tank vibrations with reference to $I_{DC} = 0$ A (light grey, right y axis). (b) Changes of even and odd vibration harmonics in comparison to the sound pressure

6.7.6 Case study on transformers impacted by DC

The general observations presented in Section 6.7.5 are tested in the field on a grid coupler transformer in service (five-limb transformer). Current measurements in the transformer's star point show a superimposed parasitic DC which is most likely caused by nearby industrial facilities (e.g. by large inverters). The transformer is operated in no-load condition. Therefore, mainly the core does provide vibrations by magnetostriction. The lower plot in Figure 6.21 shows the DC component measured in the star point. The connection to ground is closed at 11.10 and re-opened at 14.25. The positive DC is flowing from the ground connection through the star point into the transformer. The DC can be switched on and off by opening and closing the star points of other transformers in the area. Thus, DC can be forced into the surveyed transformer. Using this method, the changing DC levels at 12.22 and 13.30 are obtained. The total vibration signal power as well as the summarised even and odd harmonic components can be correlated to the measured DC, see Figure 6.21, upper plot. At the beginning of the test, DC is not present and the vibration signal mainly consists of even harmonics as expected (odd harmonic content is at approx. 3% which equals the noise level of all other frequency components). With rising DC, the entire vibration signal power as well as both even and odd harmonics increases. Given the fact that even harmonics can also be influenced by load and temperature, the total vibration level is considered for DC surveillance. Therefore, only odd harmonics are considered. They do not occur at normal operating conditions and can be correlated directly to DC as their trend in Figure 6.21

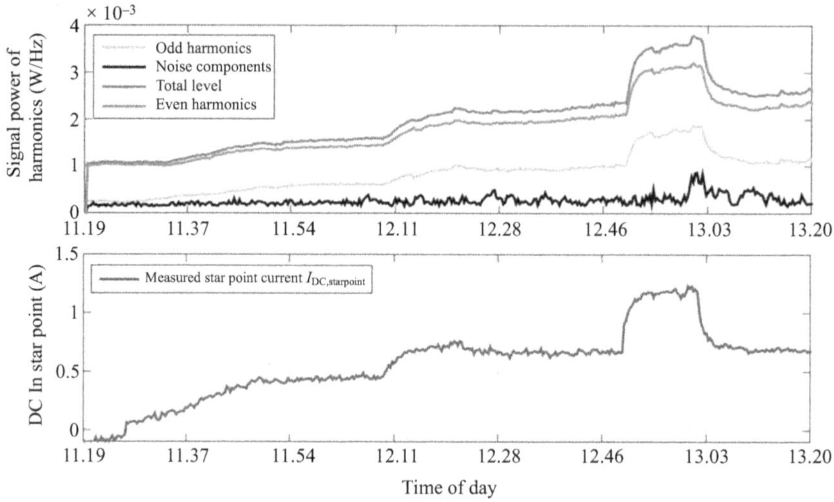

Figure 6.21 Upper plot: signal power of tank wall vibration measurement. Lower plot: DC measured in the star point of the transformer

shows. When the DC reaches its peak at 13.00, odd harmonics contribute about 25% of the entire vibration signal power.

The field test confirms vibrations being a practical, easy applicable measurement method to detect DC indirectly if star point current measurements are considered too extensive. Field test and the observations from the laboratory made in Section 6.7.6 are comparable.

6.8 Conclusion

Generally, mechanical oscillations or vibration measurements are conducted to detect two types of changes inside a power transformer: mechanical changes and DC-driven effects. An advantage of the method is its easy applicability directly on the tank wall for in-service transformers. The challenges of this technique lies within the multiple factors influencing the mechanical oscillations of transformers such as temperature, load status and the individual measurement position on the tank wall. The detailed determinations provided in this chapter can be summarised in the following facts.

Mechanical changes originated by core, windings or accessory parts can be surveyed by a continuous monitoring of the transformer. Measured signals strongly depend on the location of the sensor and the individual transformer. Hence, a general comparison to other transformer designs or other measurement positions is not usually possible. Nevertheless, long-term vibration surveillance provides trend information about changes inside the transformer if oil temperature and load factor are known and taken into account. Vibration monitoring is a time-based, comparative method and can be seen similar to frequency response analysis which uses foot printing for correlations over time. The advantage of vibration monitoring in this context is its continuous trend analysis instead of snapshot information of single measurements.

The second practical use of mechanical oscillation measurement is the DC detection. DC can be caused by GIC, interactions of AC overhead lines with nearby HVDC overhead systems or by DC components coupling from earth through grounded star points into transformers. Any DC causing a direct magnetic flux inside the core leads to an extended frequency spectrum of vibrations: The usual mechanical spectrum consists of doubled electric frequency and harmonics. DC adds signal power at half of the mechanical basic frequency (and all its multiples). The effects of DC on transformers depend on the amount of DC, its distribution on the phases and the transformer's core design. Three-limb transformers are less influenced by a DC distributed symmetrical on the phases. Vibrations of five-limb transformers always change vibration behaviour at superimposed DC.

References

[1] M. Wang, A.J. Vandermaar, K.D. Srivastava. Review of condition assessment of power transformers in service. IEEE Electrical Insulation Magazine, Volume 18, Issue 6. 2002: p. 12–25.

[2] E. Gockenbach, H. Borsi. Condition monitoring and diagnosis of power transformers. In International Symposium on Electrical Insulating Materials. 2008 September 7–11: p. 16–19.

[3] IEEE Power & Energy Society. IEEE Guide for the Interpretation of Gases Generated in Oil-Immersed Transformers, C57.104 New York: IEEE; 2009.

[4] International Electrotechnical Commission. IEC 60270 High-Voltage Test Techniques – Partial Discharge Measurements, 3. Edition Geneva, Switzerland; 2000.

[5] S. Coenen, S. Tenbohlen, T. Strehl, S. Markalous. Fundamental characteristics of UHF PD probes and the radiation behaviour of PD sources in power transformers. In International Symposium on High Voltage Engineering; 2009; Johannesburg, ZA.

[6] S. Coenen, S. Tenbohlen, S. Markalous, T. Strehl. Performance check and sensitivity verification for UHF PD measurements on power transformers. In 15th International Symposium on High Voltage Engineering; 2007; Ljubljana, Slovenia.

[7] CIGRÉ. Brochure 342 – Mechanical Condition Assessment of Transformer Windings using Frequency Response Analysis (FRA), SC A2 WG A2.26; 2008.

[8] International Electrotechnical Commission. IEC 60076-10 Power Transformers Part 10: Determination of Sound Levels; 2001.

[9] R. Henshell, P. Bennett, H. McCallion, M. Milner. Natural frequencies and mode shapes of vibration of transformer cores. Proceedings of the Institution of Electrical Engineers, Volume 112, Issue 11. 1965 November: p. 2133–2139.

[10] H.L. Rivera, J.A. Garcia-Souto, J. Sanz. Measurements of mechanical vibrations at magnetic cores of power transformers with fiber-optic interferometric intrinsic sensor. IEEE Journal of Selected Topics in Quantum Electronics, Volume 6, Issue 5. 2000: p. 788–797.

[11] P. Kung, L. Wang, M.I. Comanici. Fiber optics temperature/vibration and moisture monitoring in power transformers. Electrical Insulation Conference. 2011 June 5–8: p. 280–284.

[12] J. Shengchang, S. Ping, L. Yanming, X. Dake, C. Junling. The vibration measuring system for monitoring core and winding condition of power transformer. In Proceedings of 2001 International Symposium on Electrical Insulating Materials (ISEIM); 2001: IEEE. p. 849–852.

[13] B. García, J. Burgo, Á. Alonso. Transformer tank vibration modeling as a method of detecting winding deformations Part I: Theoretical foundation. IEEE Transactions on Power Delivery, Volume 21, Issue 1. 2006 January: p. 157–163.

[14] B. García, J. Burgo, Á. Alonso. Transformer tank vibration modeling as a method of detecting winding deformations – Part II: Experimental verification. In IEEE Transactions on Power Delivery, Volume 21, Issue 1. 2006 January. p. 164–169.

[15] S. Borucki. Diagnosis of technical condition of power transformers based on the analysis of vibroacoustic signals measured in transient operating conditions. IEEE Transactions on Power Delivery, Volume 27, Issue 2. 2012: p. 670–676.

[16] R.S. Girgis, M.S. Bernesjö, S. Thomas, J. Anger, D. Chu, H.R. Moore. Development of ultra-low-noise transformer technology. IEEE Transactions on Power Delivery, Volume 26 Issue 1. 2011: p. 228–234.

[17] G. Fasching. Materials for Electrical Engineering. Vienna: Springer Wien New York; 2005.

[18] International Electrotechnical Commission. IEC 60076-10-1 Determination of Sound Levels Application Guide; 2005-10.

[19] R. Girgis, K. Vedante. Effects of GIC on power transformers and power systems. In Transmission and Distribution Conference and Exposition. 2012; Orlando, FL.

[20] M. Heindl, M. Beltle, S. Tenbohlen, U. Sundermann, F. Schatzl. Betriebsverhalten von Leistungstransformatoren in Hybridnetzen. In VDE Smart-Grid Kongress; 2012; Stuttgart, Germany.

[21] M. Beltle, M. Schuehle, S. Tenbohlen. Influences of direct currents on power transformers caused by AC-HVDC interactions in hybrid grids. In International Symposium on High Voltage Engineering; 2015; Pilsen, Czech Republic. p. paper 300.

[22] IEEE Power & Energy Society – Power System Instrumentation and Measurements Committee. IEEE Standard 1459-2010: Definitions for the Measurement of Electric Power Quantities Under Sinusoidal, Nonsinusoidal, Balanced, or Unbalanced Conditions. New York: IEEE; 2010.

[23] M. Heindl. M. Beltle, M. Reuter, D. Schneider, S. Tenbohlen, D. Oyedokun. T. Gaunt. Investigation of GIC related effects on power transformers using modern diagnostic methods. In International Symposium on High Voltage Engineering; 2011; Hannover.

[24] Geomagnetic Disturbances: Their impact on the power grid. IEEE Power and Energy Magazine, Volume 11, Issue 4. 2013: p. 71–78.

[25] Cigré Working Group B 2.41. Cigre Brochure 583 Guide to the Conversion of Existing AC Lines to DC Operation. In; May 2014.

[26] B. Rusek, J. Wulff, K.-H. Weck, *et al.* Ohmic coupling between AC and DC circuits on hybrid overhead lines. In Cigré 2013; 2013; Auckland.

[27] R.S. Girgis, K.B. Vedante. Impact of GICs on power transformers. IEEE Electrification Magazine, Volume 3, Issue 4. 2015 December: p. 8–12.

Chapter 7
Lifecycle management of power transformers in a new energy era

Carlos Gamez[1]

7.1 Introduction

The power transformer industry was kick-started 150 years ago by the nascent electrical power industry. As towns and cities became electrified, the need to move electrical energy over long distances made the power transformer an indispensable element of the overall power generation, transmission and distribution system.

I often wonder what it would have been like to work in those embryonic stages of the electrical industry. With many discoveries and new ideas being tried on a frequent basis, it must have been an exciting time to be an electrical engineer.

As the core technologies that defined the main elements of electrical systems matured, developments in new materials and manufacturing techniques kept advancing the field. Power transformers achieved higher voltage ratings and physical sizes were reduced.

Fast forward to the present day and we face a new challenge which the previous generations did not experience. We live in a time where renewable power generation technologies, the main modes of transportation and the foundational concepts of electrical systems, are in a state of flux. The engineers that started the industry had the challenge of creating it in the first place; we have the challenge of transitioning from this model into one that is not completely defined yet.

The capacity of renewable sources of energy and reverse power flows are putting the stability of modern electrical systems to the test. Energy storage solutions are being developed and trialled, which means that our fundamental assumptions about energy generation and distribution are being redefined. Electrical vehicles seem to be gaining real momentum and their impact on electrical networks is yet to be understood and experienced. New electronic, computing and algorithmic technologies are enabling levels of instrumentation, monitoring and decision-making that previous generations could have only dreamt of.

[1]Engineers Tools, Australia

Add to this the state of an aging infrastructure and downward pressure on the price of electrical energy and you will realise we find ourselves in a very interesting period in the electrical energy industry.

As with any other period of change, our current state not only entails great challenges but also holds the promise of new opportunities. Opportunities to rethink our assumptions, learn from previous mistakes, devise novel and innovative solutions and set our civilisations up for the next 100 years.

In this chapter, I will discuss some ideas and opinions of what this evolving environment means in terms of power transformer lifecycle management. This text is not meant to be an authoritative or scientifically rigorous treatment of this subject matter but rather merely a thought-provoking exercise for actors and stakeholders involved in planning, designing, building operating and maintaining the transformers and electrical networks of the future.

7.2 A changing landscape

Many things have changed since the inception of the first electrical networks. The population of the world has grown exponentially and, consequently, our demand for energy has grown in that way too. Not only has the number of people dramatically increased but also the quality of life has significantly improved. This improvement in quality of life is directly correlated to the energy consumption per capita. A better life means higher use of energy, and thus, electricity.

In 1980, the United States of America occupied the first place in electricity generation in the world at 32.9% of the total electricity generated. China was on the fifth place with 4% [1]. In 2011, this ranking had reversed. As of 2011, China was in the first place with 21.2% of the world's electricity generation, whilst the USA was at 19.3% occupying the second place (Figure 7.1).

In just 30 years, the total amount of electricity produced by the top 20 countries in the world had tripled. The industrialisation and growth of the two most populous countries in the world, India and China, have meant an exponential increase in the amount of power generated and consumed by these economies (Figure 7.2).

According to the most recent U.S. Energy Information Administration report (the International Energy Outlook 2017 [3]), this trend shows no signs of slowing down (Figure 7.3). The projections show, 'Between 2015 and 2040, world energy consumption increases by 28% ... with more than half of the increase attributed to non-OECD Asia (including China and India), where strong economic growth drives increasing demand for energy [3]'.

On the other hand, the OECD countries grow at more modest rates compared to non-OECD countries. The IEO report estimates, 'Energy consumption in non-OECD countries increases 41% between 2015 and 2040 in contrast to a 9% increase in OECD countries [3]'.

In terms of energy production and consumption, this means that in the next two decades the world will move at two different speeds.

OECD countries like the United States, the United Kingdom, France, Australia, Japan and Germany went through their more significant growth period

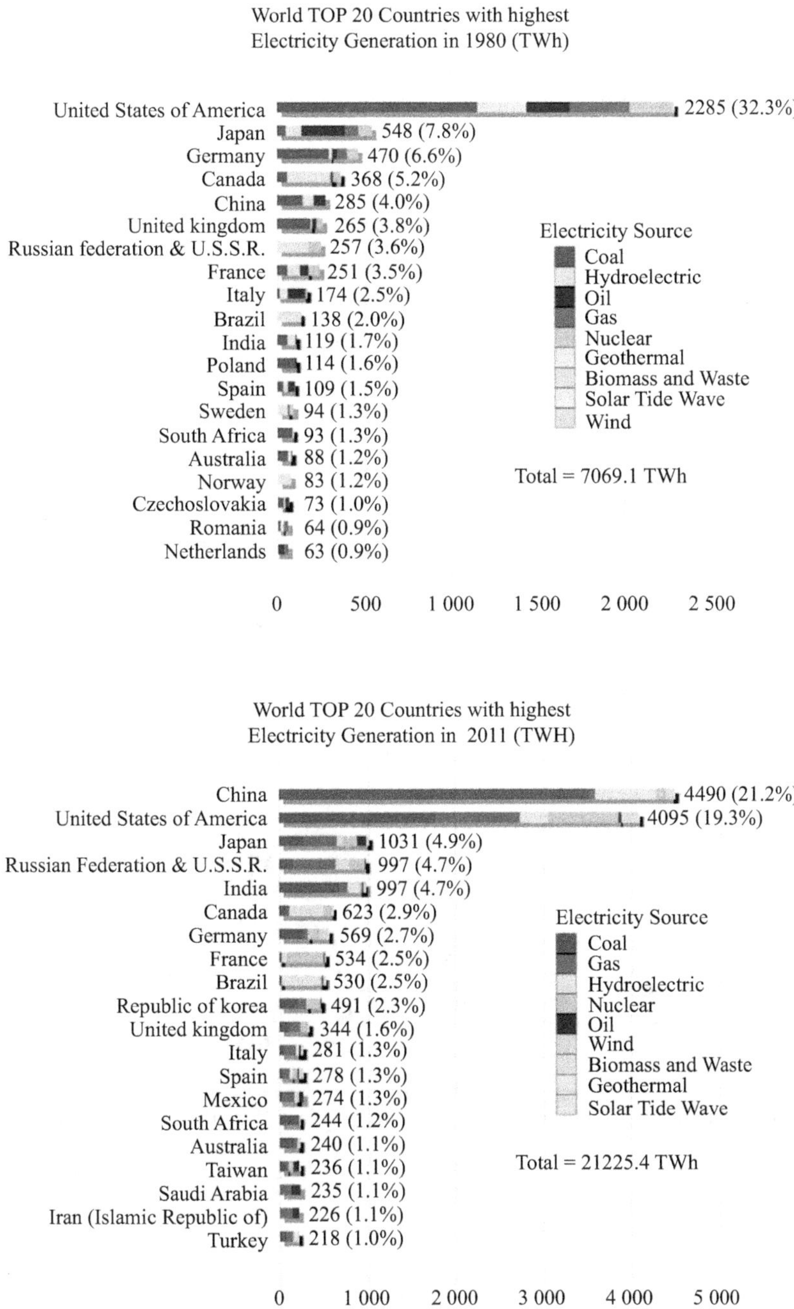

Figure 7.1 Comparison of top 20 countries with highest electricity generation – 1980 and 2011 [1]

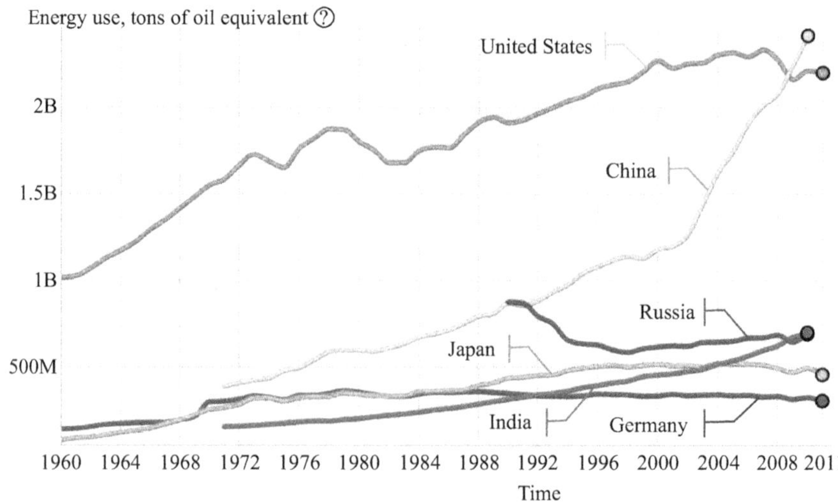

Figure 7.2 Energy use per country over time [2]

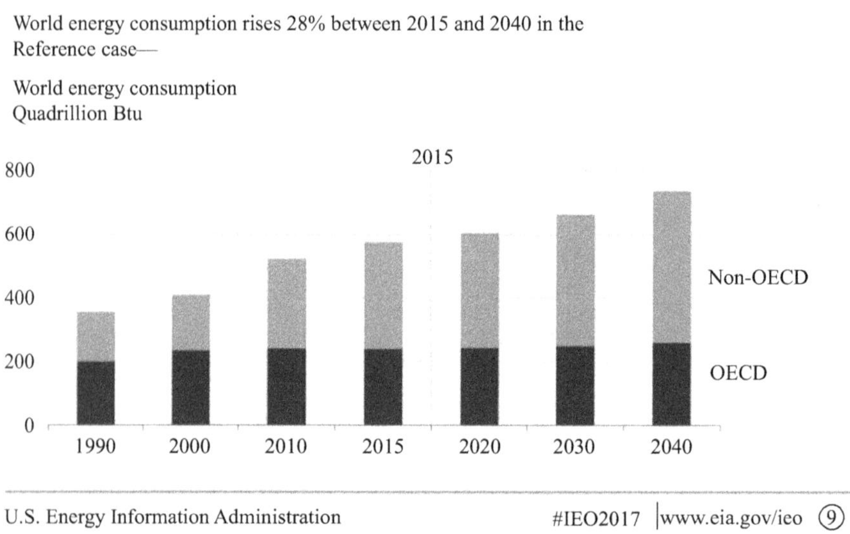

Figure 7.3 Projected energy consumption growth between 2015 and 2040 (reference case) [3]

over the last 50 years. These countries have now the challenge of low growth-rates in electricity demand and ageing networks and infrastructures.

On the other hand, non-OECD economies like India and China are, and will continue, experiencing higher rates of economic growth. These countries face different challenges in the form of growing networks and infrastructure bases.

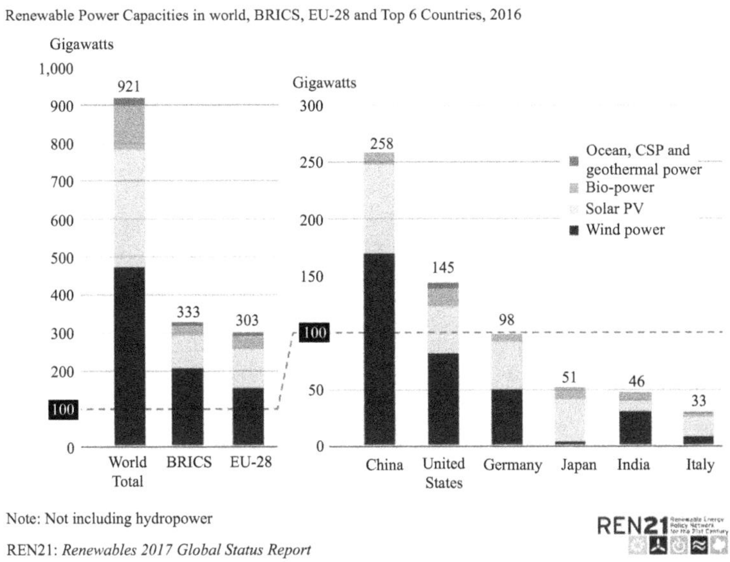

Figure 7.4 Renewable capacities as shown in REN21 2017 GSR [4]

However, these challenges are not exactly the same as those experienced by the OECD group 50 years ago. New environmental pressures and constraints (climate change, sustainability) as well as new available technologies to address these challenges (renewable energies, electric vehicles, internet) present a different landscape to economies currently undergoing expansion.

These differences in turn call for the adaptation of asset management philosophies and strategies. These strategies should be aligned to the particular context in which the assets are installed and operated.

7.2.1 Renewable energy sources

Another force that will play an increasingly important role in the shape of future networks is the penetration of renewable energy generation sources. Whilst fossil fuel generation is not going away anytime soon, it is also true that market incentives, decreasing costs, increasing efficiencies and general public appetite for renewable energy sources will continue to promote participation of renewable energies in the overall energy production market.

According to REN21 Renewables 2017 Global Status Report 'As of 2015, renewable energy provided an estimated 19.3% of global final energy consumption' [4].

Moreover, according to the same report, most of the growth in renewable energies over the last few years has occurred in the non-OECD countries, in particular China as shown in Figure 7.4.

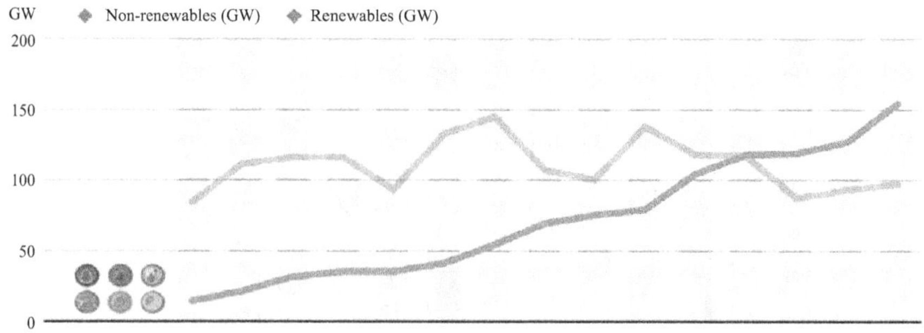

	2001	2002	2003	2004	2005	2006	2007	2008	2009	2010	2011	2012	2013	2014	2015
Renewables (GW)	14	21	31	35	35	41	54	69	75	79	104	118	119	127	154
Non-Renewables (GW)	84	111	116	116	93	133	145	107	100	138	118	117	87	93	97

Source: *IRENA, 2016b*

1 Excludes 154 GW of pure and mixed pumped storage capacity otherwise included in hydropower capacity. The bulk of this 154 *GW is pure pumped storage capacity that contains no renewable energy generation component but is instead a storage medium for grid power of any origin*

2 Including solar power and heat, wind power, hydropower, ocean energy, geothermal power and heat, and modern bioenergy.

Figure 7.5 IRENA 2017 Rethinking energy report – renewable vs. non-renewable power capacity additions between 2001 and 2015 [5]

This growth of renewable energy production is only expected to continue its current trend, which as of 2012 surpassed the addition in power capacity of non-renewable power sources. This is shown in Figure 7.5 taken from the International Renewable Energy Agency report 'REthinking Energy 2017: Accelerating the global energy transformation' [5].

Renewable generation technologies present various challenges to our traditional models of production, transmission and distribution of electrical energy. I would like to highlight two of the most important in my opinion.

1. Variability of supply
2. Decentralisation of generation

7.2.1.1 Variability of supply

There are a range of energy production technologies that are encompassed under the renewables banner. Hydropower, wind, bioenergy, solar PV, geothermal, concentrated solar and ocean energy are amongst those being developed and deployed today.

Not all these technologies work in the same manner and have varying degrees on impact on the variability of energy supply.

In hydropower for example, as long as the dams are at capacity, the power generation can be dispatched in a controlled manner by manipulating the feed to the turbines. However, technologies that depend on the availability of a natural power source such as wind or solar offer less degree of control as they are dependent on the availability of the primary source of energy to be transformed into electricity.

The premise under which most of the existing electric power infrastructure was created is relatively simple, power is generated at a power plant, transmitted for some distance and then consumed at the other end of the lines. In this paradigm, if the consumer demands more energy, the generating stations ramp up their production (by burning more fuel or placing more generators online) and pump more energy into the network. The system is kept in a constant and delicate state of balance in which the energy being consumed is being produced in real time.

Under this model, large swings in either power production or consumption are relatively rare, and the bigger and more redundant the system is, the rarer these instances are making the system more resilient. In other words, losing a large portion of the generation being injected into the system or losing a large segment of the consumer base are relatively rare events.

The directionality and consistency of power generation and consumption have been the defining features of our modern networks since their inception. Many of the engineering calculations and decisions have been made around this model. Systems are planned, designed, built, operated and maintained having these principles in mind.

As more renewable sources of energy are introduced in the market, a greater degree of variation in the available generated energy needs to be dealt with. This is not to say that operating systems with large proportion of renewable sources is impossible, but rather that new planning and operating paradigms will have to be developed and used into the future of electric energy production, transmission and consumption.

7.2.1.2 Decentralisation of generation

The second aspect of renewable energy that is worth mentioning is the decentralisation of the energy generation sources.

As described in the previous section, our traditional network model is that of large power generating stations where a primary source of energy is converted into electricity, its characteristics appropriately modified by a step-up transformer, transmitted over relatively long distances and then converted again at the receiving end in order to make it suitable for consumption.

The generating stations would be planned and built around strategic locations where the fuel source was readily available. Whilst similar approaches can be followed with some renewable technologies such as hydropower, solar and wind; it is also true that now generation can come from the consumer.

Until the advent of residential solar photovoltaic (PV) systems, never before in the history of electric networks had consumers had the ability to push energy back into the system. In some countries, such as in Australia, the number of households with rooftop PV generation is as high as 16% [6].

Distribution networks will have to adapt to reverse flows of electricity coming from a multitude of small generating stations (i.e. from each house). What would the implications be for system stability? Protection systems? Revenue? These are all questions that we have yet to answer.

7.2.2 Energy storage

Finally, given the conditions and new energy sources mentioned before, it seems fitting that new energy storage solutions are devised. If the continuity in the energy availability of these renewable sources cannot be relied upon, we will have to develop means to store this energy when it is available and consume it when it is necessary, which will not necessarily occur at the same time of the day.

Some energy storage technologies are not new, such as pumped hydro storage. However, the improvement in battery energy densities is making new energy options feasible.

Large-scale battery storage systems are currently being deployed such as in the case of South Australia which at the time of writing will be the biggest storage of its kind [7].

This represents the design, installation, operation and maintenance of assets that had never been seen before in electrical networks, at least in this scale.

The networks of the future will need to adapt to these new technologies and make the best use of these new capabilities.

7.3 Impact on asset management strategies

With all these changes in the electrical energy landscape, the question that needs to be answered from the perspective of the Asset Manager is: so what? In other words, how do we best respond to all these changes in the way we produce and consume electrical energy? What changes should we make, if any, to current strategies? With the industry in a state of flux, how much can we rely on our current forecasts and plan accordingly?

In my opinion, these questions can be answered by looking at the ways we operate, maintain and replace ageing transformers in existing electrical networks.

While the electrical network as a concept makes a transition between what it is at the moment and what it will be in the future, we will have to find ever more efficient, reliable and safe ways to operate the existing infrastructure, of which power transformers are a major component.

Also, our decision-making frameworks will have to be nimble and flexible in order to adapt to a constantly changing environment.

Most modern asset management systems have been designed using reliability centred maintenance principles [8,9]. Using reliability centered maintenance (RCM), the asset manager systematically analyses the function, failure modes, failure effects, consequences and corresponding actions that mitigate risks. With new technologies and forms to produce and distribute electrical energy, new functions and failure modes emerge in the context of networks with large populations of aged assets. This will prompt the revisit of our current assumptions, models and risk mitigation techniques in order to adapt to a new reality which has not quite settled yet.

7.3.1 Operation, maintenance and replacement of ageing assets

Traditionally, transformers have been expected to have an economic life that ranges from 30 to 50 years. While in some cases nameplate age is used as a reference for remaining life, the actual life of a particular transformer depends on a multitude of factors [10,11].

These factors include operational stresses such as loads and weather conditions play a role on how 'hard' the transformer is utilised. The amount and magnitude of abnormal events such as lighting and switching overvoltage as well as overload and through-fault current levels will also have an impact on the longevity of a particular transformer. Aspects of the design of each transformer will also have an influence on the transformer behaviour and resilience to these stresses as well as to normal operating conditions. Longevity of certain components such as bushings and tap-changers also play a role in the overall life equation.

All these elements ultimately determine how long a transformer will last. In general, the life of a transformer can end in one of the two ways:

1. Experiencing an unexpected and catastrophic failure or
2. Degrading to a point that is considered too unreliable for continued operation and thus being retired (i.e. de-commissioned).

If either of these outcomes occur in an uncontrolled fashion or earlier than necessary, it means valuable and scarce resources are being wasted or, at least, not being used in the most efficient manner possible.

This is an engineering challenge at its best. The question then becomes: how to use the available engineering knowledge and techniques to prevent failure from occurring in an uncontrolled mode or retiring a transformer too early? Both of which pose undue reliability, safety or financial risks.

In order to address the first point, we will have to apply all the collective knowledge gained about transformers so far and apply the best analytical tools in order to consistently identify and address emerging failure modes before they become life threatening for the transformer.

Addressing the second point will require questioning and, if applicable, changing of the criterion we have used so far to judge end-of-life for power transformers and the decisions it drives. In recent times, some authors have started to suggest that we look more closely at these criteria [12] and consider what would be the impact of changing it.

Knowing that a failure mode is present in a transformer is useful, but the other 50% of the equation is in what to do with this information. New techniques might have to be devised to economically repair defects that otherwise would have condemned the unit for decommissioning.

Novel field monitoring and diagnosis techniques will have to be explored in order to provide this information in an opportune and efficient manner so the right decision can be made at the right time.

Better solutions might have to be developed to replace or repair critical components (i.e. bushings and on load tap changers) whose life expectancies are somewhat misaligned with that of the core and coils of the transformer.

Obsolescence criteria might have to be revisited. We might have to investigate smarter refurbishment options for obsolete components that in the past would have condemned these components for disposal.

In summary, as it has been the case in the past, a good amount of engineering solution-finding will continue to be required in the next few decades while our systems make the transition from the old to the new model of electrical energy generation, transmission and distribution.

7.4 The advent of artificial intelligence

Artificial intelligence, and in particular its subset of machine learning (ML) techniques, have had a notable re-emergence in the past decade. This progress is the result of continued advances in our understanding of how the brain works and processes information, the accessibility to ever more powerful computers and the availability of massive amounts of training data thanks to the internet.

The authors in this field have proposed that the neurons in our brain, and in particular those in the neo-cortex, are a hierarchically structured pattern recognition machine [13,14]. These neurons use physical connections to each other in order to recognise and store spatial and temporal relationships between the signals they receive and generate.

While an in-depth exploration of ML history, evolution and current state would require its own book and is outside the scope of this chapter, a simplified definition could be stated as follows. ML is a set of algorithmic techniques that attempt to emulate how the brain processes information in order to automate analytical model building.

Rather than programming every possible variation of a problem space, ML algorithms attempt to learn from existing data (i.e. inputs and outputs) and adjust the parameters of models that describe the correlations in this data. When properly calibrated, these models allow the machines to accurately classify data or predict outcomes when given novel inputs.

There are many specific mathematical techniques and algorithms that have been tried over the last few decades in order to achieve this objective. If you read on this topic, you will come across terms like supervised and unsupervised learning, clustering, artificial neural networks, deep learning neural networks, anomaly detection, etc.

These ML techniques are currently being used in a wide range of consumer-facing products and services. From the algorithms that select what appears on your Facebook® feed to the recommendation systems used to translate a piece of text or show you targeted advertising on Google® ads. The virtual voice-activated and conversational assistant called Siri® integrated to Apple® products such as the iPhone® makes use of these technologies too. Impressive and accurate results have been obtained in machine vision, speech recognition and natural language processing problems.

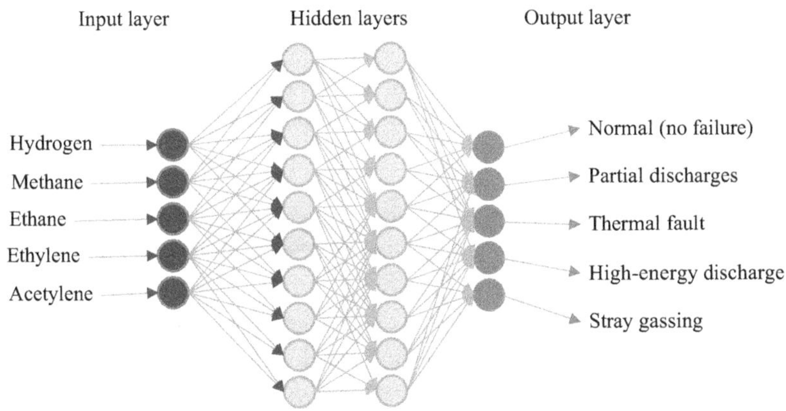

Figure 7.6 Conceptual example of neural network for DGA data analysis

The advances that have been achieved in this field have prompted enthusiasts, students, professors, engineers and companies all over the world to imagine a future in which these same technologies are used to identify specific failure modes or predict the moment in which a transformer is going to fail.

As many researchers have found, the application of these techniques to power transformer condition diagnosis is proving to be challenging. At their core, the development of ML models and algorithms depends on the calibration of the parameters of a network of artificial neurons which map complex input vectors (sometimes comprising hundreds of inputs) to a defined set of outputs. This is a process called supervised learning. A depiction of what this would look like for dissolved gas analysis for example is shown in Figure 7.6. In this case, a set of gas concentrations is used as an input in trying to identify or classify the failure mode present in the transformer.

For this technique to work, a sizable amount of examples is required in order to teach the algorithm which inputs correlate to which outputs. In classical image recognition problems, a set of images has been pre-tagged by humans and then fed to the ML algorithms. The algorithms then adjust the parameters of the neural networks in order to map the content of the images to the tagged outputs. This is an overly simplistic representation of the complexities behind these processes, but it should be sufficient to explain this point.

The amount of high-quality tagged failure mode and condition data is simply not enough for this research to provide practical outcomes. Given the lack of this data, many researchers [15–17] use failure classification techniques that are well known in the industry which their authors put together by painstakingly analysing study cases available to them and 'tagging' each case. Typical examples of these are the Duval Triangles and Pentagons, the IEC and IEEE guidelines and the various techniques that rely on gas ratios amongst others [18–22].

While dissolved gas analysis results have been heavily studied, classified and systematised to provide practical diagnosis tools, we have yet to see frameworks or tools with similar level of maturity that utilise all the other condition indicators that we have for power transformers, i.e. visual inspections (external and internal), electrical tests, etc.

This brings us to the question of: how mature is this technology in regards to its practical application to power transformers?

Many of these modern technologies can be placed in a continuum called the hype cycle.

> The hype cycle is a branded graphical presentation developed and used by the American research, advisory and information technology firm Gartner, for representing the maturity, adoption and social application of specific technologies. [23]

Although this is a very subjective and personal opinion, if I had to guess where the applicability of current artificial intelligence techniques is in regards to power transformer condition management, I would venture to say that it is somewhere between the 'Peak of Inflated Expectations' and the 'Slope of Enlightenment'. It could be argued that ML as applied to consumer-facing products has probably entered the 'Plateau of Productivity'. This has prompted very positive and hopeful expectations in regards to its application to industrial problems (Figure 7.7).

In my opinion, the main challenge we have in the power transformer industry in order to successfully develop these models is the availability of sufficient adequately classified, tagged and ordered data.

Our transformer databases have different shapes, names and degrees of quality. They are held in information silos and belong to thousands of different companies around the world.

I am an optimistic and I believe we will arrive to a future in transformer condition monitoring and diagnosis in which many of the mundane and repetitive

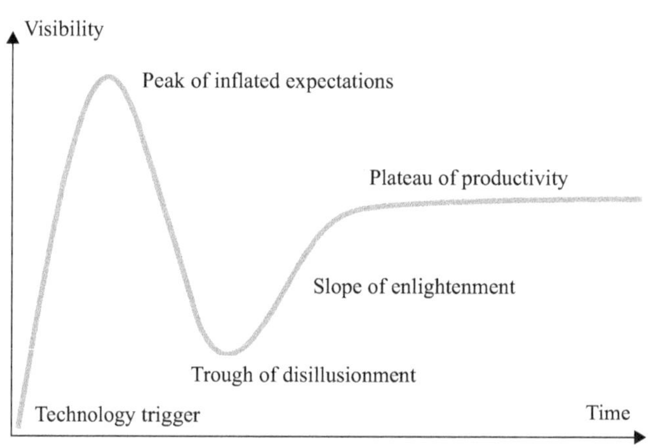

Figure 7.7 Gartner hype cycle [24]

diagnosis tasks have been automated and produce highly targeted and accurate assessments.

In this vision of the future, we have to develop practical mechanisms of cooperation that allow us to anonymously share sufficient and high-quality structured transformer condition and failure mode data. This in turn allows researchers to use this data to design and produce algorithms which help to improve the management of transformers around the world.

In this moment in time, I believe the best compromise we can make is to use all the power afforded by modern modelling tools, online monitors and computer systems to be an aid and an assistant to asset managers, subject-matter experts and decision makers.

7.5 Analysis automation as an aid to lifecycle management

If we are not yet at the point where we can just feed data into a computer and expect it to give us a rich and nuanced diagnosis with the level of quality we currently obtain from a subject matter expert, is there anything we can do with the available technologies? I think the answer is yes.

The condition assessment of power transformers, as well as all other assets in an electrical network, fits within a larger risk management framework. While the condition of the asset is important and probably half of the equation, other considerations such as consequence of failure, operational, legislative and compliance requirements are the other half. To the extent that these condition assessments are consistent, readily available and opportune, they will provide added value to the organisation owning these assets as they are fed into wider decision-making frameworks.

When the entire lifecycle of power transformers is considered, particularly in the context of large populations of existing assets, a wide range of decisions have to be made. Common questions asked during this lifecycle are

- Can a particular transformer take increased loads or do we need additional transformers?
- Do we need to replace or refurbish a transformer?
- When do we need to retire a transformer?
- Is it possible to extend the life of a transformer? And if so, what investment do we need to do it?
- How many spares do we need to keep in stock?
- What is the best maintenance task to perform on a transformer and when is the most opportune time to do so?

Having a formal methodology and framework to analyse condition and classify faults not only helps to answer these questions but also has side benefits like providing a centralised repository of asset knowledge and enabling agile condition status communication across the organisation. Remember the need for structured condition and failure information from the previous section? This formalised analysis framework would also function as a preparatory step in the ML journey.

Whilst there are many different and valid ways in which this problem could be attacked, an automated data acquisition and analysis framework could possibly be used as a solution. This framework would help realise additional value particularly in organisations where some level of automation already exists. When properly designed, the concepts and tools used to build such a system would not only be useful to manage transformers but also be extensible to other asset classes.

This system would need to fulfil two main requirements:

1. Consistently identify whether any failure mode is present in a given transformer and what the failure mode is
2. Given the current conditions of the transformer, determine the remaining life of the transformer

The following is a suggestion of conceptual components that could be used to construct this system.

7.5.1 Condition attributes

The first component has the responsibility of defining the best attributes that describe what the asset is (Descriptors), what attributes better correlate to its condition (Condition Indicators) and which attributes represent all the demands that the asset is subject to (Stress Factors).

In this context, condition is understood as the type and number of failure modes currently present (if any) on a particular transformer. These failure modes are those identified during an RCM analysis of the asset in question.

7.5.1.1 Descriptors

Descriptors do not change with condition and are normally based on the nameplate characteristics of the asset, attributes such as serial number, manufacturer, year of manufacture, etc. Additional to nameplate details, descriptors can include company specific fields like tag numbers, computerised system identifiers, etc.

7.5.1.2 Condition indicators

This category identifies all the attributes that characterise the condition of each asset. The condition indicators are ideally chosen to be as responsive to changes in the asset's condition as possible.

There are two types of condition indicators: subjective and objective.

The objective condition indicators are any attributes that can be measured and quantified without the need for human interpretation. Typical examples of objective measures are the result of electrical tests, oil analyses, age, temperature, pressures, etc.

The subjective condition indicators are those that do require a level of human interpretation in order to determine their current status. Examples of subjective indicators are visual inspections, like oil leaks, corrosion levels, etc.

In order to try to remove the subjectivity of the second type of indicators, the observations made on them can be captured as a selection from a determined set of choices rather than as an open text descriptions.

7.5.1.3 Stress factors

A stress factor identifies any element that places a demand on the asset. This demand is a contributor to the asset's loss of life and therefore affects its condition over time.

For example, the load placed on a power transformer causes a certain current to flow through its windings. This current in turn generates heat which translates in increased temperatures on the coils. Temperature has a direct correlation with the loss of life of the insulation on a transformer. Therefore, load can be considered a stress factor for a power transformer.

7.5.2 Measurements

A measurement is a specific instance of a condition attribute. These measurements should be time-stamped and associated with metadata such as name of the individual or instrument that acquired the data. Examples of measurements on a specific date and time are as follows:

- Winding or oil temperatures
- Load currents and terminal voltages
- Content of dissolved gases

In this context, condition attributes are the definitions of these parameters and measurements are the actual values of these attributes.

7.5.3 Analysis rules

An analysis rule can be seen as a black box that contains our current knowledge about the behaviour of a particular asset class, in this case, power transformers.

The inputs to this black box are the descriptors, condition indicators and stress factors and the outputs are the identified failure modes (if any) and a score that quantifies our confidence on this assessment. The complexity of these rules is wide ranging and can vary between simple functions and formulas to entire algorithms that take into account several inputs and perform more complicated processing.

Another characteristic of the analysis rules is that they are composable. This means that the output of a particular rule could be used as the input of a more complex rule or algorithm.

In general, analysis rules can be classified in three broad categories, classification rules, limit rules and algorithmic rules.

Once the desired set of rules has been defined, the data available for a selected portion of the asset base is processed through these rules. An individual subject matter expert (SME) or group of SMEs then validates the outcomes of these rules and whether they correlate well with the known current state of these assets.

As new learning is gained from failures, root cause analyses and other types of defects, the set of rules for a specific asset class and failure mode is amended to reflect the gained knowledge.

A set of rules has to be defined for each failure mode identified during an RCM analysis. For example, if 'hot internal connection' has been identified as a potential failure mode, a set of rules that utilise the appropriate Condition Attributes to determine how likely it is that a particular transformer has an active 'hot internal connection' failure mode has to be defined.

7.5.4 Implementation tool

The final component of the system is the enabler that allows the processes necessary to run this system to occur in an automated manner. This typically takes the shape of some form of software or customised programme.

The tool needs to be able to ingest the raw condition attributes data in the form of measurements, execute the analysis rules and provide the means to collate, visualise and interrogate the outcomes of this process.

In its simplest form, this system could be implemented in a software package like Microsoft® Excel. On the other hand, a custom solution could be acquired or developed to meet the specific needs of this process. In large organisations, this could mean the integration of disparate systems with specific functions and data protocols.

In general, the simplest the solution is, the more cost effective is to implement but also tends to be the least scalable or flexible.

7.6 The digital substation

As discussed throughout this chapter, the future of electrical power networks is currently in a state of flux, and it would be very hard to predict exactly how that future looks like. However, as we have also discussed, we have a good idea of what technologies and processes are likely to play an important role in that future.

The power transformer industry is comprised of an ecosystem of manufacturers, service providers, consultancies, utilities and end users. All these actors have different interest and drivers which emerge from looking at the same problems from various angles.

As technological advancements acquire more capabilities at reduced costs, they will enable new business models to become viable and realise the benefits that are required to keep moving forward.

The reader will find many definitions when searching the term 'digital substation'. In the context of this chapter, I am using digital substation as a term that encompasses the set of technical standards, communication technologies, monitoring devices, analysis algorithms, field services, software tools and business processes that would enable the execution of substation maintenance activities in one seamless electronic value chain. From capturing the data to performing the analysis and making asset management decisions in a reliable and repeatable way.

In the last few sections of this chapter, we will explore the various components or sub-systems that would enable a digital substation to become a reality.

7.6.1 Value proposition

As a first item, we need to look at the value proposition of the digital substation concept. It would not be justifiable to implement any of the solutions we have talked about if they did not provide additional value to the business that owns the assets (i.e. substations). The value proposition has to answer a very simple question: why are we doing this?

In the early days of my engineering career, I had a manager that used to say, 'This is not an engineering playground'. He was referring to the fact that we engineers are curious by nature and sometimes that curiosity and pure sense of exploration can distract us from the desired outcomes of a particular project. We get so engaged in the how we are doing something that we might lose sight of the why we are doing it.

The digital substation space is full of shiny and wonderful technologies that we could only have dreamt of a decade or two ago. However, and as much as I like to play with technology myself, we need not to lose sight of the end goal which is to create a more reliable, efficient, safer, environmentally responsible and less costly electrical power network.

The value of implementing this type of solution needs to be clearly identified and stated from the start of the project. The sense of having a clear goal and a set of deadlines helps to focus the energy of all the members of the team and gives the project a better chance to have a successful outcome.

7.6.2 Technical standards

There is a myriad of technical standards that are relevant to the concept of digitisation of electrical networks. Sometimes this is referred to as the smart grid.

In the context of the digital substation, there is a set of technical standards that are highly relevant in this area, the 'IEC 61850 – Communication networks and systems in substations' series [25].

This series of technical documents defines a set of philosophies, requirements and protocols that allow the integration and communication of intelligent electronic devices.

The IEC 61850 standard enables the interoperability of hardware from various vendors and the standardisation of software interfaces.

This standard has found wide international acceptance and all major hardware vendors offer support for it on their devices.

7.6.3 Hardware and software technologies

In large organisations, there is a wide range of hardware types and models as well as a good amount of individual software packages that make up the landscape for that particular enterprise.

Whilst there might be opportunities to build a single monolithic system that covers all the requirements, it is more likely that hardware and software will have to become highly modular and agnostic of each other.

The best analogy I can think of in this case is the Lego® brick system. Hardware and software, through the use of widely used technical standards should become agnostic and independent of each other. Each piece of hardware or software should be able to work with any other piece of hardware and software.

The end user, i.e. the organisation implementing the digital substation concept, should be able to create highly customised and valuable solutions by combining the necessary components.

It is likely that companies that own large asset bases already have enterprise management, maintenance management and custom application systems in place. Building digital substations in this environment will benefit the most when using devices and software components that can easily integrate with larger corporate systems. This ease of integration will also result in reduced implementation costs and faster iteration cycles as companies try different configurations and particular setups in search of the one that better suits their needs.

7.6.4 Business processes

Last but not least, none of the potential value created by the digital substation can be realised if the right business processes are not implemented. If the data and information is available but there is not a clear way to use it to drive decisions around the organisation, the value of this data is lost in databases and spreadsheets that do not have a practical impact on the goals that the organisation has set for itself.

The right business processes move the implementation of a digital substation from the nice engineering toy space into a core business pillar that drives critical decisions and allows the organisation to derive true value.

7.7 Summary

In this chapter, we started by exploring the changing landscape of the electrical energy industry. We examined the economic drivers and different speeds at which various regions of the world are moving. We discussed how renewable energy production technologies and energy storage solutions are likely to change the traditional assumptions and models under which the electrical networks of the present were designed and built. We also discussed how these factors are creating new demands and the requirement of new solutions in the various aspects of electrical network management.

We established that all these factors will have an impact on our current asset management strategies, in particular how we operate and maintain power transformers.

We then reviewed what possible technologies can be used to respond to these new challenges, and how artificial intelligence and in particular ML might soon be capable of aiding in the condition assessment of large fleets of assets.

We analysed how these technologies can be integrated to an end-to-end condition monitoring and analysis solution.

And finally, we looked at how all these could possibly integrate using the concept of a digital substation.

I hope that even if it is in a small measure, these ideas have sparked interesting questions on the reader's mind. Useful questions that can be taken to each of the stakeholders that play a role in this industry.

As it has happened throughout human history, it will ultimately be the people that works in these organisations, the ones that ask the right questions and create the best solutions to the particular problems they are facing and, as in the process, move the world forward another notch.

References

[1] World Bank, World Development Indicators. *The Shift Project Data Portal.* [Online] [Cited: 01 10 2017.] www.tsp-data-portal.org.

[2] World Bank, World Development Indicators. Energy Production by Country. *Gapminder*. [Online]. [Cited: 01 10 2017.] https://www.gapminder.org/tools/#_locale_id=en;&state_time_value=2010;&entities_show_geo_$in@=usa&=rus&=chn&=ind&=jpn&=deu;;;;&marker_axis/_y_which=energy/_use/_total&scaleType=linear;&color_which=country;;;&chart-type=linechart.

[3] U.S. Energy Information Administration. International Energy Outlook 2017. [Online] 14 09 2017. [Cited: 08 10 2017.] https://www.eia.gov/outlooks/ieo/pdf/0484(2017).pdf.

[4] Renewable Energy Policy Network for the 21st Century. Global Status Report. [Online] 2017. [Cited: 08 10 2017.] http://www.ren21.net/status-of-renewables/global-status-report/.

[5] International Renewable Energy Agency. IRENA REthinking Energy 2017. [Online] 2017. [Cited: 08 10 2017.] http://www.irena.org/DocumentDownloads/Publications/IRENA_REthinking_Energy_2017.pdf.

[6] Australian PV Institute. Mapping Australian Photovoltaic Installations. [Online] 2017. [Cited: 08 10 2017.] http://pv-map.apvi.org.au/historical#4/-26.67/134.12.

[7] ABC News. Tesla to supply world's biggest battery for SA, but what is it and how will it work? *ABC News*. [Online] 08 07 2017. [Cited: 08 10 2017.] http://www.abc.net.au/news/2017-07-07/what-is-tesla-big-sa-battery-and-how-will-it-work/8688992.

[8] SAE International. *Evaluation Criteria for Reliability-Centered Maintenance (RCM) Processes*. s.l.: SAE International, 2009. JA1011_200908.

[9] IEC. *Dependability management - Part 3-11: Application Guide – Reliability Centred Maintenance*. s.l.: IEC, 2009. IEC 60300-3-11:2009.

[10] Gamez, Carlos. *Power Transformer Life*. Transformers Magazine. s.l.: Merit Media Int., 2014. Vol. 1, 1,2,3.

[11] Gamez, Carlos. *Life Assessment of Power Transformers*. Sydney: TechCon Asia Pacific, 2013.

[12] Duval, Michel. *Transformers with Low Degree of Polymerisation of Paper*. Transformers Magazine, 06 10 2014, Vol. 1, 3, pp. 26–31.

[13] Hawkings, Jeff and Blakeslee, Sandra. *On Intelligence: How a New Understanding of the Brain will Lead to the Creation of Truly Intelligent Machines*. s.l.: Times Books, 2004. ISBN 0-8050-7456-2.

[14] Kurzweil, Ray. *How to Create a Mind: The Secret of Human Thought Revealed*. s.l.: Penguin Books, 2013. ISBN-10: 0143124048.
[15] Bakar, N. and Abu-Siada, Ahmed. *Fuzzy Logic Approach for Transformer Remnant Life Prediction and Asset Management Decision*. IEEE Transactions on Dielectrics and Electrical Insulation, 2016, Vol. 23, 5. 10.1109/TDEI.2016.005936.
[16] Barle, Nisha, Jha, Manoj Kumar and Qureshi, M. F. *Artificial Intelligence Based Fault Diagnosis of Power Transformer – A Probabilistic Neural Network and Interval Type-2 Support Vector Machine Approach*. International Journal of Innovative Research in Science, Engineering and Technology, 2015, Vol. 4, 1. ISSN(Online): 2319-8753.
[17] Souahlia, Seifeddine, Bacha, Khmais and Chaari, Abdelkader. *Artificial Intelligence Tools Aided-decision for Power Transformer Fault Diagnosis*. International Journal of Computer Applications (0975–8887), 2012, Vol. 38, 3, pp. 1–8.
[18] CIGRE. *DGA in Non-Mineral Oils and Load Tap Changers and Improved DGA Diagnosis Criteria*. s.l.: Working Group D1.32, 2010. 443.
[19] CIGRE. *Recent Developments in DGA Interpretation*. s.l.: Joint Task Force D1.01/A2.11, 2006. 296.
[20] Institute of Electrical and Electronics Engineers (IEEE). IEEE Guide for the Interpretation of Gases Generated in Oil-Immersed Transformers. 2008. C57.104-2008.
[21] International Electrotechnical Commission (IEC). Mineral Oil-Filled Electrical Equipment in Service – Guidance on the Interpretation of Dissolved and Free Gases Analysis. 2015. IEC 60599:2015.
[22] Duval, Michel. *A Review of Faults Detectable by Gas-in-Oil Analysis in Transformers*. IEEE Electrical Insulation Magazine, May-June 2002, Vol. 18, 3, pp. 8–17.
[23] Wikipedia. Hype Cycle. *Wikipedia*. [Online] [Cited: 21 10 2017.] https://en.wikipedia.org/wiki/Hype_cycle.
[24] Kemp, Jeremy. Gartner Research's Hype Cycle Diagram. *Wikipedia*. [Online] 27 12 2007. [Cited: 21 10 2017.] The underlying concept was conceived by Gartner, Inc. https://en.wikipedia.org/wiki/File:Gartner_Hype_Cycle.svg.
[25] IEC. Communication Networks and Systems for Power Utility Automation – ALL PARTS. s.l.: IEC, 2017. IEC 61850:2017 SER.

Further reading

[1] Gamez, Carlos. *Power Systems Assets Management in the Digital Age*. Auckland, New Zealand: Electricity Engineers' Association, 2015.
[2] Saha, Tapan K. *Review of Modern Diagnostic Techniques for Assessing Insulation Condition in Aged Transformers*. IEEE Transactions on Dielectrics and Electrical Insulation, 2003, Vol. 10, 5.

Chapter 8

Power transformer asset management and remnant life

Norazhar Abu Bakar[1]

Abstract

The quality of insulation system within power transformer reflects the overall health condition of the asset. A timely and reliable maintenance decision along with the remaining operational life estimation of the asset can be identified through the measurements of some parameters reflecting the degradation rate of the transformer dielectric insulation. However, the process is not always straightforward due to the complexity of the insulation structure and degradation process. This chapter proposes a new fuzzy-logic approach to provide a proper asset management decision and predict the remaining operational life of a power transformer based on some insulating oil tests. The developed fuzzy-logic model is validated through field data collected for various transformers of pre-known health condition and life span. The results show that the proposed model is reliable and can be facilitated to provide a timely asset management decision with less reliance on expert personnel.

8.1 Introduction

This chapter addresses developing an expert model to estimate the remnant life and asset management decision of power transformers based on routine insulating oil tests.

Statistics show that the bulk of power transformer fleets within developed countries such as the United States, the United Kingdom, and Australia are approaching or have already exceeded their design lifetimes as they were installed prior to 1985 [1]. With the current global economic crisis, the mindset within the electricity utility industry is centred on getting the most usage out of existing equipment rather than installing replacements. An ageing population of large power

[1]Faculty of Electrical Engineering, Universiti Teknikal Malaysia Melaka, Malaysia

transformers requires more reliable monitoring systems and diagnostics to detect incipient faults and to provide an efficient predictive maintenance plan that can extend the operational life of the asset and minimize the possibility of catastrophic failures [2].

Several papers investigating the correlation between some condition monitoring parameters such as furan, dissolved gases, water content, and temperature with the transformer's remnant life using mathematical-based approaches can be found in the literature [3–6]. However, due to the complexity in developing a mathematical model for such a non-linear problem, researchers have limited their models to only one or two parameters without giving attention to all the accelerating ageing factors, thereby risking making an incorrect estimation for the transformer's remnant life [7]. IEEE Std. C57.91-2011: Guide for Loading Mineral-Oil-Immersed Transformers and Step-Voltage Regulators [8] and IEC Standard 60076-7: Power transformers – Part 7: Loading guide for oil-immersed power transformers [9] proposed mathematical models to estimate a transformer's remnant life based on hot-spot temperature within the transformer. Two other parameters, water content in paper and dissolved oxygen in oil, are considered in one proposed remnant life mathematical-based model [10]. A further two parameters, 2-furfural (2-FAL) concentration in oil and the degree of polymerization (DP) of paper insulation which reflects the remaining life of solid insulation and hence a transformer, have also been found to correlate [6].

In addition to the inaccuracy of these mathematical-based models due to neglecting other key parameters that contribute to transformer ageing, the validity of measurements using the proposed parameters in these models such as hot-spot temperature and water content in paper is questionable.

To overcome the limitations of mathematical-based approaches in estimating transformer remnant life, some models based on artificial intelligence (AI) have been developed. An AI-based model using gene expression programming and dissolved gas analysis (DGA) parameters was developed to identify the criticality of power transformer and asset management [11]. Also, another study proposed a fuzzy-logic-based model using furan and carbon oxides to assess the transformer ageing condition [12].

Although more condition monitoring parameters have been considered in AI-based models, none of the proposed models considered all of the key factors affecting the transformer life, such as rise in operating temperature, oxygen level, and water content in paper. Moreover, most of the existing transformer remnant life estimation models were developed without incorporating a reliable asset management decision model. An AI-based asset management model was developed using expert machine learning and three condition monitoring parameters: DGA, polarization and depolarization current (PDC), and dielectric frequency domain spectroscopy (FDS) [13]. However, the parameters (PDC and FDS) used by this model are not regularly measured through transformer's routine inspections.

A transformer remnant life and asset decision fuzzy-logic model was proposed in [14], which considered several condition monitoring parameters such as partial discharge (PD), leakage reactance, dielectric dissipation factor, and frequency response analysis (FRA); however, the model cannot be implemented online as several input parameters can only be measured offline. Moreover, not all parameters required by the model, such as FRA, are available during routine inspection of transformers. Statistical analysis for transformers' failure-rate population is used in [15,16] to estimate the transformer remaining life based on analytical model. However, this approach is affected by various environmental conditions and maintenance procedure which reduce the accuracy of the developed estimation model.

Another approach to assess transformer condition using a health index (HI) or a score table is proposed in [17,18]. The HI-based technique is developed based on a pre-defined weighting factor for each parameter and uses historical data which calls for expert personnel. This means results from HI-based techniques can be inconsistent when developed by various experts and used by various utilities [19].

This chapter proposes a new fuzzy-logic approach that utilizes data gathered mainly from insulating oil routine assessment such as furan, DGA, interfacial tension (IFT), water content, and operating temperature in evaluating the remnant life and health condition of power transformer.

The key advantage of the proposed model in this work over previously published models is that all input parameters proposed in the model can potentially be measured online or on-site which facilitates a proper and timely maintenance action based on the model output [20–24]. The model also considers the rate of increase of key parameters that significantly affect transformer health condition, such as 2-FAL, carbon monoxide (CO), and IFT. The generation rate of furan's derivatives in transformer oil has been found to be more important in identifying paper health condition than its absolute level [3]. The CO increment rate in insulating oil along with the CO_2/CO ratio provides valuable information about paper degradation activity [25]. On the other hand, the IFT decrement rate has been found to be more effective than the oil acid number increment rate in identifying oil contamination criticality [26].

8.2 Transformer health condition

The quality of the insulation system within power transformers that comprises dielectric insulation paper and oil reflects the overall health condition of the transformer. The combination of heat (pyrolysis), moisture (hydrolysis), and air (oxidation) within operating transformers causes oil and paper decomposition which results in a number of gases that relates to the cause and effect of various faults [1]. Faults such as overheating, PD, and sustained arcing produce a range of gases such as hydrogen (H_2), methane (CH_4), acetylene (C_2H_2), ethylene (C_2H_4), and ethane (C_2H_6) that dissolve in insulating oil and can be quantified using DGA. The concentration of these gases in transformer oil can aid in identifying the criticality level of the transformer health condition [20].

Chemical products such as carbon oxides and furan derivatives are formed due to paper insulation degradation and can be detected through dissolved gas in oil and furan analyses, respectively. As transformer operational life is strongly correlated to the age of paper insulation, carbon oxides and furan in oil are widely used as indicators to predict the transformer remnant life and paper ageing criticality [5]. However, any oil treatment will eliminate carbon oxides and furan contents, hence limiting the reliability of these gases in estimating the transformer's remaining life [2]. It takes approximately 3 years for furan concentration to retain its original value before oil reclamation [27].

High operating temperature, water content in paper, and oxygen dissolved in oil accelerate the degradation of paper insulation. Studies show that the rate of ageing of a transformer with a high level of oxygen (above 16,500 ppm) and wet paper insulation is 40 times the ageing rate of a similar transformer with a low level of oxygen (below 7,000 ppm) and dry paper insulation [4]. Additionally, the transformer's remnant operational life is halved for every 6–8 °C increment of the operating temperature [8,9].

On the other hand, the quality of insulating oil influences the performance and the service life of the transformer [28]. During the oil ageing process, oil gets contaminated with dissolved decay products such as organic acids, peroxides, aldehydes, and ketone formed as a result of the chemical reaction between mineral oil molecules and oxygen dissolved in oil [29]. The acids formed due to this oxidation process attack the metal tank, forming sludge that reduces the dielectric strength of the oil. Meanwhile, the acids also attack the cellulosic chain of the paper, causing accelerating insulation degradation [17]. Sludge and contamination development in insulating oil can be identified by measuring the IFT value of the oil. Oil is considered severely contaminated and sludge is expected to increase when the IFT value drops to 22 mN/m [17].

DGA, furan analysis, water content, and IFT are normally performed during routine transformer inspection. Conventionally, these analyses are conducted in a laboratory environment. Recently, several analytical techniques have been developed which allowed these parameters be measured online or on-site. Dissolved gases in insulating oil can be measured online by using photo-acoustic spectroscopy or a hydrogen online monitoring [20].

Meanwhile, the use of UV–vis spectroscopy to measure the furan (2-FAL) concentration and the IFT value of the transformer oil, which can be performed on-site and has the potential to be implemented online, has been proposed by Abu-Siada *et al.* [21] and Bakar *et al.* [22]. Instead of using the Karl Fischer titration technique to measure water content in oil, Martin *et al.* [24] proposed the use of water-activity probes immersed in transformer oil to measure the water concentration.

Table 8.1 summarizes the parameters used for the asset management and remnant life model proposed for this study.

Table 8.1 Proposed parameters for transformer remnant life and asset management decision model

Category	Parameters
Remnant life estimation	Furan, water content, oxygen, operating temperature rise
Paper ageing criticality	Furan, carbon oxides
Faults criticality	Fault gases (CH_4, C_2H_2, C_2H_4, C_2H_6)
Contamination criticality	IFT
Relative accelerated ageing criticality	Water content, oxygen, operating temperature rise

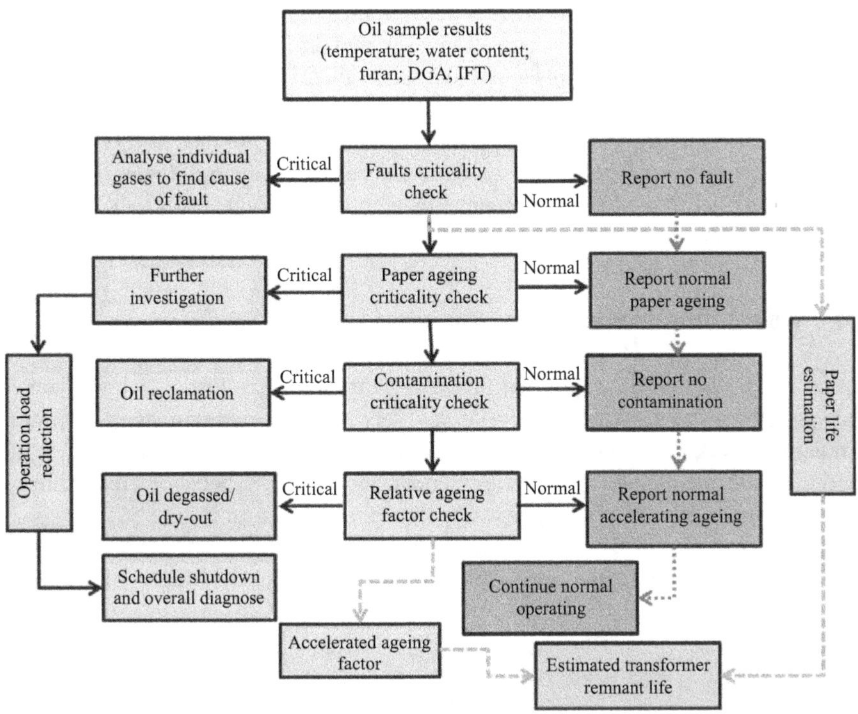

Figure 8.1 Flow chart of the proposed approach [31]

8.3 Proposed approach

An asset management decision model was designed as proposed in the flow chart of Figure 8.1. Four criticalities comprising faults criticality, paper ageing criticality, contamination criticality, and relative accelerating ageing are examined to allow a proper maintenance action on the transformer. As shown in Figure 8.1, the faults criticality is examined based on DGA data to determine the health condition of the transformer which accords with IEEE Std. C57.104 guidelines [25]. If the faults criticality check is positive, further analyses on individual dissolved gases are to be performed to find the fault causes. The condition of paper insulation is examined through the furan and carbon oxides data in accordance with the guidelines reported in [17].

If paper ageing is found to be critical, the trend of increases of the critical factor is to be monitored through observing the increment of the monitored parameters per a particular period that depends on the significance of the measured parameters, and operational load reduction is recommended. The contamination level in oil is to be examined through the IFT value, as recommended in the IEEE Std. C62-1995 [30]. If the contamination level is found to be critical, oil is recommended for reclamation.

The relative accelerated ageing factor is examined through transformer operating temperature, water content in paper, and oxygen in oil. If this factor is found to be critical in accordance with [17], oil is recommended for rehabilitation through oil degassing or dry-out. In cases where all investigated criticality factors are normal, the transformer is reported as healthy and is to be scheduled for the next routine inspection.

Paper ageing criticality is used to estimate the remaining operational life of paper insulation in accordance with the recommendation published in [17]. Accelerated ageing factor is determined by relative accelerated ageing factor criticality as presented in [29]. Transformer remnant life is obtained by integrating the estimated paper remaining life and accelerated ageing factors, as shown in Figure 8.1.

8.4 Fuzzy-logic model development

A fuzzy-logic model to implement the flow chart shown in Figure 8.1 was developed based on the pre-known correlations between paper ageing, faults, oil contamination level, and relative accelerating ageing factors with transformer health condition and oil maintenance recommendations. A fuzzy decision trees concept was constructed based on a 'top-down' method, as shown in Figure 8.2.

The root node of the model is the transformer health condition, while paper ageing, faults, contamination criticalities, and relative accelerating ageing are the

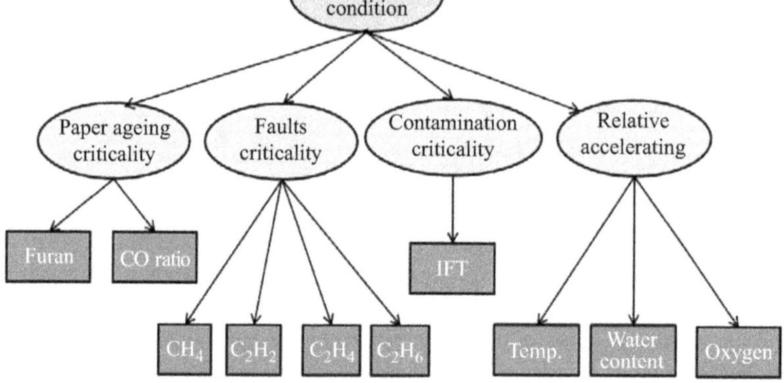

Figure 8.2 Fuzzy decision tree concept [31]

internal nodes. Furan, CO_2/CO ratio, fault gases (CH_4, C_2H_2, C_2H_4, and C_2H_6), IFT, operating temperature, water content in paper, and oxygen dissolved in oil are the leaf nodes or input variables to the model.

The proposed overall fuzzy-logic asset management decision model is shown in Figure 8.3. The development of the fuzzy sub-model for the furan factor is described in detail below. The same procedure was used to develop the fuzzy models for the other factors.

8.4.1 Furan criticality

Furan concentration (2-FAL) is generated by paper degradation and is normally used to estimate the paper remaining life. A fuzzy model was developed to estimate the paper's remaining life where furan concentration and furan generation

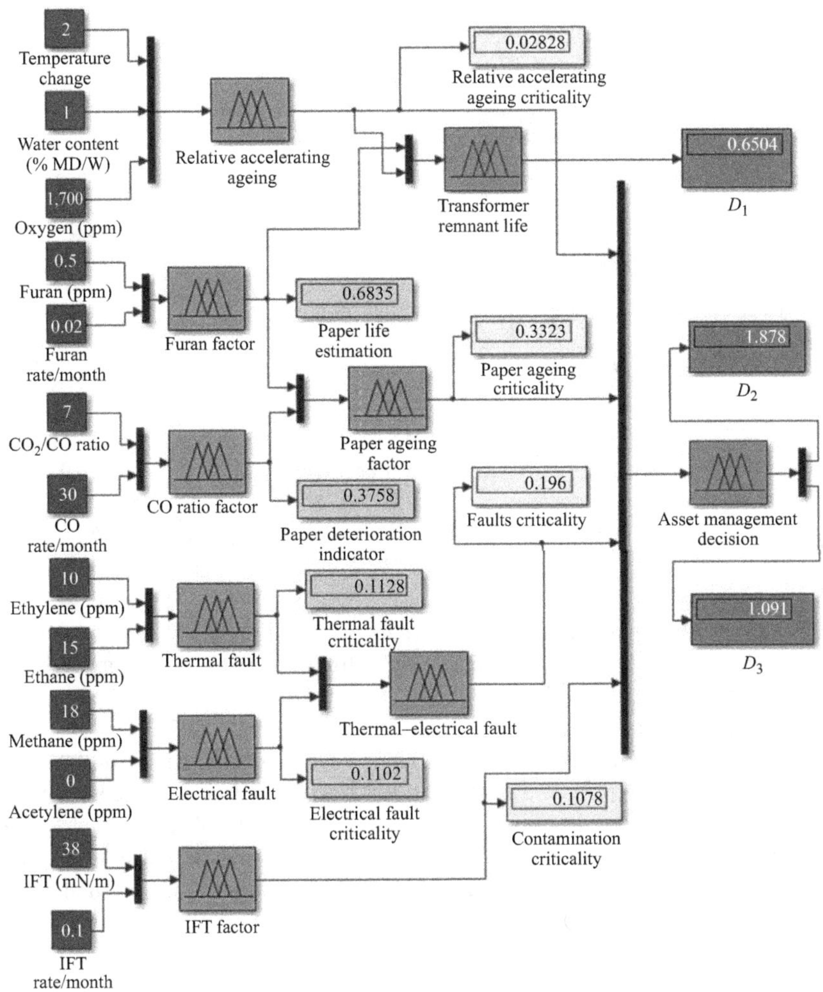

Figure 8.3 Proposed fuzzy-logic asset management decision model [31]

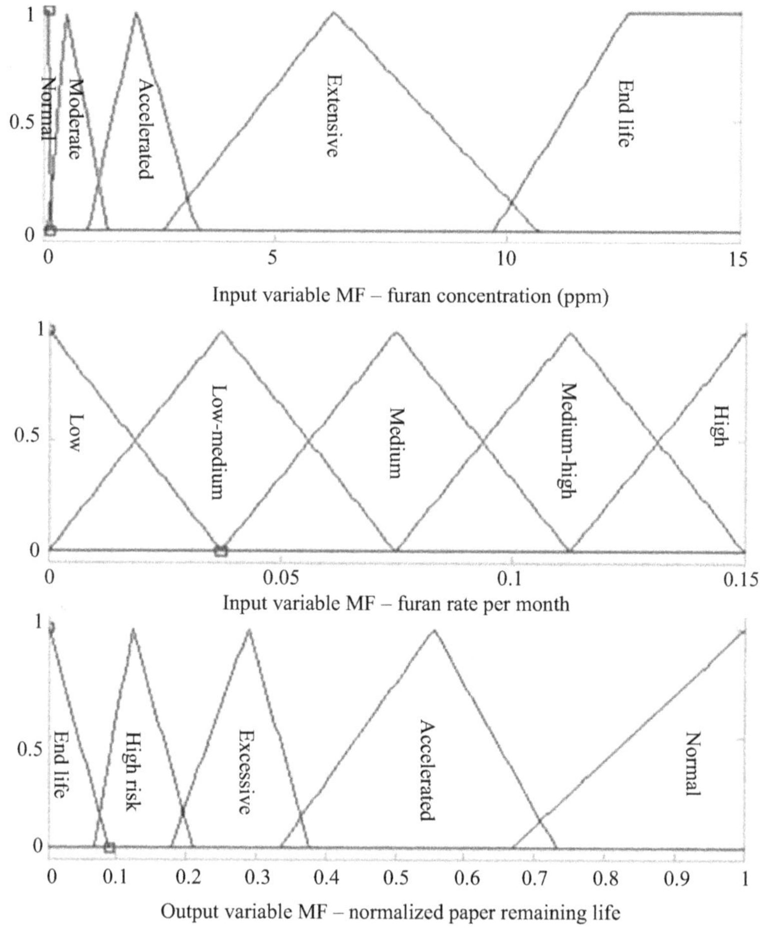

Figure 8.4 Input and output variables' MF for furan fuzzy model

rate/month are the inputs to the model in parts per million (ppm) and ppm/month, respectively, while the output variable is the estimated paper remaining life which is normalized on a scale 0–1, whereas 1 reflects new paper with 100% remaining life, and 0 reflects the end of the paper's life.

There are five membership functions (MF) based on furan concentration in oil: 'Normal', 'Moderate', 'Accelerated', 'Extensive', and 'End Life, and five sets based on furan rate/month named 'Low', 'Low-Medium', 'Medium', 'Medium-High', and 'High'. These were developed based on the correlation between furan and paper condition [17], as shown in Figure 8.4. The output is represented by five triangular MFs named 'Normal', 'Accelerated', 'Excessive', 'High Risk', and 'End Life'.

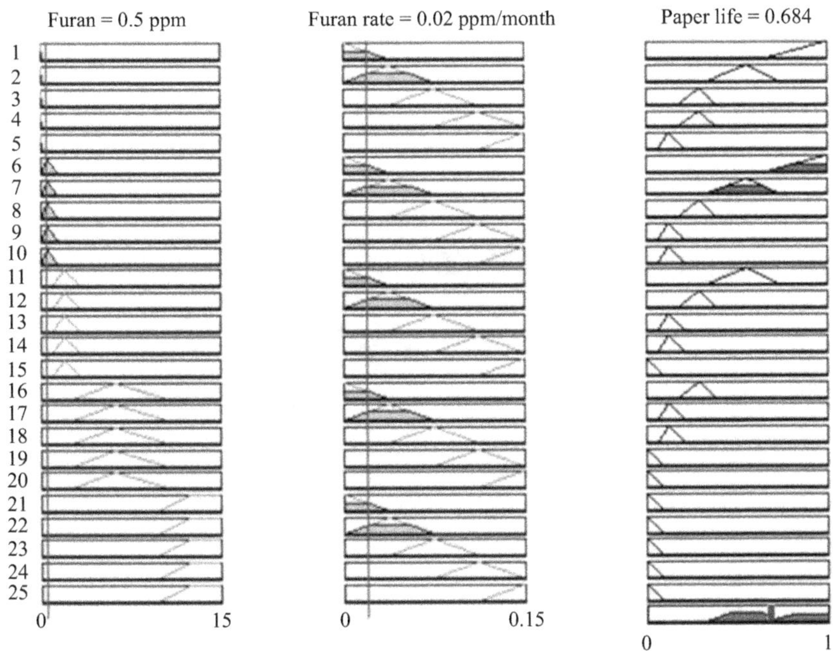

Figure 8.5 Developed fuzzy rules for paper life estimation based on furan

Fuzzy-logic rules in the form of 'IF-AND-THEN' were developed to correlate input and output parameters based on the correlation between furan concentration/ rate of increment per month and paper remaining life [17], as shown in Figure 8.5, which shows the 25 rules developed to map this correlation.

The model was tested for input parameters, 0.5 ppm furan concentration and an increment rate of 0.02 ppm/month. The model output is 0.684 i.e. the estimated remaining life of paper insulation is 68.4% of its designed life which is normally 40 years [17]. This output is consistent with the correlation of paper DP, furan, and paper life developed in [17]. The paper life estimation for any set of input data (furan and furan increment rate) can be estimated using the surface graph shown in Figure 8.6.

8.4.2 CO ratio criticality

CO and CO_2 are the main gases evolved from paper degradation. A set of fuzzy rules correlating CO_2/CO ratio and CO generation rate/month with paper deterioration was developed in accordance with IEEE Std. C57.104 recommendations [25]. MF for CO_2/CO ratio and CO generation rate/month are considered on the scale 0–15 and 0–350 ppm/month, respectively, as shown in Figure 8.7. The output variable, paper deterioration, is assumed to be in the range 0–1. Output approaching 1 indicates significant paper deterioration while output close to 0 reflects normal paper condition.

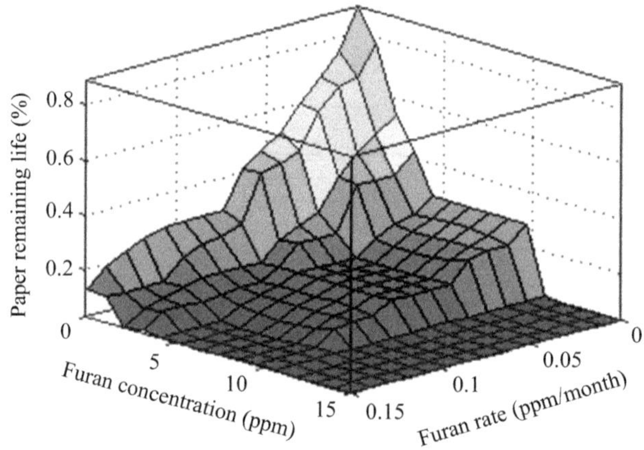

Figure 8.6 Paper life estimation surface graph

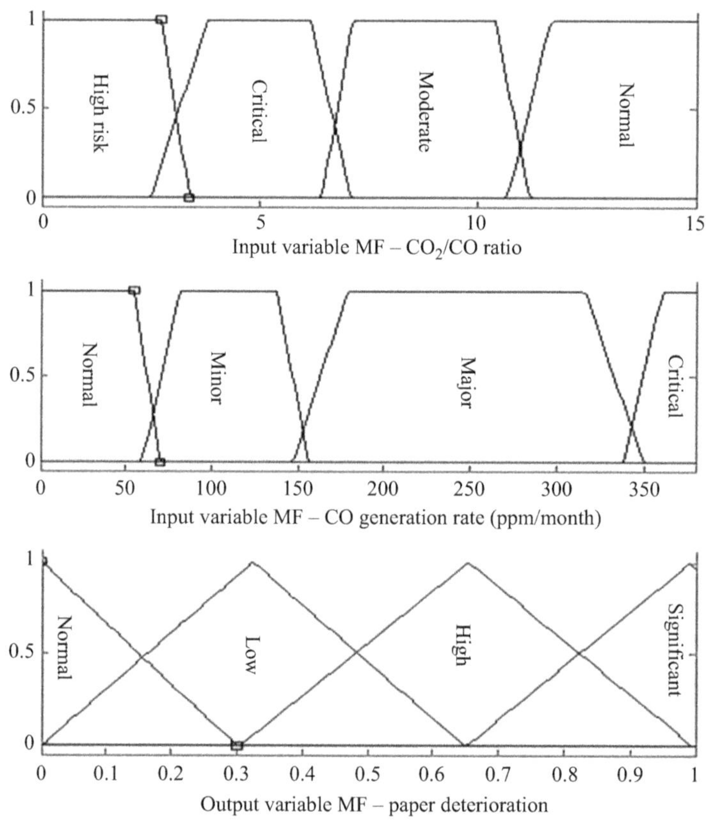

Figure 8.7 Input and output variables' MF for CO ratio fuzzy model

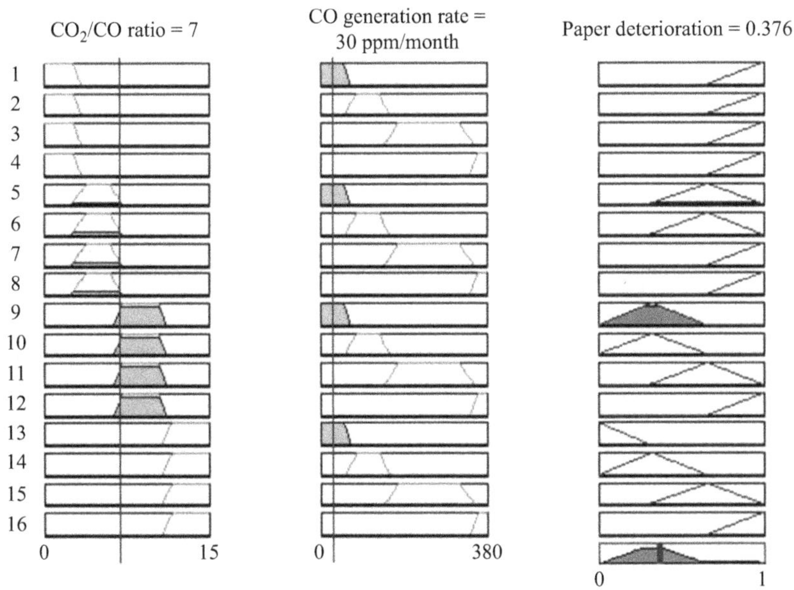

Figure 8.8 Developed fuzzy rules for paper deterioration based on CO ratio

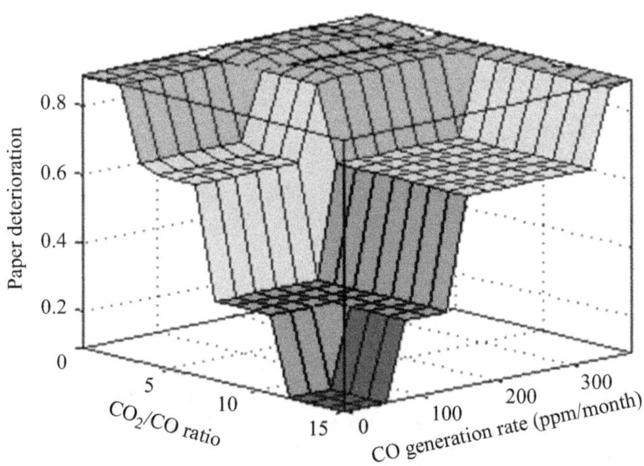

Figure 8.9 Paper deterioration surface graph

The model is tested with inputs, CO_2/CO ratio (7) and CO generation rate (30 ppm/month). The fuzzy-logic numerical output is 0.376, which corresponds to moderate deterioration paper condition as shown in Figure 8.8. This result aligns well with the recommendation in [17]. The paper deterioration for any set of input data (CO_2/CO ratio and CO generation rate) can be estimated using the surface graph shown in Figure 8.9.

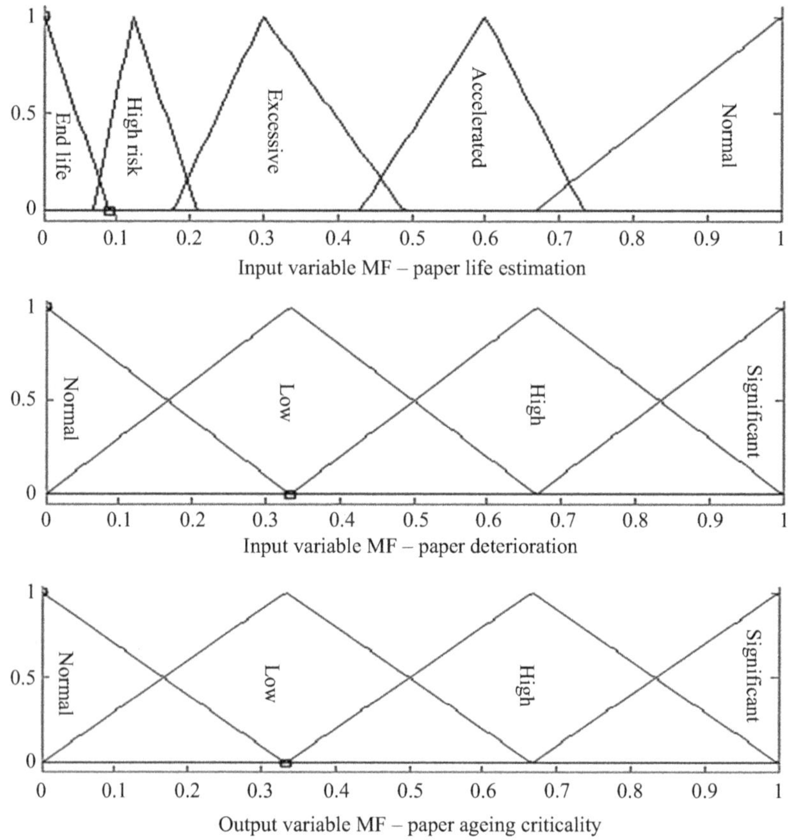

Figure 8.10 Input and output variables' MF for paper ageing criticality fuzzy model

8.4.3 Paper ageing criticality

The overall paper ageing criticality is obtained by integrating the furan and CO ratio factors fuzzy models as shown in Figure 8.3. The input to the model are the same outputs of the furan and CO ratio fuzzy models, and the output representing paper ageing criticality is assumed to be in the range of 0–1 (normal to significant criticality), as shown in Figure 8.10.

A set of fuzzy rules that relate the input and the output variables for overall paper ageing model is developed in accordance with FIST 3-31 guidelines, as shown in Figure 8.11 [17]. The model is tested with output values obtained from furan and CO ratio fuzzy models: 0.684 and 0.376, respectively. A paper ageing criticality of 0.332 is resulted. This value is corresponding to moderate paper ageing activity.

The overall paper ageing criticality for any set of input variable can be estimated through the 3D surface graph shown in Figure 8.12.

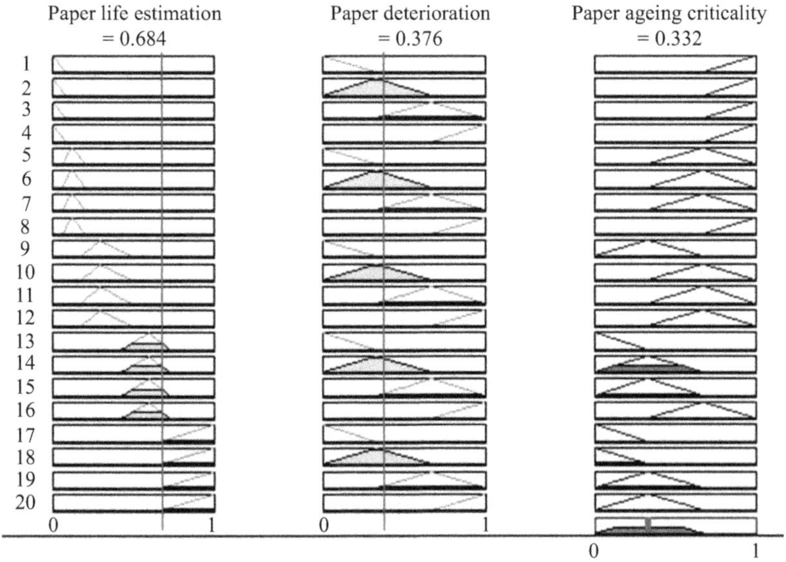

Figure 8.11 Developed fuzzy rules for paper ageing criticality

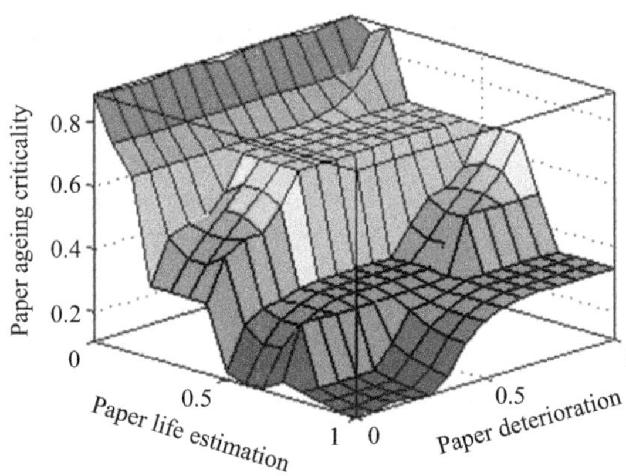

Figure 8.12 Paper ageing criticality surface graph

8.4.4 Relative accelerating ageing criticality

A fuzzy-logic model was developed using operating temperature rise, oxygen in oil, and water content in paper insulation parameters as inputs to the model where the output is the accelerating ageing factor based on the correlations of these factors with paper ageing [29].

The MF of operating temperature rise, oxygen, and water content are considered on the scale 0–30 °C, 0–23,000 ppm, and 0%–4.5% moisture by dry weight

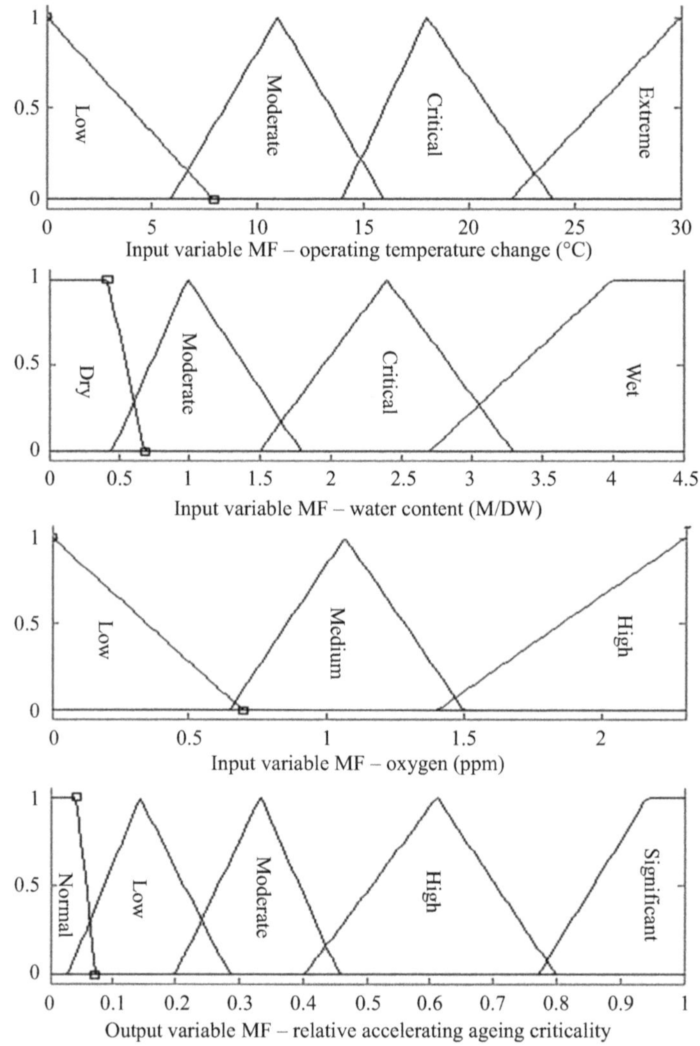

Figure 8.13 Input and output variables' MF for relative accelerating ageing fuzzy model

(M/DW) of paper insulation, respectively, as shown in Figure 8.13. The output of the model is considered on the range 0–1 (normal to significant criticality).

A set of fuzzy rules which correlate inputs and output variables is shown in Figure 8.14. The model is tested with inputs, operating temperature rise (2 °C), water content (1 M/DW), and oxygen (1,700 ppm). The fuzzy-logic numerical output is 0.028, which corresponds to normal accelerating ageing factor. The relative accelerating ageing criticality for any set of input variable can be estimated through the 3D surface graph shown in Figure 8.15.

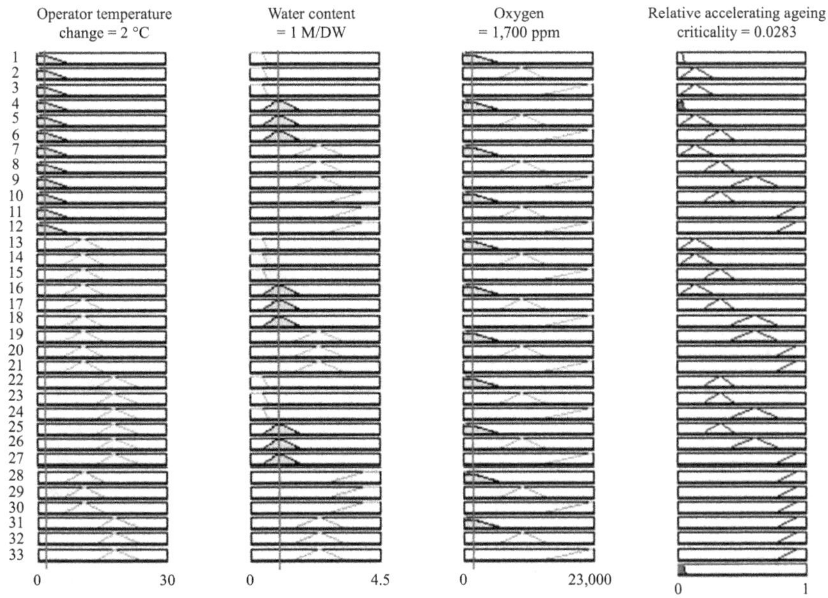

Figure 8.14 Developed fuzzy rules for relative accelerating ageing

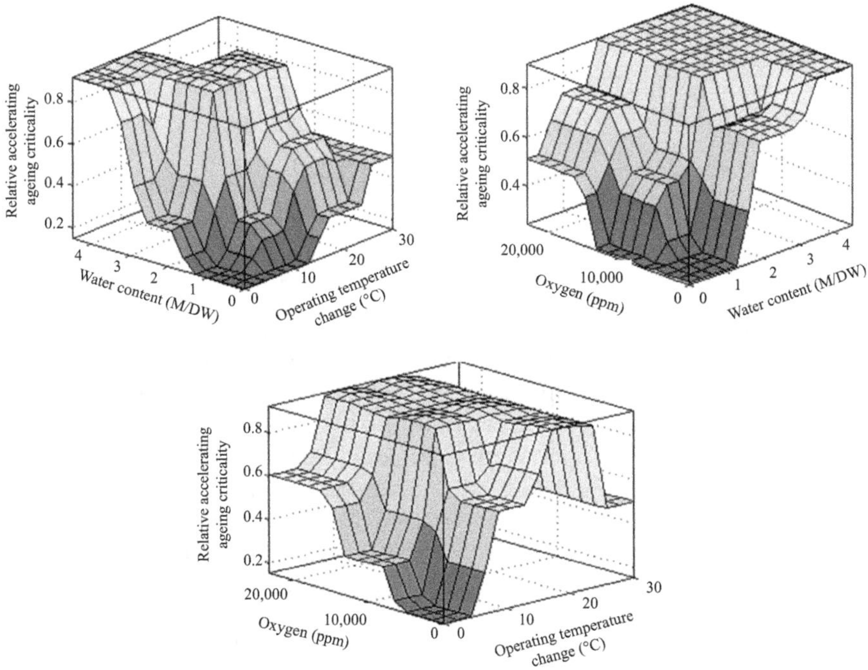

Figure 8.15 Relative accelerating ageing surface graph

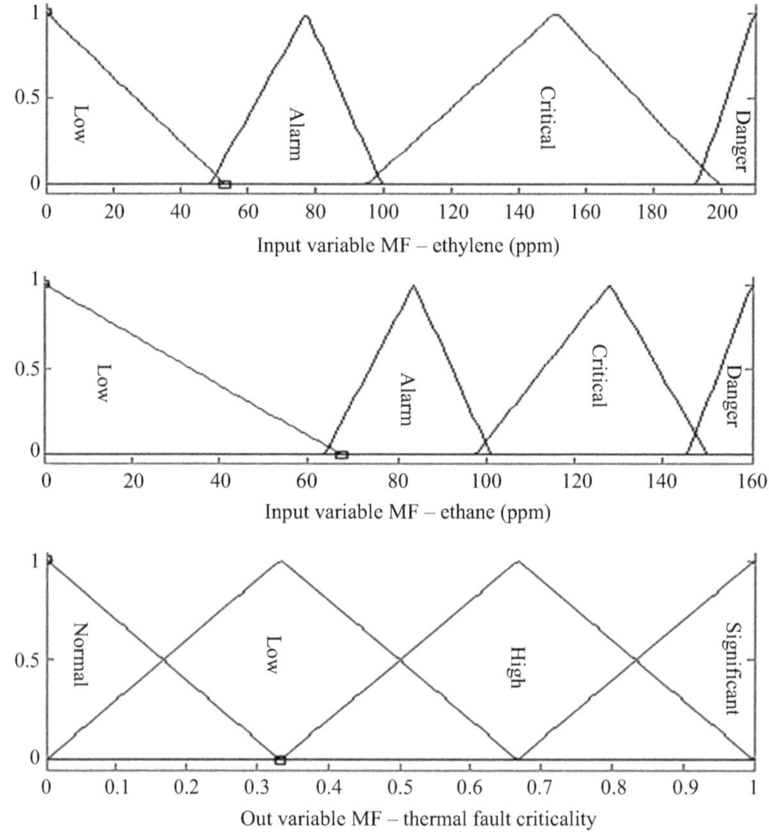

Figure 8.16 Input and output variables' MF for thermal fault fuzzy model

8.4.5 *Thermal fault criticality*

C_2H_4 and C_2H_6 are the main gases generated by oil decomposition due to thermal stress. A thermal fault fuzzy model is developed to assess the oil thermal criticality where C_2H_4 and C_2H_6 concentrations in ppm are the inputs to the model, while the output represents the thermal fault criticality.

Input variable MF are considered on the scale 0–210 and 0–160 ppm for C_2H_4 and C_2H_6, respectively. The output MF are measured on the scale 0–1 (normal to significant criticality) as shown in Figure 8.16. The fuzzy rules for thermal fault model are developed in accordance with IEEE Std. C57.104 recommendation [25] as shown in Figure 8.17. The model is tested with inputs, C_2H_4 (10 ppm) and C_2H_6 (15 ppm). The fuzzy-logic numerical output is 0.1128, which corresponds to less significant thermal fault condition. This result agrees with the IEEE Std. C57.104 guidelines [25]. The thermal fault criticality for any set of input variable can be estimated through the 3D surface graph shown in Figure 8.18.

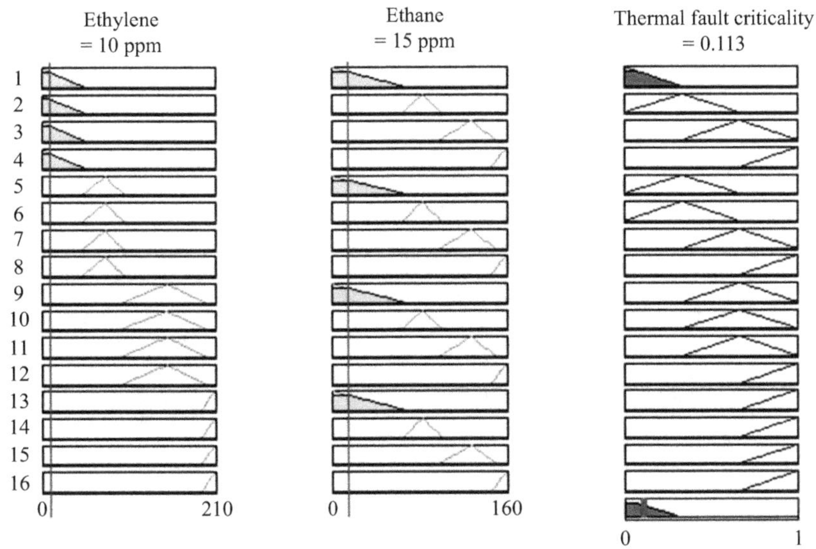

Figure 8.17 Developed fuzzy rules for thermal fault criticality based on ethylene and ethane

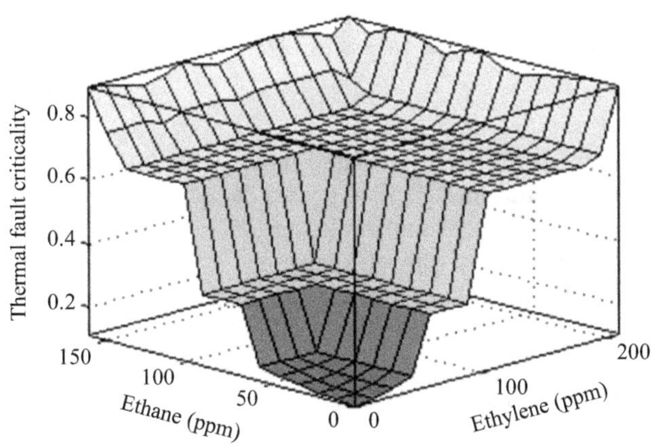

Figure 8.18 Thermal fault surface graph

8.4.6 Electrical fault criticality

CH_4 and C_2H_2 are the main gases generated by oil decomposition due to PD or arcing faults. An electrical fault fuzzy model was developed to assess the electric fault criticality where CH_4 and C_2H_2 are the inputs to the model in ppm, and the output is the electrical fault criticality. Input variable MF were considered on the scale 0–1,100 and 0–90 ppm for CH_4 and C_2H_2, respectively, as shown in

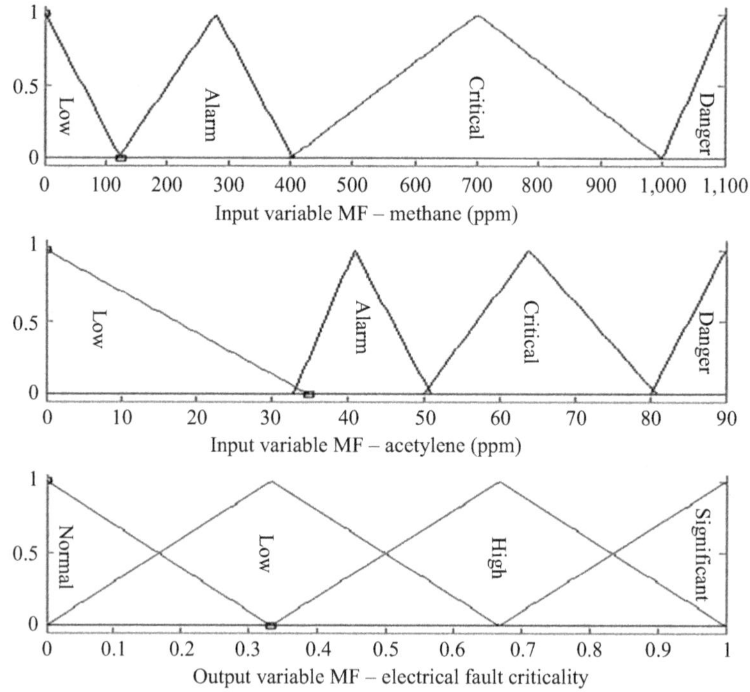

Figure 8.19 Input and output variables' MF for electrical fault fuzzy model

Figure 8.19. The output MF were measured on the scale 0–1 (normal to significant criticality). The fuzzy rules for electrical fault model were developed in accordance with the IEEE Std. C57.104 recommendation [25] as shown in Figure 8.20. The model was tested with inputs, CH_4 (18 ppm) and C_2H_2 (0 ppm). The fuzzy-logic numerical output is 0.1102, which corresponds to a less significant electrical fault condition. This result is consistent with the IEEE Std. C57.104 guidelines [25]. The electrical fault criticality for any set of input variable can be estimated through the 3D surface graph shown in Figure 8.21.

8.4.7 Overall thermal–electrical fault criticality

The overall thermal–electrical fault criticality can be assessed by integrating thermal and electrical fault criticalities fuzzy models. The inputs to the overall thermal–electrical fault criticality model are the outputs of the two aforementioned models while the output representing the overall fault criticality is assumed to be in the range of 0–1 (normal to significant criticality), as shown in Figure 8.22.

A set of fuzzy rules developed to correlate between inputs and output variables as shown in Figure 8.23. The model was tested with the output values obtained from thermal and electrical fault fuzzy models, 0.1128 and 0.1102, respectively. The final output of the fuzzy-logic model is 0.1078, which corresponds with a less

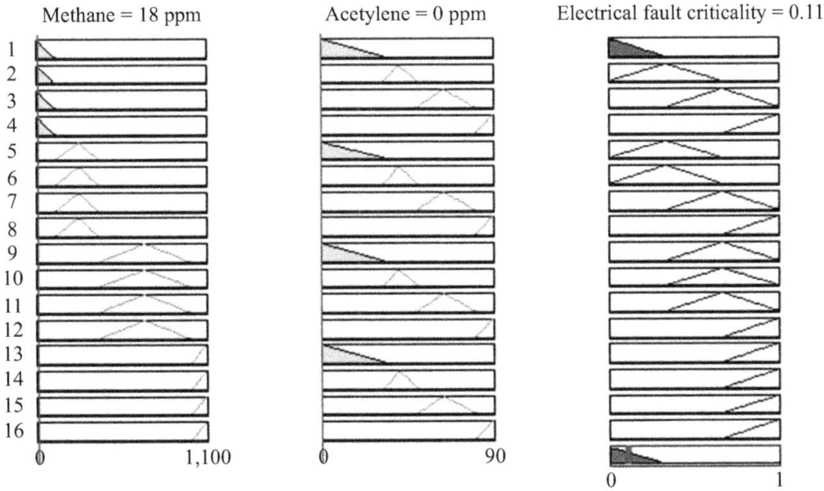

Figure 8.20 Developed fuzzy rules for electrical fault criticality based on methane and acetylene

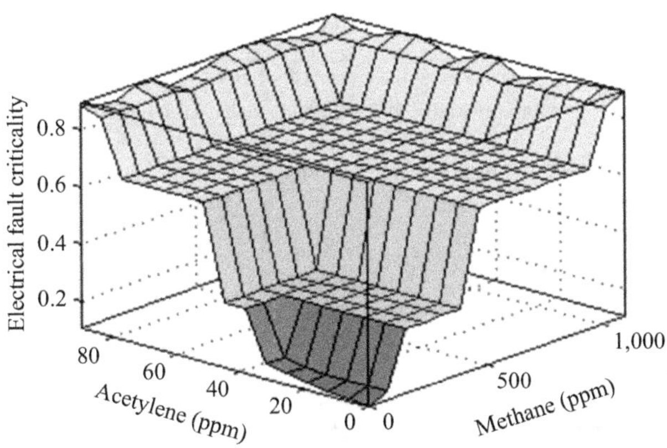

Figure 8.21 Electrical fault surface graph

significant thermal–electrical fault activity. This output is consistent with the IEEE Std. C57.104 as all fault gases are well below the designated minimum threshold fault limit [25]. The overall thermal–electrical fault criticality for any set of input variable can be estimated through the 3D surface graph shown in Figure 8.24.

8.4.8 IFT criticality

A fuzzy-logic model was developed using IFT and IFT decrement rate as inputs to the model where output is the IFT criticality that reflects the contaminations

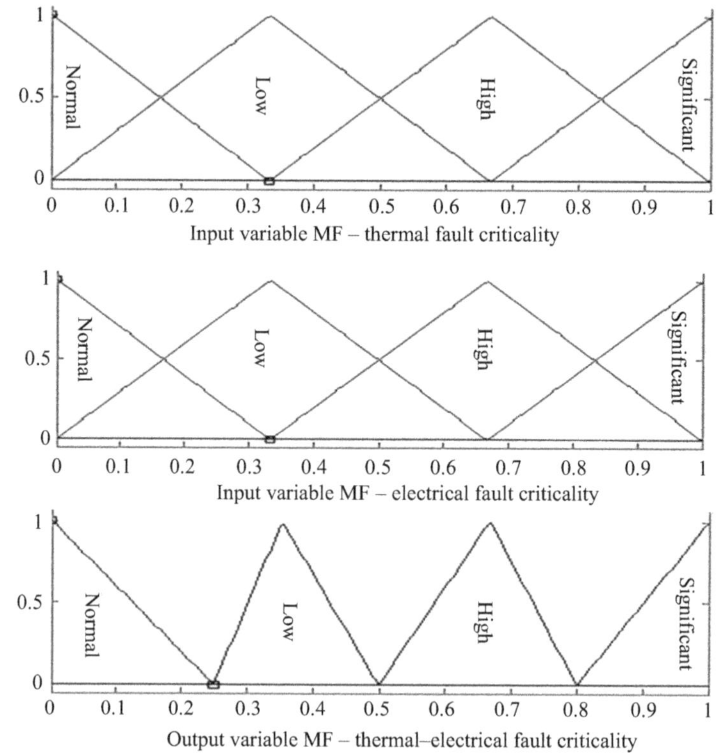

Figure 8.22 Input and output variables' MF for overall thermal–electrical fault fuzzy model

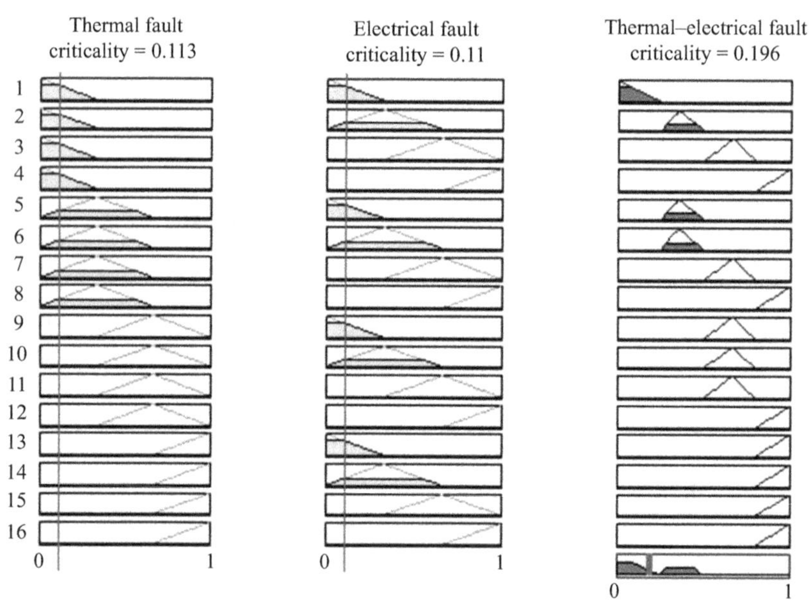

Figure 8.23 Developed fuzzy rules for overall thermal–electrical fault criticality

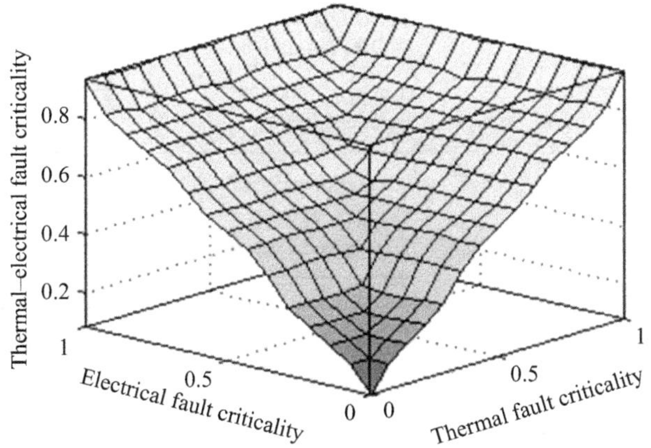

Figure 8.24 Overall thermal–electrical fault surface graph

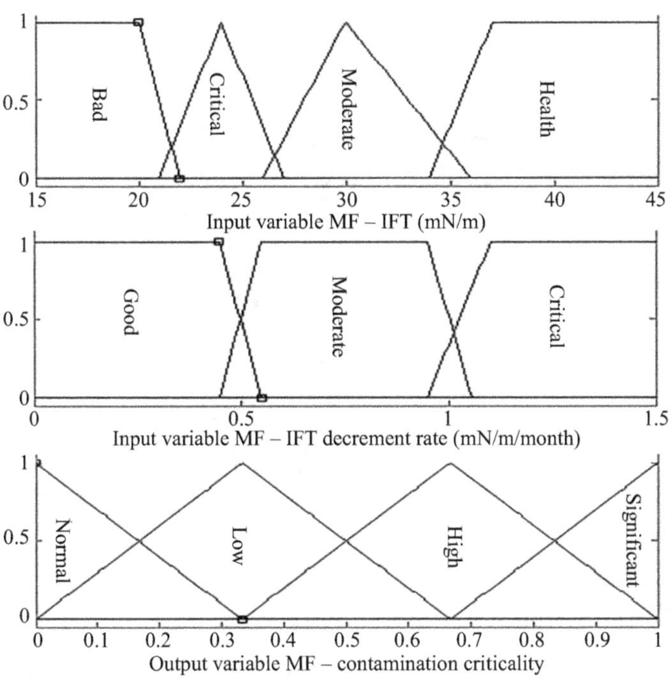

Figure 8.25 Input and output variables' MF for IFT criticality fuzzy model

in transformer oil. MF for IFT and IFT decrement rate are considered on the scale 15–45 and 0–1.5 mN/m/month, respectively, while the IFT criticality output MF are considered on the scale 0 (no contamination) to 1 (significant contamination) as shown in Figure 8.25.

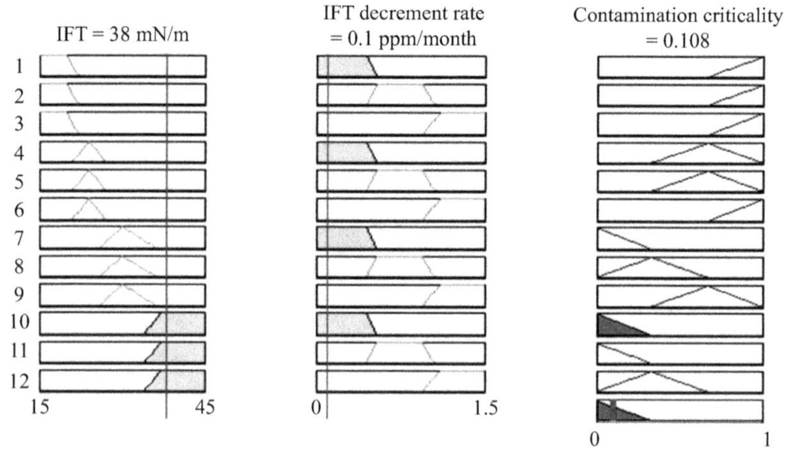

Figure 8.26 Developed fuzzy rules for contamination criticality based on IFT

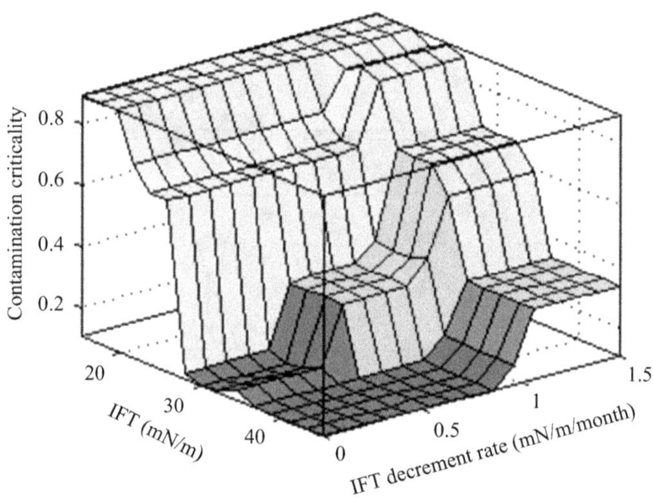

Figure 8.27 Contamination criticality surface graph

A set of fuzzy rules relating the input and output variables for the IFT criticality model was developed in accordance with the guidelines reported in FIST 3-31 [17], as shown in Figure 8.26. The IFT criticality fuzzy model is tested with inputs, IFT (38 mN/m) and IFT decrement rate (0.1 mN/m/month). The fuzzy-logic model output is 0.107, which corresponds to the minor contamination level detected in the oil. The IFT criticality for any set of input variable can be estimated through the 3D surface graph shown in Figure 8.27.

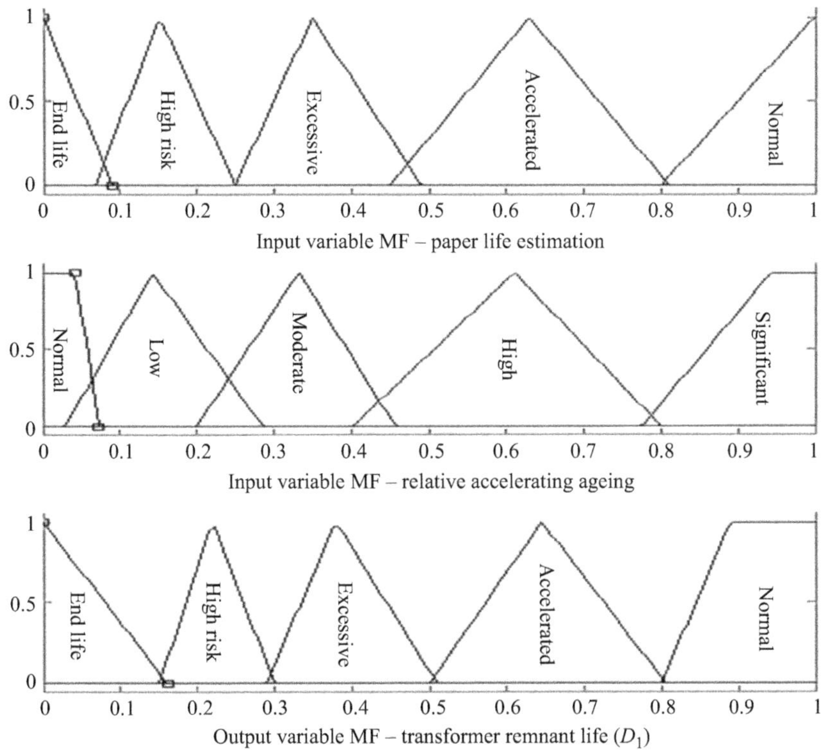

Figure 8.28 Input and output variables' MF for transformer remnant life (D_1) fuzzy model

8.4.9 Remnant life estimation

The transformer remnant life can be estimated by integrating the furan and relative accelerating ageing criticalities, as shown previously in Figure 8.3. The inputs to the model are the same outputs of the furan and relative accelerating ageing fuzzy models and the MF of the output which represents the transformer remnant life (D_1) are normalized on a scale 0–1, whereas 1 reflects a new transformer with 100% remaining life, and 0 reflects the end of the transformer's life, as shown in Figure 8.28.

A set of fuzzy rules that relate the input and output variables for the model was developed in accordance with the published recommendations [4], as shown in Figure 8.29. The model was tested with the output values obtained from furan and relative accelerating ageing fuzzy models, 0.684 and 0.028, respectively. The fuzzy model output is 0.65, which corresponds to 65% of transformer remaining life. The transformer remnant life for any set of input variables can be estimated through the 3D surface graph shown in Figure 8.30.

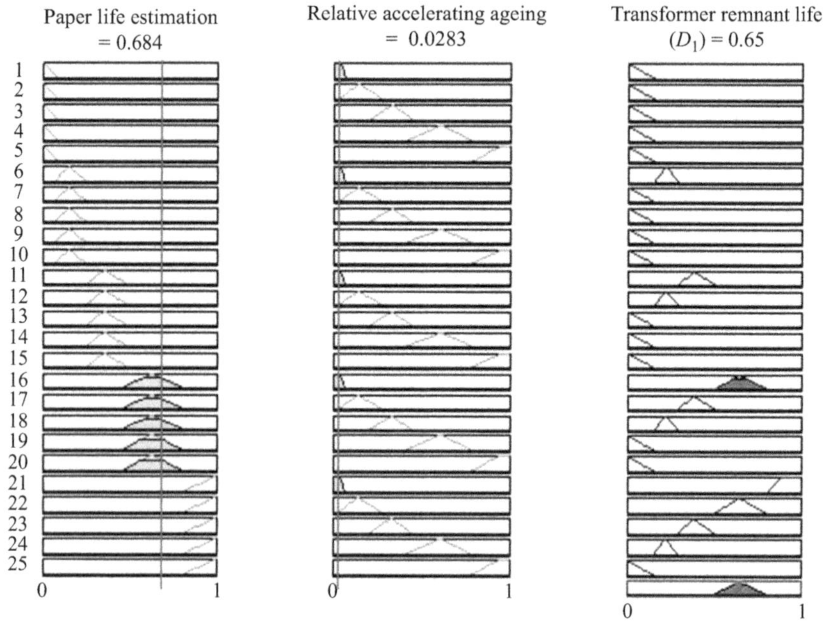

Figure 8.29 Developed fuzzy rules for transformer remnant life estimation (D_1)

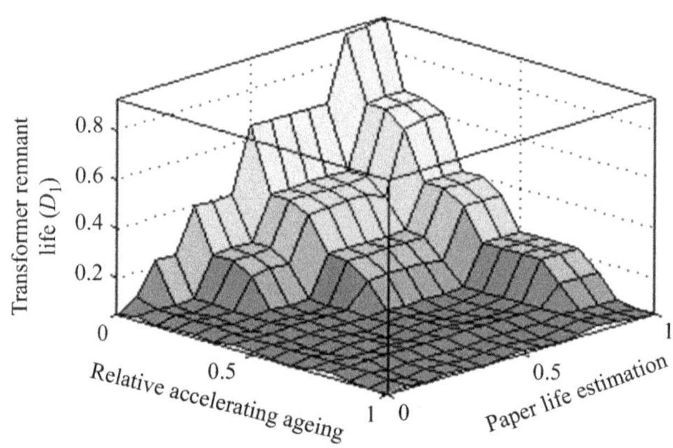

Figure 8.30 Transformer remnant life (D_1) surface graph

8.4.10 Asset management model

An asset management decision model was developed by integrating paper ageing, faults, contamination, and relative accelerated ageing criticalities, as shown in Figure 8.3.

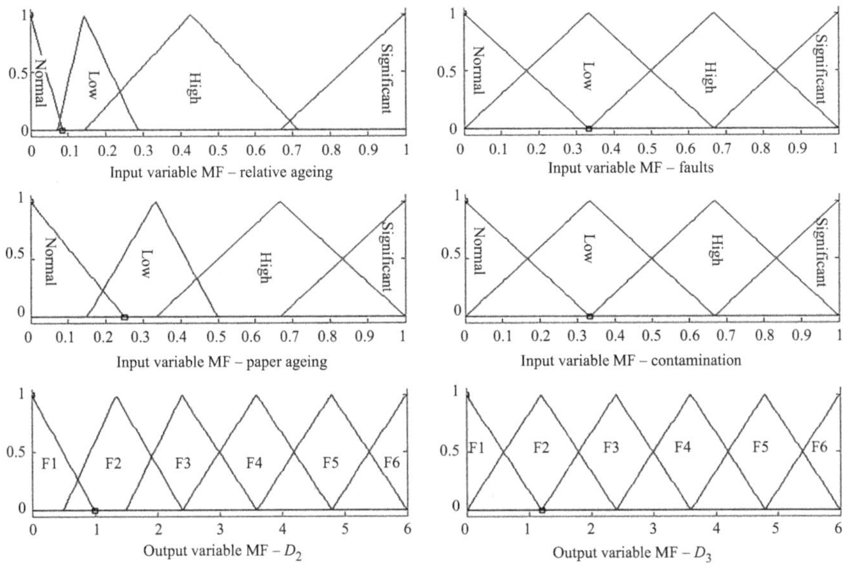

Figure 8.31 Input and output variables' MF for asset management decision fuzzy model

Inputs to the model are the outputs of the mentioned models. Two output decisions are made based on these factors: transformer operation condition (D_2) and recommendation for oil maintenance (D_3). Both output decisions are mapped on a scale 0–6 as shown in Figure 8.31. Fuzzy rules developed to correlate between input and output variables are shown in Figure 8.32.

The overall asset management decision fuzzy model was tested with the output values obtained from the four aforementioned models, paper ageing (0.332), faults (0.196), contamination (0.107), and relative accelerated ageing criticalities (0.028). The fuzzy model outputs are 1.88 and 1.09 for D_2 and D_3, respectively. The asset management decision, D_2 and D_3, for any set of input variable can be estimated through the 3D surface graph shown in Figures 8.33 and 8.34, respectively.

Eighteen transformer asset management decision codes that are associated with the combination of D_2 and D_3 were developed to represent the criticality of the transformer along with asset management decision. Table 8.2 provides the proposed asset management decision code corresponding to various numerical numbers resulted from the asset management fuzzy-logic model (D_2 and D_3). The overall results of the asset management model decisions in Figure 8.3 are consistent with the code 4 shown in Table 8.2, which corresponds to a transformer in good condition.

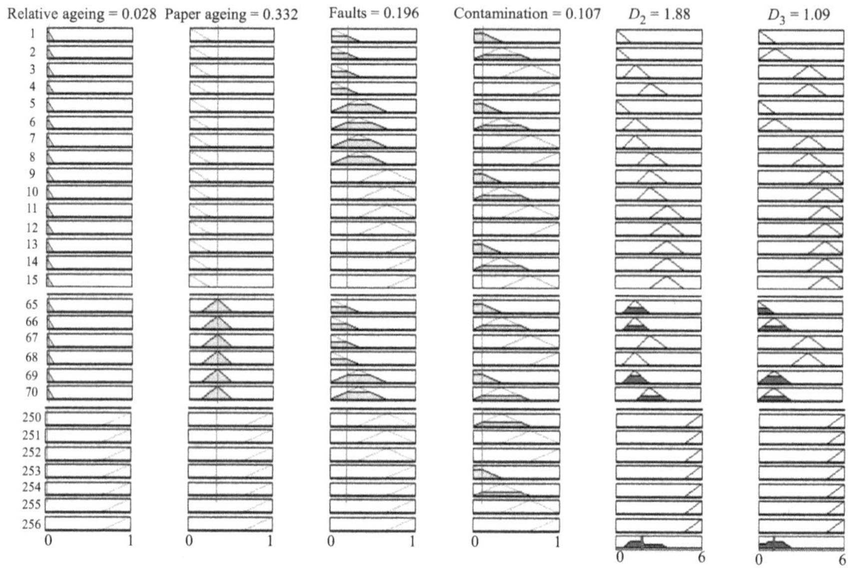

Figure 8.32 Developed fuzzy rules for asset management decision (D_2 and D_3)

8.5 Case study on pre-known condition of power transformer

To examine the accuracy of the proposed model, insulating oil analyses of 16 power transformers of different rating and life span with pre-known health conditions from previously published research papers [19,32] along with some collected field data were tested using the proposed model shown in Figure 8.3 to compare the model's output with the actual health conditions of these transformers that have been identified by experienced health and asset management utility team based on transformer regular measured condition monitoring parameters. Clarification of the actual health conditions of the investigated transformers are given in Table 8.3, while Table 8.4 shows the overall model output for 16 oil samples. Results obtained from the proposed fuzzy-logic model were compared with the actual condition of the transformer, as shown in Table 8.4.

The model outputs (D_2,D_3), for the first two transformers, were less than 1, (0.98,0.97) for sample 1 and (0.9,0.9) for sample 2 which correspond to code 1 in Table 8.2. This code indicates that all investigated critical factors are within normal limits and the transformer condition is considered 'very good'. This result is consistent with the guidelines for the IEEE Std. 57.106 [33] and FIST 3-31 [17] as all input parameters are within safety limits. It also matches with the actual condition of the two transformers.

The model output for samples 3 and 4 reveals code 4 which indicates one or more of the investigated critical factors were over normal limits. The very low CO_2/CO ratio may influence the decision for sample 3. On the other hand, sample 4

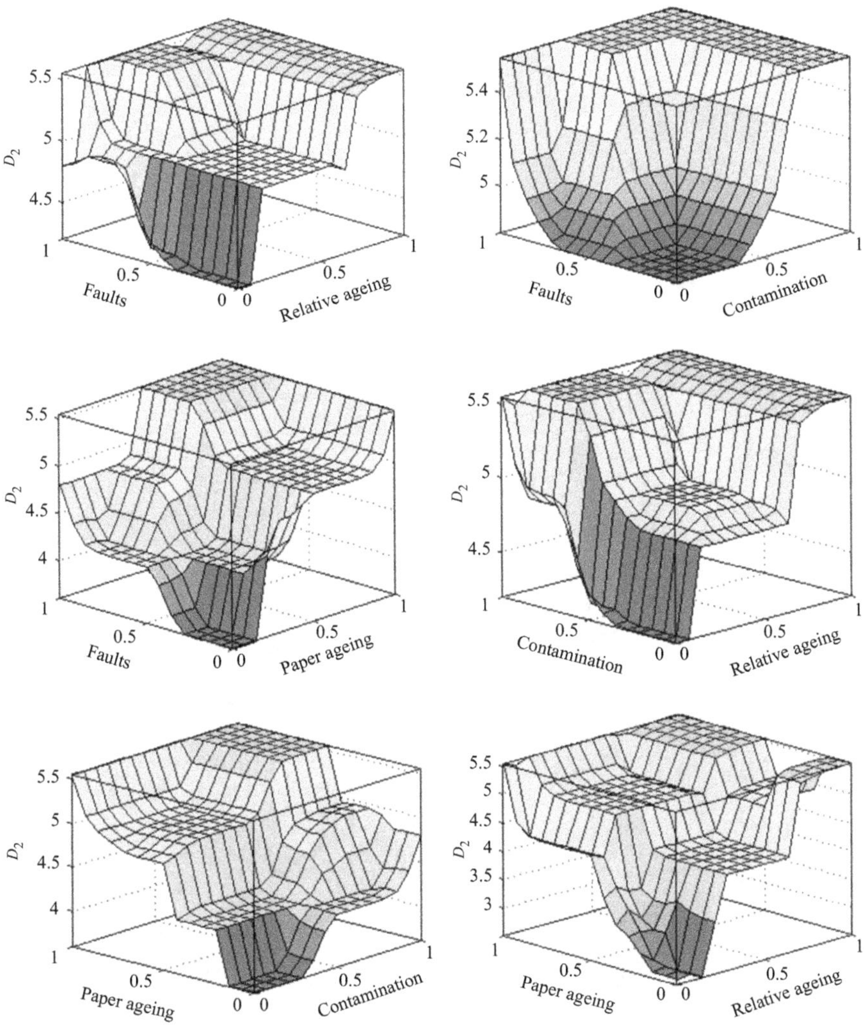

Figure 8.33 Asset management decision (D_2) surface graph

which was taken from a 22-year-old transformer indicates 2-FAL more than 0.1 ppm, revealing incipient paper ageing acceleration. Exercise caution monitoring on overall factors is recommended.

As stated in Table 8.3, samples 5 and 6 are obtained from aged transformers with very contaminated insulating oil. The model outputs for both samples reflect code 6, which indicates either paper degradation is accelerated or contamination level has entered the critical zone. Oil reclamation is recommended.

The model outputs for sample 7 is (2.6, 1.9) which are corresponding to code 7. This code reflects paper accelerated ageing with moderate faults factor criticality as

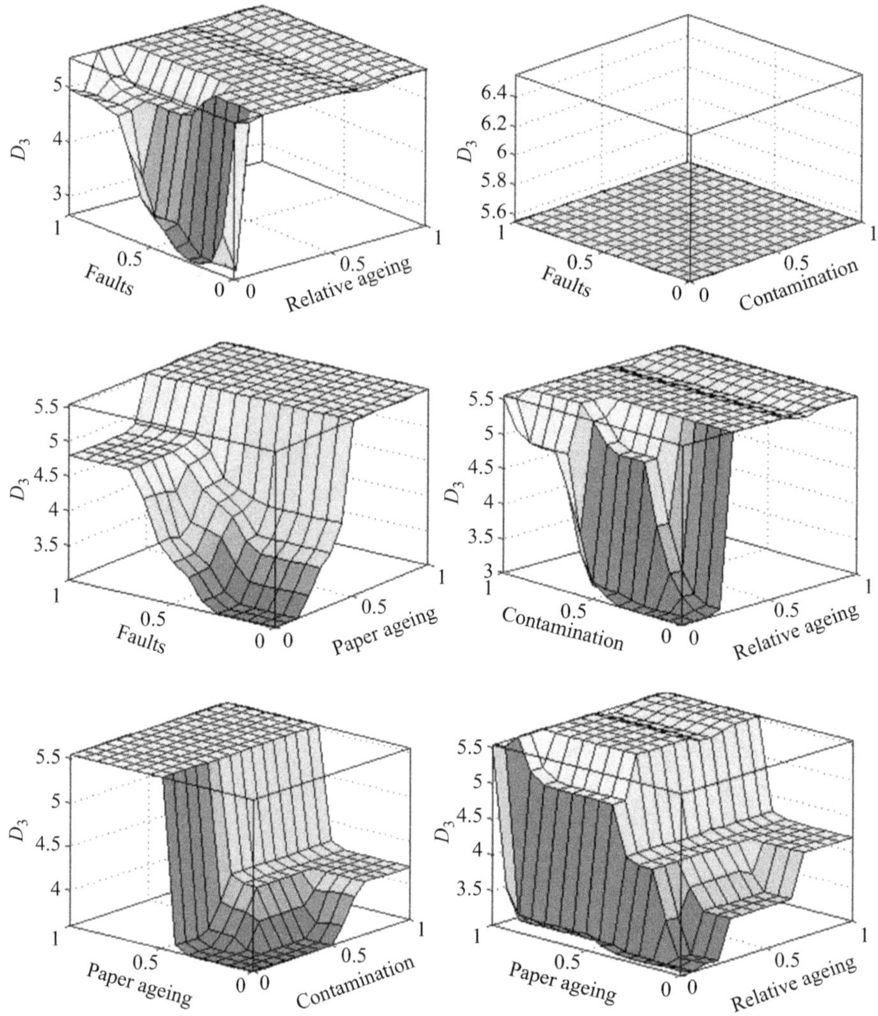

Figure 8.34 Asset management decision (D_3) surface graph

C_2H_6 concentration exceeds its safety limit. Specific trend monitoring on critical factor is required, and predictive maintenance is recommended.

Paper insulation of transformer 8 is reported to be nearly wet and moderately contaminated. The fuzzy output model for sample 8 (2.7, 3.5) is corresponding to code 9 which requires specific trend monitoring on critical factor. Code 9 reveals accelerated paper ageing and either relative accelerating ageing and/or oil contamination is critical. Oil reclamation is recommended.

Transformers 9 and 10 are found to be faulty where the model output results is (3.6, 4.8) which is corresponding to code 12. The model results agree with the

Table 8.2 Proposed transformer asset management decision code

Code no.	Model output		Trans. condition	Recommended decision
	D_2	D_3		
1	$0 \le D_2 < 1$	$0 \le D_3 < 1$	Very good	Continue normal operation
2	$0 \le D_2 < 1$	$1 \le D_3 < 2$	Good	Continue normal operation
3	$1 \le D_2 < 2$	$0 \le D_3 < 1$	Good	Continue normal operation
4	$1 \le D_2 < 2$	$1 \le D_3 < 2$	Good	Exercise caution. Monitor the trend of all factors
5	$1 \le D_2 < 2$	$2 \le D_3 < 3$	Average	Exercise caution. Monitor the trend of all factors. Degassing/dry-out recommended
6	$1 \le D_2 < 2$	$3 \le D_3 < 4$	Average	Exercise caution. Monitor the trend of all factors. Reclamation recommended
7	$2 \le D_2 < 3$	$1 \le D_3 < 2$	Average	Exercise extra caution. Specific trend monitoring on critical factor. Plan for predictive maintenance
8	$2 \le D_2 < 3$	$2 \le D_3 < 3$	Average	Exercise extra caution. Specific trend monitoring on critical factor. Degassing/dry-out recommended
9	$2 \le D_2 < 3$	$3 \le D_3 < 4$	Average	Exercise extra caution. Specific trend monitoring on critical factor. Reclamation recommended
10	$3 \le D_2 < 4$	$2 \le D_3 < 3$	Poor	Exercise extreme caution. Specific trend monitoring on critical factor with restricted load operation. Degassing/dry-out recommended
11	$3 \le D_2 < 4$	$3 \le D_3 < 4$	Poor	Exercise extreme caution. Specific trend monitoring on critical factor with restricted load operation. Reclamation recommended
12	$3 \le D_2 < 4$	$4 \le D_3 < 5$	Poor	Exercise extreme caution. Fault detected. Analyse individual gases to find cause of fault. Schedule for total condition monitoring diagnoses
13	$4 \le D_2 < 5$	$2 \le D_3 < 3$	Bad	Exercise extreme caution. Specific trend monitoring on critical factor. Reduce load operation to below 70%. Plan outage. Degassing/dry-out recommended
14	$4 \le D_2 < 5$	$3 \le D_3 < 4$	Bad	Exercise extreme caution. Specific trend monitoring on critical factor. Reduce load operation to below 70%. Plan outage. Reclamation recommended
15	$4 \le D_2 < 5$	$4 \le D_3 < 5$	Bad	Exercise extreme caution. Fault detected. Analyse individual gases to find cause of fault. Urgent total condition monitoring diagnoses required

(Continues)

Table 8.2 (Continued)

Code no.	Model output		Trans. condition	Recommended decision
	D_2	D_3		
16	$4 \leq D_2 < 5$	$5 \leq D_3 < 6$	Bad	Exercise extreme caution. Specific trend monitoring on critical factor. Reduce load operation to below 60%. Plan outage. Consider removal from service
17	$5 \leq D_2 < 6$	$4 \leq D_3 < 5$	Very bad	Exercise extreme caution. Fault detected. Analyse individual gases to find cause of fault. Immediate action required. Consider removal from service
18	$5 \leq D_2 < 6$	$5 \leq D_3 < 6$	Very bad	Exercise extreme caution. Reduce load operation to below 50%. Immediate action required. Consider removal from service

actual condition of the two transformers as presented in Table 8.3, whereas transformer 9 is detected with arcing in oil and transformer 10 is detected with tap changer failure. Further investigation on individual dissolved gases is recommended to find the cause of fault.

The fuzzy model output for sample 11 coincides with code 11, which indicates accelerated paper ageing with high criticalities of relative accelerating ageing and contamination. Transformer 11 is reported to have wet paper insulation and highly contaminated insulating oil. Extreme caution monitoring on related factors with restricted load operation is necessary and oil reclamation is recommended.

Fuzzy output for sample 12 coincides with code 10 which indicates significant criticality of relative accelerated ageing factor. Oil is recommended for degassing/dry-out depending on water content and oxygen level.

The model output for sample 13 (4.3, 3.6) was associated with code 14 which indicates significant transformer criticality. A transformer under this code requires specific trend monitoring on critical factors with load reduction below 70%. Oil reclamation and plan for outage are recommended for this transformer.

Code 15 was the fuzzy model output for sample 14 which reflects high thermal fault and excessive paper degradation. A transformer in this category is considered highly critical. Further assessment on individual dissolved gases is recommended to find the cause of fault. Immediate overall transformer condition diagnoses are suggested.

Samples 15 and 16 were associated with code 18 of the fuzzy model output. With extreme paper degradation and highly contaminated insulating oil, transformers 15 and 16 are considered to be in a high-risk condition. A transformer under this code should be taken out of service immediately to prevent any potential catastrophic failure.

Table 8.3 Pre-known condition of 16 transformers

Sample	Real age (years)	Trans. status	Description
1	4	Good	New and healthy transformer
2	2	Good	New and healthy transformer
3	18	Good	Healthy transformer, but CO exceed limits
4	22	Good	Aged and healthy transformer
5	35	Average	Aged transformer with average paper degradation. Oil is contaminated
6	28	Average	Aged transformer with average paper degradation. Oil is contaminated
7	30	Average	Aged transformer with average paper degradation. No faults detected, but ethane exceeds safety limits
8	23	Average	Aged transformer with accelerated paper degradation. Paper is nearly wet and oil is moderately contaminated
9	38	Poor	Faulty transformer (arcing in oil). Average paper degradation
10	25	Poor	Faulty transformer (Failed tap changer). Normal paper degradation. Acetylene level is very high. Oil is contaminated
11	42	Poor	Aged transformer with average paper degradation. High oxygen concentration. Oil is too contaminated
12	49	Poor	Aged transformer with average paper degradation. Paper is wet and highly risk for failure
13	56	Bad	Aged transformer with accelerated paper degradation. Paper is wet and highly risk for failure. Oil is very contaminated
14	50	Bad	Faulty transformer (Thermal faults exceeding 700 °C). Excessive paper degradation. Very high oxygen concentration
15	63	Bad	Aged transformer with excessive paper degradation. Oil is contaminated
16	36	Bad	Aged transformer with excessive paper degradation. Paper is nearly wet and oil is contaminated

The overall results of the transformer criticality measures proposed by this model are highly consistent with the actual condition of the investigated 16 power transformers.

8.6 Conclusion

This chapter has addressed developing an expert model to estimate the remnant life and asset management decision of power transformers based on routine insulating oil tests.

A new fuzzy-logic approach to predicting the remnant life of power transformers and to provide a proper asset management decision tool based on insulating

Table 8.4 Fuzzy-logic model output with criticality code proposed

Sample	Temp. rise	Water content	Oxygen	Furan	Furan rate	CO_2/CO ratio*	CO rate	Ethylene	Ethane	Methane	Acetylene	IFT	IFT rate	Model output (D_1) (%)	Model output (D_2)	Model output (D_3)	Code apply	Model condition	Actual condition
1	1	0.6	500	<0.01	0	18	5	2	3	4	0	40	0.2	91	0.98	0.97	1	VG	G
2	1	0.45	180	<0.01	0	10	0	0	0	0	0	42	0.1	93	0.9	0.9	1	VG	G
3	1	0.8	800	<0.01	0	2	22	2	4	22	0	40	0.2	89	1.8	1.2	4	G	G
4	1	0.45	1,050	0.23	0.01	10	5	9	1	1	0	40	0.2	67	1.8	1.2	4	G	G
5	2	1	1,510	0.61	0.01	10.6	10	2	2	3	3	16	0.8	63	1.87	3.6	6	A	A
6	2	0.8	1,200	0.25	0.01	10	5	3	1	2	0	20	0.5	65	1.8	3.6	6	A	A
7	2	0.85	158	0.12	0.02	10	5	7	90	23	0	32	0.5	63	2.6	1.9	7	A	A
8	1	2	1,817	1.05	0.05	11	8	31	9	29	0	28	0.8	38	2.7	3.5	9	A	A
9	2	0.9	1,515	0.3	0.01	10	10	458	14	286	884	32	0.4	63	3.6	4.8	12	P	P
10	2	1	780	0.11	0.01	10	10	13	1	3	73	25	0.5	63	3.6	4.8	12	P	P
11	2	2.6	3,250	0.24	0.01	10	10	2	0	2	0	18	0.5	38	3.7	3.6	11	P	P
12	2	3	4,715	0.29	0.01	10	10	1	0	1	0	28	0.5	32	3.2	2.1	10	P	P
13	2	3	5,112	0.99	0.05	4.9	10	8	7	34	0	14	1	31	4.3	3.6	14	B	B
14	2	0.85	30,280	1.38	0.05	10	5	150	18	55	22	41	0.2	22	4.8	4.8	15	B	B
15	2	1.1	2,680	4.46	0.1	5.8	20	7	10	39	0	20	0.5	19	5.2	5.5	18	VB	B
16	2	1.8	1,700	5.53	0.1	3	40	15	18	22	0	19	0.5	7	5.6	5.6	18	VB	B

*CO_2/CO ratio only applicable when the accumulated values of CO_2 and CO exceeds 2,500 and 350 ppm, respectively. Else, CO_2/CO ratio is set as 10.
(very good = VG; good = G; average = A; poor = P; bad = B)

oil routine inspection was developed. All input parameters used in the proposed model can be measured on site and/or online. Eighteen codes which reflect the overall fuzzy-logic output have been developed to cover all criticality conditions of an operating power transformer and recommend an appropriate asset management decision based on the proposed model. Accuracy of the overall model was assessed through historical data of 16 transformers of different ages, ratings, and pre-known operating conditions. Results of the model were found to be consistent with real transformer conditions. The development of the model is expected to be beneficial for quick and reliable insulation condition assessment and for assisting inexperienced engineers in proposing a proper management decision based on current transformer condition.

Acknowledgement

It is acknowledged that most of the work in this chapter is part of the author thesis, 'A New Technique to Detect Loss of Insulation Life in Power Transformers' from Curtin University, Australia.

References

[1] M. Arshad and S. M. Islam, "Significance of cellulose power transformer condition assessment," *IEEE Transactions on Dielectrics and Electrical Insulation*, vol. 18, pp. 1591–1598, 2011.

[2] "IEEE Guide for the Evaluation and Reconditioning of Liquid Immersed Power Transformers," in *IEEE Std C57.140-2006,* pp. c1–67, 2007.

[3] A. M. Emsley, X. Xiao, R. J. Heywood, and M. Ali, "Degradation of cellulosic insulation in power transformers. Part 2: Formation of furan products in insulating oil," *IEE Proceedings on Science, Measurement and Technology*, vol. 147, pp. 110–114, 2000.

[4] L. E. Lundgaard, W. Hansen, D. Linhjell, and T. J. Painter, "Aging of oil-impregnated paper in power transformers," *IEEE Transactions on Power Delivery*, vol. 19, pp. 230–239, 2004.

[5] N. A. Bakar, A. Abu-Siada, and S. Islam, "A Review on Chemical Diagnosis Techniques for Transformer Paper Insulation Degradation," in *Australasian Universities Power Engineering Conference, AUPEC 2013*, 2013.

[6] L. Cheim, "Furan analysis for liquid power transformers," *IEEE Electrical Insulation Magazine*, vol. 28, p. 8, 2012.

[7] D. Martin, C. Yi, C. Ekanayake, M. Hui, and T. Saha, "An updated model to determine the life remaining of transformer insulation," *IEEE Transactions on Power Delivery*, vol. 30, pp. 395–402, 2015.

[8] "IEEE Guide for Loading Mineral-Oil-Immersed Transformers and Step-Voltage Regulators," in *IEEE Std C57.91-2011*, pp. 1–123, 2012.

[9] "Power Transformers – Part 7: Loading guide for Oil-Immersed Power Transformers," in *IEC 60076-7:2005 Std,* 2005.

[10] A. M. Emsley, R. J. Heywood, M. Ali, and X. Xiao, "Degradation of cellulosic insulation in power transformers 4. Effects of ageing on the tensile strength of paper," *IEE Proceedings on Science, Measurement and Technology*, vol. 147, pp. 285–290, 2000.

[11] A. Abu-Siada and S. Islam, "A new approach to identify power transformer criticality and asset management decision based on dissolved gas-in-oil analysis," *IEEE Transactions on Dielectrics and Electrical Insulation*, vol. 19, pp. 1007–1012, 2012.

[12] Z. Qian and Z. Yan, "Fuzzy synthetic method for life assessment of power transformer," *IEE Proceedings on Science, Measurement and Technology*, vol. 151, pp. 175–180, 2004.

[13] T. K. Saha and P. Purkait, "Investigation of an expert system for the condition assessment of transformer insulation based on dielectric response measurements," *IEEE Transactions on Power Delivery*, vol. 19, pp. 1127–1134, 2004.

[14] M. Arshad, S. Islam, and A. Khaliq, "Fuzzy logic approach in power transformers management and decision making," *IEEE Transactions on Dielectrics and Electrical Insulation*, vol. 21, pp. 2343–2354, 2014.

[15] L. Chmura, P. H. F. Morshuis, J. J. Smit, and A. Janssen, "Life-data analysis for condition assessment of high-voltage assets," *IEEE Electrical Insulation Magazine*, vol. 31, pp. 33–43, 2015.

[16] R. A. Jongen, P. H. F. Morshuis, E. Gulski, J. J. Smit, J. Maksymiuk, and A. L. J. Janssen, "Application of statistical methods for making maintenance decisions within power utilities," *IEEE Electrical Insulation Magazine*, vol. 22, pp. 24–35, 2006.

[17] D. Hydroelectric Research and Technical Services Group, Colorado, "Facilities, Illustrations, Standards and Techniques; Transformer Diagnostic," *US Department of Interior Bureau of Reclamation*, vol. 3–31, pp. 1–62, June 2003.

[18] A. Jahromi, R. Piercy, S. Cress, J. Service, and W. Fan, "An approach to power transformer asset management using health index," *IEEE Electrical Insulation Magazine*, vol. 25, pp. 20–34, 2009.

[19] A. E. B. Abu-Elanien, M. M. A. Salama, and M. Ibrahim, "Calculation of a health index for oil-immersed transformers rated under 69 kV using fuzzy logic," *IEEE Transactions on Power Delivery*, vol. 27, pp. 2029–2036, 2012.

[20] N. A. Bakar, A. Abu-Siada, and S. Islam, "A review of dissolved gas analysis measurement and interpretation techniques," *IEEE Electrical Insulation Magazine*, vol. 30, pp. 39–49, 2014.

[21] A. Abu-Siada, S. P. Lai, and S. M. Islam, "A novel fuzzy-logic approach for furan estimation in transformer oil," *IEEE Transactions on Power Delivery*, vol. 27, pp. 469–474, 2012.

[22] N. A. Bakar, A. Abu-Siada, S. Islam, and M. F. El-Naggar, "A new technique to measure interfacial tension of transformer oil using UV–vis

spectroscopy," *IEEE Transactions on Dielectrics and Electrical Insulation*, vol. 22, pp. 1275–1282, 2015.

[23] M. Zhixin and W. Jingyu, "Detection of dissolved gas in oil-insulated electrical apparatus by photoacoustic spectroscopy," *IEEE Electrical Insulation Magazine*, vol. 31, pp. 7–14, 2015.

[24] D. Martin, C. Perkasa, and N. Lelekakis, "Measuring paper water content of transformers: a new approach using cellulose isotherms in nonequilibrium conditions," *IEEE Transactions on Power Delivery*, vol. 28, pp. 1433–1439, 2013.

[25] "IEEE Guide for the Interpretation of Gases Generated in Oil-Immersed Transformers – Redline," in *IEEE Std C57.104-2008* pp. 1–45, 2009.

[26] M. Kohtoh, S. Kaneko, S. Okabe, and T. Amimoto, "Aging effect on electrical characteristics of insulating oil in field transformer," *IEEE Transactions on Dielectrics and Electrical Insulation*, vol. 16, pp. 1698–1706, 2009.

[27] D. Martin, T. Saha, T. Gray, and K. Wyper, "Determining water in transformer paper insulation: effect of measuring oil water activity at two different locations," *IEEE Electrical Insulation Magazine*, vol. 31, pp. 18–25, 2015.

[28] M. Wang, A. J. Vandermaar, and K. D. Srivastava, "Review of condition assessment of power transformers in service," *IEEE Electrical Insulation Magazine*, vol. 18, pp. 12–25, 2002.

[29] N. Lelekakis, J. Wijaya, D. Martin, and D. Susa, "The effect of acid accumulation in power-transformer oil on the aging rate of paper insulation," *IEEE Electrical Insulation Magazine*, vol. 30, pp. 19–26, 2014.

[30] "IEEE Guide for Diagnostic Field Testing of Electric Power Apparatus – Part 1: Oil Filled Power Transformers, Regulators, and Reactors," in *IEEE Std 62-1995*, pp. 1–64, 1995.

[31] N. A. Bakar and A. Abu-Siada, "Fuzzy logic approach for transformer remnant life prediction and asset management decision," *IEEE Transactions on Dielectrics and Electrical Insulation*, vol. 23, pp. 3199–3280, 2016.

[32] A. Abu-Siada, S. Hmood, and S. Islam, "A new fuzzy logic approach for consistent interpretation of dissolved gas-in-oil analysis," *IEEE Transactions on Dielectrics and Electrical Insulation*, vol. 20, pp. 2343–2349, Dec 2013.

[33] "IEEE Guide for Acceptance and Maintenance of Insulating Oil in Equipment," in *IEEE Std C57.106-2002 (Revision of IEEE Std C57.106-1991)*, pp. 0_1–27, 2002.

Index

www.ingramcontent.com/pod-product-compliance
Lightning Source LLC
LaVergne TN
LVHW011913200926
848484LV00002B/613
9781785612541